Library of Congress Cataloging-in-Publication Data

Lefohn, Allen S.
Surface-level ozone exposures and their effects on vegetation/Allen S. Lefohn.
    p. cm.
  Includes bibliographical references and index.
  ISBN 0-87371-169-6
  1. Plants, Effect of atmospheric ozone on.   I. Title.
QK751.L43 1991
581.5'222—dc20
                                                               91-33463
                                                               CIP

LEWIS PUBLISHERS, INC.
121 South Main Street, Chelsea, MI 48118

PRINTED IN THE UNITED STATES OF AMERICA
0 1 2 3 4 5 6 7 8 9

Surface Level Ozone Exposures and their Effects on Vegetation

Allen S. Lefohn

THE EDITOR

 Dr. Allen S. Lefohn, president and founder of A.S.L. & Associates, received his Ph.D. in physical chemistry from the University of California at Berkeley. Over the past 22 years, Dr. Lefohn's research has involved the development of exposure-response relationships describing the effects of ozone and sulfur dioxide on vegetation, the analysis of air quality data for developing dose-response surrogates for assessing the effects of air pollutants on vegetation, the characterization of European and North American air quality and wet deposition data for assessing possible reasons for forest decline, the development of field surveys to assess the effects of SO_2 on agricultural crops, and the analysis of acidic deposition wetfall and effects data to better assess the potential for biological effects. Dr. Lefohn has focused his research efforts on better understanding (1) the quantification and relationship between pollutant exposure and naturally occurring processes and (2) the possible effects of air pollutants on the ecosystem.

He is currently an organizer of and participant in the United Nations World Meteorological Organization (WMO) Expert Committee to evaluate surface-level ozone exposures and trends at remote locations in the world. He was the lead author of the National Acid Precipitation Assessment Program (NAPAP) State-of-Science Report No. 7: Air Quality Measurements and Characterizations for Vegetation Effects Research and has served as a senior scientist to the U.S. EPA National Crop Loss Assessment Network (NCLAN) program.

Dr. Lefohn has published over 70 peer-reviewed papers and 80 technical reports, edited 4 books, presented oral papers, and participated in a number of television and radio interviews and panel presentations. He is an Executive Editor of the prestigious scientific journal *Atmospheric Environment* and is an Adjunct Instructor of Environmental Engineering at Montana Tech in Butte, Montana. He currently resides with his family in Clancy, Montana.

ACKNOWLEDGMENTS

The authors wish to acknowledge the many research institutions, both in the government and in the private sector, that have provided financial assistance over the years for a better understanding of the formation, transport, fate, and effects of surface-level ozone on vegetation. Although many of our research efforts have been published in the open literature, we believe that bringing together our collective thoughts on this important subject under one cover will provide students, instructors, active researchers, and policymakers with the knowledge they require to develop and use research results in the most defensible and appropriate manner.

The editor wishes to acknowledge the following staff members of A.S.L. & Associates: Mrs. Janell Foley for her assistance in organizing the material and Ms. Phyllis Lefohn for editing and proofing the chapters.

THE CONTRIBUTORS

Dr. William L. Chameides
School of Geophysical Sciences
Georgia Tech
Atlanta, GA 30332
(404) 894-3893

Dr. Arthur H. Chappelka
108 M Whitesmith Hall
Auburn University
Auburn, AL 36849
(205) 844-1047

Dr. Boris I. Chevone
Department of Plant Pathology
and Physiology
Virginia Tech
Blacksburg, VA 24060
(703) 231-6530

Dr. Sagar V. Krupa
Department of Plant Pathology
495 Borlang Hall
1991 Buford Circle
University of Minnesota
St. Paul, MN 55108
(612) 625-9291

Dr. Allen S. Lefohn
A.S.L. & Associates
111 North Last Chance Gulch, Ste 4a
Helena, MT 59601
(406) 443-3389

Dr. James P. Lodge, Jr.
801 Circle Drive
Boulder, CO 80302
(303) 449-7712

Dr. William J. Manning
Department of Plant Pathology
Fernald Hall
University of Massachusetts
Amherst, MA 01003
(413) 545-2289

Dr. Victor C. Runeckles
Department of Plant Science
University of British Columbia
Vancouver, B.C. V6T 1Z4
Canada
(604) 822-6829

CONTENTS

CHAPTER 1

Introduction

Allen S. Lefohn, A.S.L. & Associates, Helena, MT

1.1 SURFACE-LEVEL OZONE AND ITS IMPORTANCE

The existence of an allotrope of oxygen called ozone was discovered in the mid 19th century. Although the vast majority of ozone in the atmosphere is located at stratospheric altitudes, in some locations ozone levels in the troposphere are high enough to affect plant and animal life. Ozone was first identified as a significant, phytotoxic, gaseous air pollutant in Southern California half a century ago. It has progressively become the major air pollutant in many parts of the world. Although the build-up of tropospheric ozone is typical of urban areas with high automobile densities, its ready transport to nonurban, rural, and pristine areas can result in adverse effects on the growth of crops, forest trees, and other natural vegetation.

Numerous investigators have studied the adverse effects of ozone on vegetation growth and productivity. Over the past decade several general reviews of the effects of air pollutants on vegetation have appeared that discuss biochemical, metabolic, and physiological consequences of exposure to ozone. Building upon these reviews, the authors of the chapters of this book have summarized the most current research results so that the reader is provided with a clear understanding of (1) what is known about the formation, fate, and effects of surface-level ozone on vegetation, and (2) the considerations that must be addressed when one establishes an air quality standard to protect vegetation.

Based on evidence presented in the literature and discussed in this book, it is clear that the present form of the standard may be inadequate to protect

1

vegetation from ozone exposure. Several of the important issues that require resolution when establishing a standard to protect vegetation are discussed in this book. These issues are

- The distinction between exposure to ambient conditions and dose per se
- The dynamic nature of pollutant concentrations in ambient air (i.e., concentration, fluctuations, and episodicity)
- The selection of suitable measures for defining the "dose" term in dose-response relationships
- The establishment of toxicant-response relationships
- Temporal, environmental, genetic, and developmental influences on such response
- The interactive effects of concurrent and sequential exposure to other stresses, including those caused by other pollutants

The book begins by discussing tropospheric ozone, its formation and fate. The reader is then provided with information that summarizes the characterization of ambient concentrations of ozone, as well as the rationale for selecting ozone exposure indices that bridge the gap between exposure and effects. The experimental methodology for studying the effects of ozone on crops and trees is presented. Chapter 5 discusses the uptake of ozone by vegetation and the important links among exposure, uptake, and dose. Chapters 6 and 7 focus on crop and tree responses to ozone, including the effects of ozone on plant biochemistry and metabolism, the physiology of plant cells and tissues, plant physiology and growth, and exposure-yield response relationships.

The first seven chapters of the book provide a strong platform for better understanding the basic issues relating ozone exposure and vegetation effects. Using this platform of information, Chapter 8 discusses ozone standards and their relevance to protecting vegetation.

In the U. S., the present ozone primary (health effects) and secondary (welfare effects) national ambient air quality standards (NAAQS) are both set at an hourly average concentration of 0.12 ppm, not to be exceeded on more than one day per year. In Chapter 8, it is pointed out that the current form of the standard is believed to be inadequate to protect vegetation. Because there is no requirement that the primary and secondary standards be identical, the chapter focuses on several exposure indices that may be more appropriate for protecting vegetation. In addition, future research required to (1) link exposure with experimental results and (2) develop a secondary standard to protect vegetation is discussed.

Although much is known about surface-level ozone, there are many uncertainties in our understanding of both the physical processes associated with the cycling of tropospheric ozone, as well as the effects of surface-level ozone on vegetation. Recognizing these uncertainties, the authors discuss future research directions as well as provide insight to those interested in better understanding the processes associated with ozone formation and the subsequent reactions and effects that result from ozone exposures. It is the authors' goal that the research information summarized in this book will provide students, instructors, active researchers, and policymakers with the knowledge they require to make the decisions necessary to develop and use research results in the most defensible and appropriate manner.

CHAPTER 2

Tropospheric Ozone: Formation and Fate

William L. Chameides, School of Geophysical Sciences,
Georgia Institute of Technology, Atlanta, GA

James P. Lodge, Boulder, CO

2.1   INTRODUCTION

The discovery of the existence of an allotrope of oxygen called ozone (from the Greek word meaning "smelly") is generally attributed to Schoenbein, de la Rive, Houzeaum, and Soret, working in the mid 19th century (Nicolet, 1979). These scientists established that ozone contains three atoms of oxygen (i.e., O_3), is produced by electrical discharges, and is ubiquitous in the earth's atmosphere. Subsequent investigations by Chappuis (1880), Hartley (1881a), and Huggins and Huggins (1890) established the ultraviolet and visible spectroscopy of the ozone molecule. On the basis of this spectroscopy, Hartley (1881b) was able to deduce that most of the atmosphere's O_3 was located well above the earth's surface and that absorption by O_3 molecules at these higher altitudes prevented solar radiation in the ultraviolet region from penetrating to the earth's surface.

In the early part of this century, the basic photochemical mechanism by which ozone is maintained in the stratosphere was first identified by Chapman (1930). In this reaction sequence, often referred to as the Chapman mechanism, ozone is produced by the ultraviolet photolysis of molecular oxygen at wavelengths less than 242 nm:

$$(R1) \; O_2 + h\nu \rightarrow O + O$$

followed by the three-body reaction between atomic and molecular oxygen:

$$(R2) \; O + O_2(+M) \rightarrow O_3(+M)$$

In (R1) and all other photolytic reactions, the symbol hν denotes a quantum of radiation of appropriate energy: in this case ultraviolet. In (R2) and all other three-body reactions, M is used to refer to an atmospheric molecule, such as N_2 and O_2, which takes part in the reaction and acts to stabilize the molecule produced in the reaction. Ozone loss in the Chapman mechanism is accomplished via

$$(R3) \; O_3 + h\nu \rightarrow O + O_2$$

and

$$(R4) \; O + O_3 \rightarrow 2O_2$$

While the Chapman mechanism was able to successfully explain many of the gross, qualitative features of stratospheric ozone, it became apparent upon closer examination that (R3) and (R4) were inadequate to remove all the O_3 generated by (R1) and (R2). Subsequent studies established the importance of additional gas-phase reactions in which hydrogen oxides, nitrogen oxides, and halogens catalyze O_3 loss (Hampson, 1964; Crutzen, 1970; Johnston, 1971; Stolarski and Cicerone, 1974; Molina and Rowland, 1974). More recently it has been discovered that heterogenous reactions on ice particles can also play an important role by facilitating the catalytic destruction of ozone by chlorine; these reactions are now recognized as responsible for the "ozone hole" observed during the spring over Antarctica (Solomon, 1988).

While the vast majority of the ozone (O_3) in the atmosphere is located at stratospheric altitudes where UV photons dissociate O_2 and trigger the production of O_3 via the Chapman mechanism, ozone levels in the troposphere are still sufficiently large to pose a threat to many forms of plant and animal life. However, in spite of ozone's toxic properties and the fact that its presence in the atmosphere was discovered over a century ago, the processes that control the formation and fate of tropospheric O_3 have been only recently elucidated. While many uncertainties remain in our understanding of the cycle of tropospheric ozone, it has become apparent that tropospheric O_3 is influenced by a highly complex set of photochemical reactions that can be profoundly perturbed by human activities. In this chapter the processes that control the levels of tropospheric O_3 in pristine and polluted environments are reviewed.

2.2   TRANSPORT THEORY OF TROPOSPHERIC OZONE

While initial speculations on the main source of atmospheric O_3 focused on production during thunderstorms (Nicolet, 1979), this theory gave way to the so-called transport theory for tropospheric ozone, once the general features of the O_3 distribution in the atmosphere became established. Because the ultraviolet photons necessary to drive the photochemical reactions in the Chapman mechanism do not penetrate into the troposphere, it was assumed that O_3 was chemically inert in the troposphere and could therefore be treated as a tracer of atmospheric motions. Since tropospheric ozone concentrations generally increase with altitude (see Chapter 3), implying a source from the upper atmosphere, it was concluded that O_3 is produced in the stratosphere, mixed down into the troposphere by tropospheric-stratospheric exchange processes, and eventually lost from the atmosphere by dry deposition at the earth's surface (cf. Junge, 1962).

Many of the general features of the O_3 distribution in the troposphere and its variability appear to support the transport theory for O_3. Numerous investigators have shown that tropospheric ozone levels often tend to be positively correlated with tracers of stratospheric air and negatively correlated with tracers of lower tropospheric air. These correlations have been found to exist on time scales ranging from hours to seasons, and on spatial scales ranging from the mesoscale to the global scale. The tracers that have been studied include [7]Be, a radionuclide that is primarily produced in the stratosphere by cosmic rays; potential vorticity, a dynamical property of air that is typically one or two orders of magnitude larger in the stratosphere than in the troposphere; and water vapor, which is most abundant in the atmospheric boundary layer (Herring, 1966; Danielsen, 1968, Fabian et al., 1968; Danielsen and Mohnen, 1977; Singh et al., 1978; Dutkiewicz and Husain, 1979; Routhier and Davis, 1980).

In addition to the qualitative evidence cited above, a more quantitative examination of the tropospheric O_3 budget also appears to lend credence to the transport theory as described below.

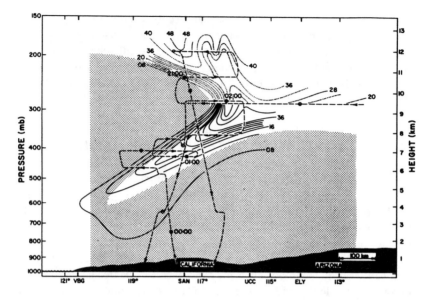

Figure 2.1.   Cross sections of O_3 concentrations (pphmv) during the 0000 GMT March 13, 1978, tropopause-folding event, as inferred from aircraft observations. The thin dashed lines are the Sabreliner flight tracks, the dotted lines are the O_3 analyses from the upper flight track, and the solid lines are the O_3 analyses from the lower flight track. The boundary between the stippled and clear areas indicates the tropopause (after Shapiro, 1980).

2.2.1   Stratospheric-Tropospheric Exchange

The primary mechanism by which air from the stratosphere is injected into the troposphere appears to involve large-scale eddies in the jet stream region. During cyclogenesis the boundary between the troposphere and stratosphere can become deformed vertically downward into and below the core of the jet stream, forming a folded structure. The term "tropopause fold" was coined by Reed and Danielsen (1959) to describe this phenomenon. Within the fold, stratospheric air containing high levels of O_3 and associated stratospheric tracers can penetrate into the troposphere (see Figure 2.1). Provided that the fold is cut off from the stratosphere or that the high O_3 levels within the fold are dissipated by turbulent mixing before the fold is returned to the stratosphere (Shapiro, 1980), this process can represent a sizeable source of O_3 to the troposphere.

The rate at which O_3 is injected into the troposphere via stratospheric-tropospheric exchange processes has been estimated by a number of investigators, including Cunnold et al. (1975), Danielsen and Mohnen (1977), Nastrom and Belmont (1977), Gidel and Shapiro (1980), and Mahlman et al. (1980). These estimates range from a high of about 7 to 9×10^{10} molecules $cm^{-2} s^{-1}$ for the Northern Hemisphere, inferred from observed ozone levels in folds and the number of folding events in the Northern Hemisphere, to a low of 5×10^{10} and 2.5×10^{10} molecules cm^{-2} s^{-1} for the Northern and Southern Hemispheres, respectively, obtained from general circulation model calculations.

2.2.2 Dry Deposition

The rate at which atmospheric trace pollutants, including O_3, are removed by attachment and chemical reaction with the Earth's surface is generally found to be linearly proportional to the species' concentration. The proportionality constant has dimensions of velocity and is referred to as the deposition velocity (V_d). Thus, for O_3 the loss via dry deposition at the earth's surface is

$$\phi_0\left(O_3\right) = V_d \, C_0\left(O_3\right) \qquad (2.1)$$

where $\phi_0(O_3)$ is the downward flux of O_3 to the earth's surface and $C_0(O_3)$ is the ambient concentration of O_3 at some reference height near the surface.

The parameter V_d is a complicated function of a variety of factors, including the meteorological conditions in the surface boundary layer, the physical properties of the surface itself, and the nature of the chemical interaction between O_3 and the surface that ultimately causes the O_3 removal (cf. Baldocchi et al., 1987). Conceptually, the deposition velocity can be thought of as an effective sedimentation velocity that would yield the same removal rate as that of a species that actually settled out with a velocity of V_d, rather than being brought to the surface by turbulence and diffusion. Because of the large variation in the physical and chemical properties of the Earth's surface, V_d can vary widely as a function of the surface type (cf. Aldaz, 1969; Galbally and Roy; 1980; Lenschow et al., 1982). For instance, over the ocean, O_3 deposition appears to be controlled by the rate at which it reacts with halides, such as iodide, within the laminar surface layer (Garland et al., 1980); seawater values measured for V_d typically range from 0.02 to 0.1 cm s^{-1}. Over a forested area or a crop canopy during the daylight hours, on the other hand, O_3 can diffuse through the stomata of the leaves and react with a variety of substances such as ascorbic acid and olefinic compounds produced by the mesophyllic cells (cf. Chameides, 1989; Hewitt et al., 1990) or with the cell material itself (Pell and Weissberger, 1976); as a result the rate-limiting processes controlling O_3 deposition to a vegetative surface during the day are largely micrometeorological, and much larger values of V_d, ranging from 0.2 to 1.0 cm s^{-1}, are typically observed.

By combining observed values for V_d as a function of the surface type with the distributions of these surface types over the globe, it is possible to derive a value for the rate of ozone destruction at the Earth's surface as a function of latitude and longitude. As summarized in Table 2.1, an analysis of this type

Table 2.1. Estimates of Ozone Dry Deposition Rates By Hemisphere

	Deposition rate (10^{10} molecules cm^{-2} s^{-1})	
Investigator	Northern hemisphere	Southern hemisphere
Aldaz (1969)	19	9
Fabian and Junge (1970)	6	3
Fabian and Pruchniewicz (1977)	7	4
Fishman and Crutzen (1978)	13	4
Galbally and Roy (1980)	10	7

implies an average global O_3 deposition rate of about 5 to 10×10^{10} molecules cm^{-2} s^{-1}. Because of the greater concentration of continents in the north (with their relatively large deposition velocities) and the larger levels of O_3 in the north (see Chapter 1), the Northern Hemispheric deposition rate is estimated to be about 1.5 to 3 times larger than that of the Southern Hemisphere.

Comparison of the estimated rates of O_3 injection into the troposphere from the stratosphere and the O_3 deposition to the earth's surface indicate an approximate balance; i.e., both processes appear to have average rates of about 5 to 10 $\times 10^{10}$ molecules cm^{-2} s^{-1}. The fact that these two processes are in approximate balance has been cited as further evidence in favor of the transport theory of tropospheric ozone; because the O_3 loss via deposition at the Earth's surface can be balanced by the source from stratospheric-tropospheric exchange, it has been argued that no other sources or sinks are needed to explain the abundance or distribution of O_3 in the troposphere (cf. Fabian and Pruchniewicz, 1977). Indeed, support for this argument can be found in the calculations of Levy et al. (1985), in which many of the general features of the tropospheric O_3 distribution were successfully reproduced from simulations, using a General Circulation Model in which O_3 was assumed to be an inert tropospheric tracer (see Figure 2.2).

In spite of the successes of the transport theory, however, it became apparent that tropospheric O_3 can also be produced photochemically, at least in polluted areas, when Haagen-Smit (1952) demonstrated that urban photochemical smog is characterized by extremely high levels of O_3. Soon after Haagen-Smit's pioneering work on urban photochemical smog, Frenkiel (1955) proposed that photochemical sources of O_3 could also be significant in the remote troposphere. However, it was not until the 1960s and early 1970s, when the role of free radicals in tropospheric photochemistry was uncovered (e.g., Levy, 1971), that the detailed photochemical mechanisms responsible for the generation of tropospheric O_3 were identified (cf. Leighton, 1961; Heicklen, 1968; Westberg and Cohen, 1969; Altshuller and Bufalini, 1971; Crutzen, 1973; Chameides and Walker, 1973). In the next section the key photochemical reactions involved in the production and destruction of tropospheric O_3 are discussed.

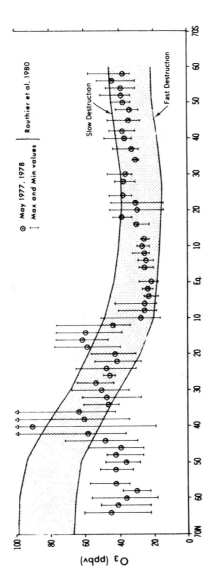

Figure 2.2A. Instantaneous observation by Routhier et al., 1980, of O_3 made in the middle troposphere compared to 500 mbar monthly and zonally averaged O_3 levels calculated with a General Circulation Model in which tropospheric O_3 is treated as an inert tracer. The "slow-destruction" and "fast-destruction" lines indicate model results for different parameterizations for dry deposition at the Earth's surface (after Levy et al., 1985).

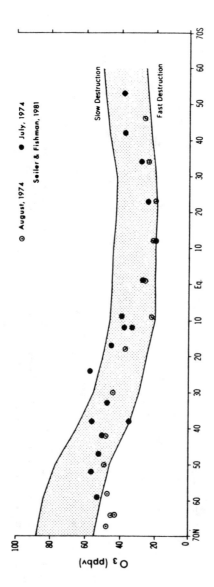

Figure 2.2B. Instantaneous observation by Seiler and Fishman, 1981, of O_3 made in the middle troposphere compared to 500 mbar monthly and zonally averaged O_3 levels calculated with a General Circulation Model in which tropospheric O_3 is treated as an inert tracer. The "slow-destruction" and "fast-destruction" lines indicate model results for different parameterizations for dry deposition at the Earth's surface (after Levy et al., 1985).

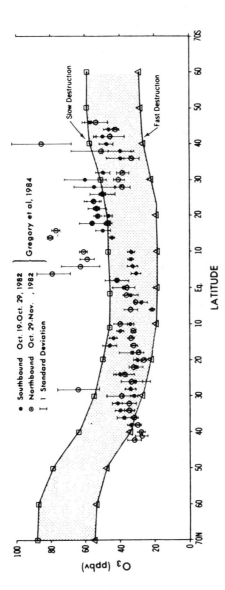

Figure 2.2C. Instantaneous observation by Gregory et al., 1984, of O_3 made in the middle troposphere compared to 500 mbar monthly and zonally averaged O_3 levels calculated with a General Circulation Model in which tropospheric O_3 is treated as an inert tracer. The "slow-destruction" and "fast-destruction" lines indicate model results for different parameterizations for dry deposition at the Earth's surface (after Levy et al., 1985).

2.3　OZONE PHOTOCHEMISTRY

While uncertainties remain in our understanding of tropospheric photochemistry, it has been reasonably well established that O_3 can be photochemically generated in both the polluted and unpolluted troposphere and that the same basic chemical mechanism is responsible for O_3 generation in both kinds of regions. This mechanism, often referred to as the "photochemical smog" reactions, involves the oxidation of hydrocarbons and carbon monoxide in the presence of nitrogen oxides (NO_x) and sunlight (cf. Calvert et al., 1972; Seinfeld, 1989). It is also interesting to note that in the unperturbed atmosphere, these reactions serve to degrade organic compounds to carbon dioxide and water vapor and, thus, are part of the carbon cycle.

Like much of the photochemistry of the troposphere, the production of tropospheric O_3 requires the presence of OH radicals. In the remote troposphere, primary production of OH radicals occurs from the near-UV photolysis of O_3 itself via (R5) and (R6) (Levy, 1971):

$$(R5)\ O_3 + hv \rightarrow O\left(^1D\right) + O_2 \left(\lambda \le 320\ nm\right)$$

$$(R6)\ O\left(^1D\right) + H_2O \rightarrow 2OH$$

In polluted atmospheres, primary production of radicals can also occur from the direct emissions of aldehydes (RCHO, where R = H, CH_3, etc.) and nitrous acid (HONO), followed by their photolysis (Calvert, 1976; Rodgers and Davis, 1989).

Once OH radicals have been generated, photochemical production of O_3 is possible in the troposphere by converting these OH radicals into peroxy radicals in the presence of nitrogen oxides. In the remote, marine troposphere where there are relatively few nonmethane hydrocarbons, O_3 production is often triggered by the oxidation of carbon monoxide (CO); i.e.,

$$(R7)\ CO + OH \rightarrow CO_2 + H$$

$$(R8)\ H + O_2(+M) \rightarrow HO_2(+M)$$

$$(R9)\ HO_2 + NO \rightarrow OH + NO_2$$

$$(R10)\ NO_2 + hv \rightarrow NO + O(\lambda \le 420\ nm)$$

$$(R2)\ O + O_2(+M) \rightarrow O_3(+M)$$

$$\text{Net: } CO + 2O_2 + hv \rightarrow CO_2 + O_3$$

The oxidation of methane (CH_4) can also be an important source of tropospheric ozone:

$$(R11) \ CH_4 + OH \rightarrow CH_3 + H_2O$$

$$(R12) \ CH_3 + O_2(+M) \rightarrow CH_3O_2(+M)$$

$$(R13) \ CH_3O_2 + NO \rightarrow CH_3O + NO_2$$

$$(R14) \ CH_3O + O_2 \rightarrow HCHO + HO_2$$

$$(R9) \ HO_2 + NO \rightarrow OH + NO_2$$

$$2 \times \ (R10) \ NO_2 + hv \rightarrow NO + O$$

$$2 \times \ (R2) \ O + O_2 \ (+M) \ \rightarrow O_3 \ (+M)$$

$$\text{Net: } CH_4 + 4O_2 + 2hv \rightarrow HCHO + H_2O + 2O_3$$

The generation of formaldehyde (HCHO) from the above oxidation sequence can also result in the production of O_3 via

$$(R14) \ HCHO + hv \rightarrow H + HCO \ (\lambda \leq 330 \, nm)$$

and

$$(R15) \ HCO + hv \rightarrow H + CO \ (\lambda \leq 360 \, nm)$$

followed by reactions (R8), (R9), (R10), and (R2).

Similar to the methane oxidation sequence, oxidation of nonmethane hydrocarbons can also result in O_3 generation via reaction sequences such as

$$(R16) \ RH + OH \rightarrow R + H_2O$$

$$(R17) \ R + O_2 \ (+M) \rightarrow RO_2$$

$$(R18) \ RO_2 + NO \rightarrow RO + NO_2$$

$$(R19) \ RO + O_2 \rightarrow HO_2 + R'CHO$$

$$(R9) \ HO_2 + NO \rightarrow OH + NO_2$$

$$2 \times (R10) \ NO_2 + hv \rightarrow NO + O$$

$$2 \times (R2) \ O + O_2 \ (+M) \rightarrow O_3 \ (+M)$$

$$Net: \ RH + 4O_2 + 2hv \rightarrow RCHO + H_2O + 2O_3$$

Note that in the above notation, RH is used to denote a hydrocarbon with R representing a hydrocarbon chain, such as CH_3CH_2 in the case of ethane, and R' is used to denote a chain with one fewer C atom. Similar to the photolysis of HCHO, additional O_3 molecules can be generated from the photolysis of R'CHO produced from (R19).

Note that peroxy radicals are a key intermediate leading to O_3 generation in all the above reaction sequences. However, the production of peroxy radicals in the atmosphere does not guarantee that O_3 will be produced; in order for O_3 to be produced, the peroxy radicals must first react with NO. When the NO_x (NO + NO_2) levels are relatively small, the peroxy radicals can react with other constituents besides NO, and as a result, O_3 is not produced. Under these conditions, photochemistry can actually result in a net loss of O_3 from (R5) and (R6), reactions of O_3 with olefinic hydrocarbons such as propylene and isoprene, and reaction with HO_2 radicals via

$$(R20) \ HO_2 + O_3 \rightarrow OH + 2O_2$$

Given the complexities of the photochemical smog reactions, it is perhaps not surprising that the relationship between ozone photochemistry and the concentrations of NO_x and hydrocarbons can be quite nonlinear. This nonlinearity is illustrated in Figure 2.3, where model-calculated ozone production rates are plotted as a function of assumed NO_x and propylene concentrations, for conditions typical of the tropospheric boundary layer with a nominal background ozone concentration of 25 ppbv. (In these calculations, propylene has been used solely for illustrative purposes; similar results are obtained when other hydrocarbon species, or mixes of hydrocarbon species, are used.) A number of important generalizations can be drawn from the results:

1. The rate of ozone photochemical production increases markedly as one

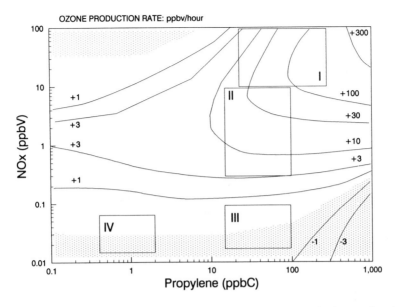

Figure 2.3.   Model-calculated net rates of ozone photochemical production as a function of assumed NO_x and propylene concentration. The calculations were carried out for summertime conditions at mid-latitudes: zenith angle = 40°; temperature = 290 K; dew point = 285 K; O_3 = 20 ppbv; CO = 120 ppbv; and pressure = 1013 mb. The shaded area in the figure denotes the concentration regime for which a net loss of ozone is calculated. The rectangles indicate the ranges of NO_x and propylene concentrations that most closely correspond to the NO_x and total hydrocarbon OH-reactivity typically observed in: (I) Urban and suburban areas; (II) Rural areas in the eastern U.S.; (III) The rain forests in Brazil; and (IV) The remote marine boundary layer. (Note the actual ozone concentration at any given location and time is a function of local meteorological conditions, ozone transport processes, and dry deposition rates as well as the photochemical production rate.)

moves from relatively pristine environments, characterized by NO_x concentrations of 100 pptv or less, to more polluted environments with higher concentrations of NO_x and hydrocarbons.

2. In "low-NO_x" environments (i.e., NO_x < 400 pptv), NO_x plays the central role in controlling ozone photochemistry. As NO_x concentrations vary from 10 to 400 pptv, the photochemistry rapidly increases and shifts from representing a net sink to representing a net source of ozone. By comparison, increasing propylene under "low-NO_x" conditions tends to increase the rate of ozone destruction due to the reaction between ozone and propylene.

3. While the ozone concentration at any given location and time is determined by a combination of meteorological and photochemical factors, the fact that the calculated ozone production is close to 0 for conditions typical of pristine environments suggests that natural levels of NO_x and hydrocarbons are only capable of generating about 20 to 30 ppbv of ozone in the boundary layer.

4. In contrast to "low-NO_x" environments, hydrocarbons can significantly enhance ozone production in more polluted regions where $NO_x > 400$ pptv. In areas with extremely high NO_x concentrations, characteristic of some urban areas, the rate of ozone production can actually become depressed by additional NO_x and, essentially, solely limited by the availability of hydrocarbons.

An additional complication in ozone photochemistry not accounted for in Figure 2.3 arises from the possible effects of halogenated compounds, that may catalyze ozone destruction in the troposphere (Chameides and Davis, 1980; Barrie et al., 1988). It has been proposed, for instance, that bromine-containing free radicals, produced in the atmosphere from the photochemical degradation of marine-biogenic bromoform, may cause large losses of ozone during the early spring in the arctic troposphere.

In the next sections we examine how the photochemical processes discussed above ultimately impact the budget and concentrations of ozone in both remote and polluted regions of the troposphere.

2.4   OZONE BUDGET IN REMOTE, PRISTINE TROPOSPHERE

While O_3 photochemistry is driven by the oxidation of CH_4 and CO over much of the remote troposphere, the oxidation of isoprene and other natural hydrocarbons can dominate in the boundary layer overlying forested regions, where biogenic emissions from trees can be quite significant (Rasmussen, 1972; Zimmerman, 1979, 1980; Lamb et al., 1987; Trainer et al., 1987; Jacob and Wofsy, 1988; Zimmerman et al., 1988). A major uncertainty in O_3 photochemistry in the remote troposphere is associated with our relatively poor understanding of the remote tropospheric nitrogen oxide budget. In the first place, there is a great sparsity of reliable NO_x measurements in the remote troposphere. Furthermore, in those remote areas where measurements have been made, the NO_x levels often fall in the critical 30 to 100-pptv range where, as illustrated in Figure 2.3, ozone photochemistry rapidly switches from supplying a net sink of O_3 to supplying a net O_3 source (cf. McFarland et al., 1979; Kley et al., 1981; Davis et al., 1987; Ridley et al., 1987; Chameides et al., 1987). As a result, the effect of photochemical processes on O_3 in the remote troposphere, and its importance relative to transport processes, remain uncertain.

However, in spite of these uncertainties, there is a growing body of evidence indicating that photochemical reactions comprise a significant fraction of the ozone budget in many regions of the remote troposphere. For instance, although the transport theory predicts that remote tropospheric O_3 should be maximum in the spring when stratospheric-tropospheric exchange is at a maximum, tropospheric O_3 reaches its peak in many remote locations of the Northern Hemisphere in the summer; the summertime peak would appear to suggest the presence of a substratospheric, photochemical source (Fishman et al., 1979; Logan, 1985; Levy et al., 1985).

Fishman and Crutzen (1978) and Fishman et al. (1979) found evidence for a photochemical source of tropospheric O_3 on the basis of the O_3 budgets in the Northern and Southern Hemispheres. Citing calculations that indicate that O_3 loss at the Earth's surface in the Northern Hemisphere is three times that in the Southern Hemisphere (see Table 2.1) and the fact that the rate of stratospheric-tropospheric exchange of O_3 in the Northern Hemisphere is, at most, twice that in the Southern Hemisphere, these authors reasoned that a photochemical source of O_3 must exist in the Northern Hemisphere to maintain the relatively high average O_3 levels found there. They further argued that this Northern Hemispheric O_3 source is caused by the enhanced levels of CO that result from anthropogenic emissions. Observations indicating that O_3 in the remote troposphere can at times be positively correlated with CO (see Figure 2.4) represent additional evidence in support of this argument (Fishman et al., 1980; Chameides et al., 1987).

While a major argument in favor of the transport theory is the fact that tropospheric O_3 levels increase with altitude, Liu et al. (1980) have pointed out that such a distribution is also consistent with a photochemical source of tropospheric O_3. Since NO levels in the remote troposphere are highest in the

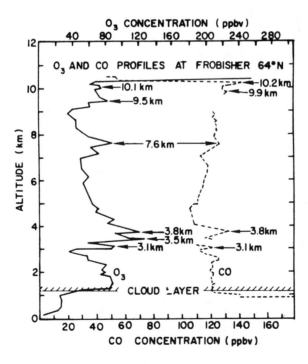

Figure 2.4.   Vertical profiles of O_3 and CO measured during descent into Frobisher, Canada (64° N), July 25, 1974. Arrows indicate significant altitudes at which anomalously high and/or low concentrations of these gases were observed (after Fishman et al., 1980).

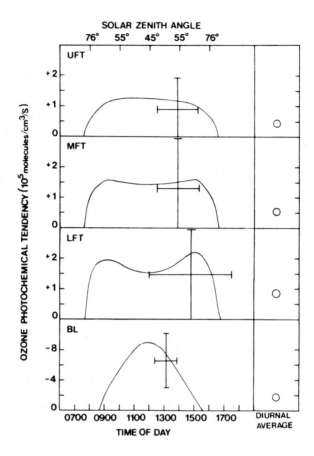

Figure 2.5.   Model-calculated net O_3 photochemical production rate as a function of time of day and solar zenith angle for the boundary layer (BL), lower free troposphere (LFT), middle free troposphere (MFT), and upper free troposphere (UFT). All results based on data gathered over the eastern and central North Pacific Ocean during the NASA GTE/CITE 1 fall 1983 field campaign. The solid lines indicate the results of a time-dependent model based on mean concentrations observed for H_2O, O_3, CO, and NO; the vertical and horizontal lines indicate the averages and standard deviations obtained from averaging the instantaneous simulations of the real-time data; and the open circles on the far right indicate the net production rates obtained by averaging the time-dependent model results (after Chameides et al., 1987).

upper troposphere (Kley et al., 1981), one might expect, on the basis of photochemical theory, that O_3 levels would be highest in the upper troposphere as well. A diagnostic analysis of what is probably one of the most extensive data sets available for studying O_3 photochemistry in the remote troposphere tends to support the arguments of Liu et al. (1980). This data set was obtained during the NASA Global Tropospheric Experiment CITE-1 airborne sampling campaign in the fall of 1983 (Beck et al., 1987). During this campaign, high-resolution measurements of O_3, NO, CO, and H_2O were made over the eastern and central North Pacific Ocean at altitudes ranging from the surface to 10 km. Integration of the model-calculated O_3 production and destruction rates based on these data (see Figure 2.5), yielded a net column loss of O_3 between the surface and 2 km of 3×10^{10} molecules $cm^{-2} s^{-1}$ and a net column production of O_3 between 2 and 10 km of 5×10^{10} molecules $cm^{-2} s^{-1}$ (Chameides et al., 1987). There are two facets of these calculations worth noting: (1) ozone photochemical production occurs primarily in the mid and upper troposphere, as proposed by Liu et al. (1980); and (2) the calculated photochemical column loss and source rates are roughly of the same order of magnitude as the column rates cited above for stratospheric-tropospheric exchange and O_3 deposition. Thus, it would appear that, while transport processes are certainly important to the budget of remote tropospheric O_3, photochemical processes are also important and cannot be ignored.

2.5 OZONE PRECURSOR RELATIONSHIPS IN THE POLLUTED TROPOSPHERE

In polluted regions of the lower troposphere and boundary layer, there is little doubt that photochemical processes cause the vast majority of summertime air pollution episodes characterized by high levels of O_3. Numerous studies using smog chambers have confirmed that O_3 levels well in excess of the 120 ppbv National Ambient Air Quality Standard for O_3 can be generated from irradiated mixtures of air containing hydrocarbons and nitrogen oxides at concentrations typical of regions directly affected by anthropogenic emissions (Jeffries et al., 1976; Carter et al., 1979; Leone et al., 1985).

However, while the role of hydrocarbons and nitrogen oxides as O_3 precursors in urban and rural O_3 episodes appears to be well established, the exact relationship between precursor levels and O_3 generation is highly complex and not yet completely understood. No doubt, it is at least partially because of our incomplete understanding of this complex relationship that efforts in the U.S. over the past decade to develop and implement an effective strategy for controlling O_3 through reductions in specific anthropogenic emissions have had very limited success (Chock and Heuss, 1987; Friedman et al., 1988; Lindsey et al., 1989).

The complexities inherent in trying to develop a strategy for controlling O_3 through hydrocarbon and nitrogen oxide emission reductions are illustrated in Figure 2.6. In this figure, isopleths of peak O_3 concentration, calculated with an air quality model, are indicated as a function of O_3 precursor levels. (Figures of this kind were originally referred to as "Haagen-Smit's" in deference to Hagen-Smit's pioneering work in photochemical smog. More recently they have come to be referred to as "EKMA Diagrams" because they are produced by the Empirical Kinetic Modeling Approach model (commonly used in the U.S. to develop emission control strategies for urban O_3 abatement). Note that the calculated effect on O_3 of reductions in VOC (volatile organic compounds; i.e., hydrocarbons and other organic compounds) and nitrogen oxides can change considerably as the ambient levels of VOC and nitrogen oxides change. Thus, in

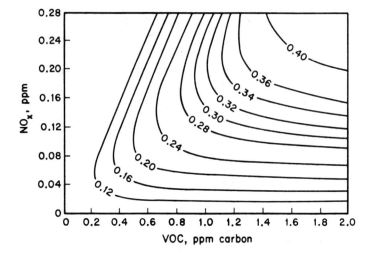

Figure 2.6.   Peak O_3 concentrations in ppmv as a function of initial organic (VOC) and NO_x levels, as generated using the U.S. Environmental Protection Agency's EKMA method under standard default conditions (after Seinfeld, 1989).

order to develop an effective strategy for O_3 control, it is first necessary to accurately define the ambient precursor levels and their reactivity in the region of interest.

In the case of hydrocarbons, this is not a simple task. There are typically hundreds of different organic compounds in the polluted troposphere (cf. Calvert et al., 1972; Seinfeld, 1989). While these compounds generally fall into several distinct classes (e.g., alkanes, alkenes, carbonyls, alcohols, carboxylic acids, and aromatics), reactivity toward OH radicals (and O_3) can vary widely from one compound to another. As a result, the concentration of each organic species needs to be accurately known in order to determine exactly where a given site is located on a diagram like Figure 2.6. Unfortunately, because speciated measurements of ambient hydrocarbon levels are time consuming and expensive, these data are relatively sparse. Another problem arises from the fact that, because of time and computer memory limitations, air quality models cannot simulate the photochemistry of each and every individual volatile organic compound found in the polluted atmosphere; as a result, models simulating urban or regional air quality must adopt approximate chemical mechanisms in which individual organics are lumped together on the basis of their class or bonding or in which a small number of specific organic compounds are chosen as surrogates to represent the oxidation of the entire suite of atmospheric volatile organic compounds (Whitten et al., 1980; Lurmann et al., 1986, 1987; Stockwell, 1986; Gery et al., 1988).

The control of O_3 may be further complicated by the nonlinearities that appear to be inherent in the generation of O_3 from hydrocarbons and nitrogen oxides. This nonlinearity is illustrated in Figure 2.7 from Lin et al. (1988), where the O_3 production efficiency (i.e., the net rate of O_3 production divided by the rate of loss of NO_x) tends to increase as the NO_x levels decrease from tens of ppbv to a few

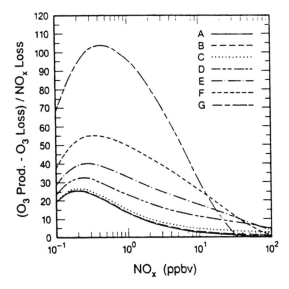

Figure 2.7.   Ozone production efficiency as a function of NO_x concentration for various NMHC/NO_x ratios (in units of ppbv of NMHC as C atoms divided by ppbv of NO_x): curve A, 0.3; curve B, 1; curve C, 4; curve D, 23.4; curve E, 50; curve F, 100; curve G, 300 (after Lin et al., 1988).

tenths of a ppbv. (A similar effect was obtained by Lin et al. (1989) for the net rate of O_3 production divided by the hydrocarbon concentration.) The calculations imply that, because the efficiency of O_3 generation increases as precursor levels decrease, areas with dispersed anthropogenic emissions can experience as much, or more, O_3 as those with concentrated emissions, and that transport processes, by acting to disperse precursors, can further exacerbate the problem. These results perhaps explain why rural areas, with relatively low anthropogenic emission rates compared to urban centers, can experience O_3 episodes as intense as those encountered in cities, even when transport from urban areas does not occur.

A potentially even more severe complication in the O_3 control problem concerns the likelihood that natural hydrocarbons emitted from trees can play a significant role in the generation of high O_3 levels in both rural and urban settings (Trainer et al., 1987; Chameides et al., 1988). Estimates based on measured hydrocarbon emission rates from individual trees and land-use patterns indicate that emissions of biogenic hydrocarbons from trees are comparable to, or larger than, anthropogenic hydrocarbon emissions in the U.S. (Lamb et al., 1987). Surprisingly, biogenic emission rates are also relatively high even in major metropolitan areas, especially those in the southern U.S. where high temperatures and significant tracts of forested land combine to favor natural emissions (see Table 2.2). Because biogenic hydrocarbons such as isoprene and α-pinene are highly reactive (Lurmann et al., 1986), they are very effective precursors of O_3. The model calculations of Trainer et al. (1987) and Chameides et al. (1988) indicate that, provided sufficient levels of nitrogen oxides are present to catalyze

Table 2.2. Natural Hydrocarbon Emission Rates From Selected Urban Areas

Urban area	Emission rate[a]	Investigator
Atlanta	30–65	Westberg and Lamb (1985)
		Chameides et al. (1988)
Tampa/St. Pete.	32	Zimmerman (1979)
Houston	28	Zimmerman (1980)
Baton Rouge	30–50	Chameides (unpubl. manusc.)
Los Angeles	19	Winer et al. (1983)

[a] Emission rates, given in kg km^{-2} day^{-1}, are for a typical summer day and have been spatially averaged over the metropolitan area. For comparison, the anthropogenic emission rate of hydrocarbons over the Atlanta metropolitan area is estimated to be about 30 kg km^{-2} day^{-1}.

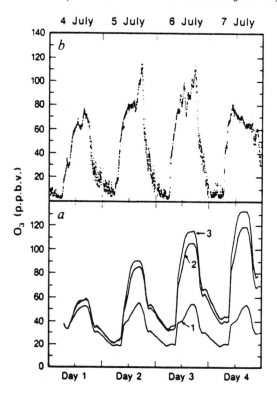

Figure 2.8. (a) Model calculated variations in O_3 during a 4-day period for: (1) CO, CH_4 oxidation only; (2) CO, CH_4, and isoprene oxidation; and (3) CO, CH_4, isoprene, and anthropogenic NMHC oxidation. (b) Observed variation in O_3 at Scotia, PA, July 4–7, 1986 (after Trainer et al., 1987).

the photochemical smog reactions, the oxidation of these natural hydrocarbons can lead to significant levels of O_3 in both the rural and urban atmosphere (see Figures 2.8 and 2.9). Given the ubiquitous presence of reactive hydrocarbons in the atmosphere due to these natural emissions, the calculations suggest that more

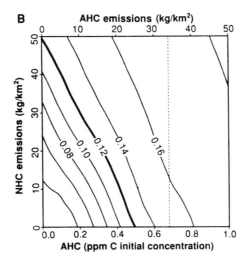

Figure 2.9.   Isopleths of the maximum hourly O_3 levels (in ppmv) calculated for June 4, 1983, in Atlanta, GA, as a function of anthropogenic hydrocarbon (AHC) and natural hydrocarbon (NHC) initial concentrations and daily emission rates. The vertical dashed line indicates the result obtained when AHC is held at present-day levels (after Chameides et al.,1988).

effective control of urban O_3 levels in the U.S. might be obtained by switching from the current strategy that focuses on reductions in hydrocarbon emissions to one that focuses on nitrogen oxide as well as hydrocarbon emission reductions. (In contrast to hydrocarbons, it appears that natural emissions of NO in the U.S. are quite small when compared to anthropogenic emissions (e.g., Logan, 1983).

2.6   CONCLUSION

While many uncertainties remain in our understanding of atmospheric photochemical processes, it is now well established that significant production and destruction of tropospheric O_3 can occur as a result of photochemical reactions. In regions directly influenced by anthropogenic emissions, the relatively high levels of nitrogen oxides catalyze O_3 photochemical generation from the oxidation of anthropogenic and natural hydrocarbons. During the summer months this mechanism can result in frequent air pollution episodes (i.e., photochemical smog) characterized by O_3 levels sufficiently high to threaten human health and harm agricultural crops and forests. The control of O_3 during these air pollution episodes is a complex problem; recent research concerning the role of natural hydrocarbons suggests that an O_3 control strategy that includes nitrogen oxide emission controls may prove to be effective.

In the remote troposphere far removed from anthropogenic influences, O_3 photochemistry can be globally significant and can result in either O_3 production or destruction, depending upon the local levels of NO. As a result, O_3 photochemical production in remote areas appears to be greatest in the upper troposphere where NO levels are high, while O_3 photochemical destruction appears most intense in the remote marine boundary layer where NO levels are typically below 10 pptv. In light of data indicating that O_3 concentrations are on the rise throughout large portions of the troposphere (cf. Logan, 1985; Volz and Kley, 1988), it is somewhat disturbing to note that, based on our current understanding of the atmosphere, a small shift in global NO levels can trigger a major change in the tropospheric O_3 balance in favor of significantly enhanced rates of O_3 photochemical production. The possibility cannot be discounted that rising anthropogenic emission rates of nitrogen oxides have already perturbed global NO levels and are, at least in part, responsible for the apparent increase in tropospheric O_3 concentrations (Isaaksen and Hov, 1987).

REFERENCES

Aldaz, L. (1969) Flux measurements of atmospheric ozone over land and water. *J. Geophys. Res.* **74**, 6942–6946.

Altshuller, A. P. (1983) Review: natural volatile organic substances and their effect on air quality in the United States. *Atmos. Environ.* **17**, 2131–2145.

Altshuller, A. P. and Bufalini, J. J. (1971) Photochemical aspects of air pollution; a review. *Environ. Sci. Technol.* **5**, 39–64.

Baldocchi, D. D., Hicks, B. B., and Camara, P. (1987) A canopy stomatal resistance model for gaseous deposition to vegetated surfaces. *Atmos. Environ.* **21**, 91–101.

Barrie, L. A., Bottenheim, J. W., Schnell, R. C., Crutzen, P. J., and Rasmussen, R. A., (1988) Ozone destruction and photochemical reactions at polar sunrise in the lower Arctic atmosphere. *Nature* **334**, 138–141.

Beck, S. M., Bendural, R. J., McDougal, D. S., Hoell, J. M., Jr., Gregory, G. L., Curfman, H. J., Jr., Davis, D. D., Bradshaw, J. D., Rodgers, M. O., Wang, C. C., Davis, L. I., Campbell, M. C., Torres, A. L., Carroll, M. A., Ridley, B. A., Sachse, G. A., Hill, G. F., Condon, E. P., and Rasmussen, R. A. (1987) Operational overview of NASA GTE/CITE-1 airborne instrument intercomparisons: carbon monoxide, nitric oxide and hydroxyl instrumentation. *J. Geophys. Res.* **92**, 1977–1985.

Calvert, J. G. (1976) Hydrocarbon involvement in photochemical smog formation in Los Angeles atmosphere. *Environ. Sci. Technol.* **10**, 256–262.

Calvert, J. G., Demerjian, K., and McQuigg, R. D. (1972) Photolysis of formaldehyde as a hydrogen atom source in the lower atmosphere. *Science* **175**, 751–752.

Carter, W. P. L., Lloyd, A. C., Sprung, J. L., and Pitts, J. N., Jr. (1979) Computer modeling of smog chamber data: progress in validation of a detailed mechanism for the photooxidation of propene and n-butane in photochemical smog. *Int. J. Chem. Kinet.* **11**, 45–103.

Chameides, W. L. (1989) The chemistry of ozone deposition to plant leaves: the role of ascorbic acid. *Environ. Sci. Technol.* **23**, 595–600.

Chameides, W. L. and Davis, D. D. (1980) Iodine: its possible role in tropospheric photochemistry. *J. Geophys. Res.* **85**, 7383–7398.

Chameides, W. L. and Walker, J. C. G. (1973) A photochemical theory of tropospheric ozone. *J. Geophys. Res.* **78**, 8751–8760.

Chameides, W. L., Davis, D. D., Rodgers, M. O., Bradshaw, J. D., Sandholm, S., Sachse, C., Hill, G., Gregory, G. L., and Rasmussen, R. A. (1987) Net ozone photochemical production over the eastern and central North Pacific as inferred from GTE/CITE 1 observations during fall 1987. *J. Geophys. Res.* **92**, 2131–2152.

Chameides, W. L., Lindsay, R. W., Richardson, J. L., and Kiang C. S. (1988) The role of biogenic hydrocarbons in urban photochemical smog: Atlanta as a case study. *Science* **241**, 1473–1475.

Chapman, S. (1930) A theory of upper atmospheric ozone. *Mem. Roy. Meteorol. Soc.* **3**, 103–125.

Chappuis, J. (1880) Sur le spectre d'absorption de l'ozone. *C. R. Acad. Sci. Paris* **91**, 985-986.

Chock, D. P. and Heuss, J. M. (1987) Urban ozone and its precursors. *Environ. Sci. Technol.* **21**, 1146–1153.

Crutzen, P. J. (1970) The influence of nitrogen oxide on the atmospheric ozone content. *Q. J. R. Meteorol. Soc.* **96**, 320–327.

Crutzen, P. J. (1973) A discussion of the chemistry of some minor constituents in the stratosphere and troposphere. *PAGEOPH* **106–108**, 1385–1399.

Cunnold, D., Alyea, F. N., Phillips, N., and Prinn, R. (1975) A three-dimensional dynamical-chemical model of atmospheric ozone. *J. Atmos. Sci.* **32**, 170-194.

Danielsen, E. F. (1968) Stratospheric-tropospheric exchange based on radioactivity, ozone and potential vorticity. *J. Atmos. Sci.* **25**, 502–518.

Danielsen, E. F. and Mohnen, V. A. (1977) Project DUSTORM Report: ozone transport and meteorological analyses of tropopause folding. *J. Geophys. Res.* **82**, 5867–5877.

Davis, D. D., Bradshaw, J. D., Rodgers, M. O., Sandholm, S. T., and Kesheng S. (1987) Free tropospheric and boundary layer measurements of NO over the central and eastern North Pacific Ocean. *J. Geophys. Res.* **92**, 2049–2070.

Dutkiewicz, B. A. and Husain, L. (1979) Determination of stratospheric ozone at ground level using Be/ozone ratios. *Geophys. Res. Lett.* **6**, 171–174.

Fabian, P. and Junge, C. (1970) Global rate of ozone destruction at the earth's surface. *Meteor. Geophys. Biokim.* **19**, 161–172.

Fabian, P. and Pruchniewicz, P. G. (1977) Meridional distribution of ozone in the troposphere and its seasonal variation. *J. Geophys. Res.* **82**, 2063–2073.

Fabian, P., Libby, W. F., and Palmer, C. E. (1968) Stratospheric residence time and interhemispheric mixing of strontium 90 from fallout in rain. *J. Geophys. Res.* **73**, 3611–3616.

Fishman, J. and Crutzen, P. J. (1978) The origin of ozone of the troposphere. *Nature* **274**, 855–858.

Fishman, J., Seiler, W., and Haagenson, P. (1980) Simultaneous presence of O_3 and CO bands in the troposphere. *Tellus* **32**, 456–463.

Fishman, J., Solomon S., and Crutzen, P. J. (1979) Observational and theoretical evidence in support of a significant *in situ* photochemical source of tropospheric ozone. *Tellus* **31**, 432–446.

Frenkiel, F. N. (1955) Ozone theory in the troposphere. *J. Chem. Phys.* **23**, 2440.

Friedman, R. M., Milford, J., Rapoport, R., Szabo, N., Harrison, K., and Van Aller, S. (1988) Urban ozone and the clean air act: problems and proposals for change. *Staff Paper from OTA's Assessment of New Clean Air Act Issues* April 1988.

Galbally, I. F. and Roy, C. R. (1980) Destruction of ozone at the Earth's surface. *Q. J. R. Meteorol. Soc.* **106**, 599–620.

Garland, J. A., Elzerman, A. W., and Penkett, S. A. (1980) The mechanism of dry deposition of ozone to seawater surfaces. *J. Geophys. Res.* **85**, 7488–7492.

Gery, M. W., Whitten, G. Z., and Killus, J. P. (1988) Development and testing of the CMB-IV for urban and regional modeling. EPA/600/3–88/012 (March).

Gidel, L. T. and Shapiro, M. A. (1980) General circulation model estimates of the net vertical flux of ozone in the lower stratosphere and the implications for the tropospheric ozone budget. *J. Geophys. Res.* **85**, 4049–4058.

Haagen-Smit, A. J. (1952) Chemistry and physiology of Los Angeles smog. *Ind. Eng. Chem.* **44**, 1342–1346.

Hampson, J. (1964) Photochemical behavior of the ozone layer. *Tech Note, Can. Armament Res. Dev. Estab.* TN 1627/64, Valcartier, Canada.

Hartley, W. N. (1881a) On the absorption spectrum of ozone. *J. Chem. Soc.* **39**, 57–60.

Hartley, W. N. (1881b) On the absorption of solar rays by atmospheric ozone. *J. Chem. Soc.* **39**, 111–128.

Heicklen, J. (1968) Gas-phase reactions of alkylperoxy and alkoxy radicals. *Advan. Chem. Ser.* **76**, 23–39.

Herring, W. (1966) Ozone and atmospheric transport processes. *Tellus* **18**, 329–336.

Hewitt, C. N., Kok, G. L., and Fall, R. (1990) Hydroperoxides in plants exposed to ozone mediate air pollution damage to alkene emitters. *Nature* **344**, 56–58.

Huggins, W. and Huggins, Mrs. (1890) On a new group of lines in the photographic spectrum of Sirius. *Proc. R. Soc.* London, **48**, 216–220.

Isaaksen, I. S. A. and Hov φ. (1987) Calculation of trends in the tropospheric concentration of O_3, OH, CO, CH_4, and NO_x. *Tellus* **39B**, 271–285.

Jacob, D. J. and Wofsy, S. C. (1988) Photochemistry of biogenic emissions over the Amazon forest. *J. Geophys. Res.* **93**, 1477–1486.

Jeffries, H. E., Kamens, R., Sexton, K. G., and Gerhardt, A. A. (1976) Outdoor smog chamber studies: light effects relative to indoor chambers. *Environ. Sci. Technol.* **10**, 1006–1011.

Johnston, H. S. (1971) Reduction of stratospheric ozone by nitrogen oxide catalysts from supersonic transport exhaust. *Science* **173**, 517–522.

Junge, C. E. (1962) Global ozone budget and exchange between stratosphere and troposphere. *Tellus* **14**, 363–377.

Kley, D., Drummond, J. W., McFarland, M., and Liu, S. C. (1981) Tropospheric profiles of NO_x. *J. Geophys. Res.* **86**, 3153–3161.

Lamb, B., Guenther, A., Gay, D., and Westberg H. (1987) A national inventory of biogenic hydrocarbon emissions. *Atmos. Environ.* **21**, 1695–1705.

Leighton, P. A. (1961) *Photochemistry of Air Pollution.* Academic Press, New York.

Lenschow, D. H., Pearson, R., Jr., and Stankov, B. B. (1982) Measurements of ozone vertical flux to ocean and forest. *J. Geophys. Res.* **87**, 8833–8837.

Leone, J. A., Flagan, R. C., Grosjean, D., and Seinfeld, J. H. (1985) An outdoor smog chamber and modeling study of toluene — NO_x photooxidation. *Int. J. Chem. Kinet.* **17**, 177–216.

Levy, H. II (1971) Normal atmosphere: large radical and formaldehyde concentrations predicted. *Science* **173**, 141–143.

Levy, H. II, Mahlman, J. D., Moxim W. J., and Liu, S. C. (1985) Tropospheric ozone: the role of transport. *J. Geophys. Res.* **90**, 3753–3780.

Lin, X., Trainer, M., and Liu, S. C. (1988) On the nonlinearity of the tropospheric ozone production. *J. Geophys. Res.* **93**, 15879–15888.

Lindsay, R. W., Richardson, J. L., and Chameides, W. L. (1989) Ozone trends in Atlanta, Georgia: have emission controls been effective? *JAPCA* **39**, 40–43.

Liu, S. C., Kley, D., McFarland, M., Mahlman, J. D., and Levy, H., II. (1980) On the origin of tropospheric ozone. *J. Geophys. Res.* **85**, 7546–7552.

Logan, J. A. (1983) Nitrogen oxides in the troposphere: global and regional budgets. *J. Geophys. Res.* **88**, 10785.

Logan, J. A. (1985) Tropospheric ozone: seasonal behavior, trends, and anthropogenic influence. *J. Geophys. Res.* **90**, 10463–10482.

Lurmann, F. W., Lloyd, A. C., and Atkinson, R. (1986) A chemical mechanism for use in long-range transport/acid deposition computer modeling. *J. Geophys. Res.* **91**, 10905–10936.

Lurmann, F. W., Carter, W. P. L., and Coyner, L. A. (1987) A surrogate species chemical reaction mechanism for urban-state air quality simulation models. Report EPA/600/3-88/014.

Mahlman, J. D., Levy, H., II, and Moxim, W. J. (1980) Three-dimensional tracer structure and behavior as simulated in two ozone precursor experiments. *J. Atmos. Sci.* **37**, 655–685.

McFarland, M. C., Kley, D., Drummond, J. W., Schmeltekopf, A. L., and Winkler, R. H. (1979) Nitric oxide measurements in the Equatorial Pacific Region. *Geophys. Res. Lett.* **6**, 605–608.

Molina, M. J. and Rowland, F. S. (1974) Stratospheric sink for chlorofluoromethanes: chlorine-atom catalyzed destruction of ozone. *Nature* **249**, 810–812.

Nastrom, G. D. and Belmont, A. D. (1977) The variability and transport of ozone near the tropopause. *Trans. Am. Geophys. Union* **58**, 461.

Nicolet, M. (1979) The first years of the study of atmospheric ozone. In *Proceedings of the NATO Advanced Study Institute on Atmospheric Ozone: Its Variation and Human Influences* (eds., Nicolet, M. and Aiken, A. C.). Report No. FAA-EE-80-20, U.S. Department of Transportation, Washington, D.C., 1–6.

Pell, E. J. and Weissberger, W. C. (1976) Histopathological characterization of ozone injury to soybean foliage. *Phytopathology* **66**, 856–861.

Rasmussen, R. A. (1972) What do the hydrocarbons from trees contribute to air pollution? *J. Air. Pollut. Control Assoc.* **22**, 537–543.

Reed, R. J. and Danielsen, E. F. (1959) Fronts in the vicinity of the tropopause. *Arch. Meteorol. Geophys. Biokl.* SA, B11, 1–17.

Ridley, B. A., Carroll, M. A., and Gregory, G. L. (1987) Measurements of nitric oxide in the boundary layer and free troposphere over the Pacific Ocean. *J. Geophys. Res.* **92**, 2025–2048.

Rodgers, M. O. and Davis, D. D. (1989) A UV photofragmentation/laser-induced fluorescence sensor for the atmospheric detection of HONO. *Environ. Sci. Technol.* **23**, 1106–1115.

Routhier, F. and Davis, D. D. (1980) Free tropospheric and boundary layer airborne measurements of H_2O over the latitude range of 58°S to 70°N: comparison with simultaneous high resolution ozone measurements. *J. Geophys. Res.* **85**, 7293–7306.

Seinfeld, J. (1989) Urban air pollution: state of the science. *Science* **243**, 745–752.

Shapiro, M. A. (1980) Turbulent mixing within tropopause folds as a mechanism for the exchange of chemical constituents between the stratosphere and troposphere. *J. Atmos. Sci.* **37**, 994–1004.

Singh, H. B., Ludwig, F. L., and Johnson, W. B. (1978) Tropospheric ozone: concentration and variabilities in clean remote atmospheres. *Atmos. Environ.* **12**, 2185–2196.

Solomon, S. (1988) The mystery of the Antarctic ozone "hole." *Revs. Geophys.* **26**, 131–148.

Stockwell, W. R. (1986) A homogeneous gas phase mechanism for use in a regional acid deposition model. *Atmos. Environ.* **20**, 1615–1632.

Stolarski, R. S. and Cicerone, R. J. (1974) Stratospheric chlorine: a possible sink for ozone. *Can J. Chem.* **52**, 1610–1615.

Trainer, M., Williams, E. J., Parrish, D. D., Buhr, M. P., Allwine, E. J., Westberg, H. H., Fehsenfeld, F. C., and Liu, S. C. (1987) Models and observations of the impact of natural hydrocarbons on rural ozone. *Nature* **329**, 705–707.

Volz, A. and Kley, D. (1988) Evaluation of the Montsouris series of ozone measurements made in the nineteenth century. *Nature* **332**, 240–242.

Westberg, H. and Lamb, B. (1985) Ozone production and transport in the Atlanta, Georgia region. EPA/600/S3-85/013, U.S. Environmental Agency, Research Triangle Park, NC.

Westberg, K. and Cohen, N. (1969) The chemical kinetics of photochemical smog as analyzed by computer. Report No. ATR-70(8107)-1, Aerospace Corporation, El Segundo, CA.

Whitten, G. Z., Hugo, H., and Killus, J. P. (1980) The carbon-bond mechanism: a condensed kinetic mechanism for photochemical smog. *Environ. Sci. Technol.* **14**, 690–700.

Winer, A. M., Fitz, D. R., Miller, P. R., Atkinson, R., Brown, D. E., Carter, W. P. L., Dodd, M. C., Johnson, C. W., Myers, M. A., Neisess, K., Poe, M. P., and Stephens, E. R. (1983) Investigation of the role of natural hydrocarbons in photochemical smog formation in California. Final Report A0-056-32, California Air Resources Board, Statewide Air Pollution Research Center, University of California, Riverside, CA.

Zimmerman, P. R. (1979) Determination of emission rates of hydrocarbons from indigenous species of vegetation in the Tampa/St. Petersburg, Florida area. EPA 904/9-77-028, U.S. Environmental Protection Agency, Atlanta, GA, 1979a.

Zimmerman P. R. (1980) Natural sources of ozone in Houston: natural organics. In *Proc. of Specialty Conference on Ozone/Oxidants — Interactions With the Total Environment* (ed., Frederick, E. R.). Air Pollution Control Association, Pittsburgh, 299–310.

Zimmerman, P. R., Greenberg, J. P., and Westberg, C. E. (1988) Measurement of atmospheric hydrocarbons and biogenic emission fluxes in the Amazon boundary layer. *J. Geophys. Res.* **93**, 1407–1416.

CHAPTER 3

The Characterization of Ambient Ozone Exposures

Allen S. Lefohn, A.S.L. & Associates, Helena, MT

3.1 THE RATIONALE FOR SELECTING EXPOSURE INDICES

This chapter focuses on characterizing ozone exposures received by forested and agricultural regions. Before describing these exposures, it is important to develop the rationale for mathematically summarizing exposure indices that serve as surrogates for dose. This section will focus on many of the issues raised in the literature concerning the selection of appropriate exposure indices.

Hourly averaged ozone concentration information is available for several forested and agricultural areas in the U.S. (Lefohn et al., 1990a). For almost 70 years air pollution specialists have explored alternative mathematical approaches for summarizing ambient air quality information in biologically meaningful forms that can serve as a surrogate for dose for both human health and vegetation effects purposes. Unfortunately, it is difficult to obtain a clear definition of "biologically meaningful." For the purposes of this chapter, I will focus on the historical uses of exposure indices and how they have been applied to exposure-response equations.

Historically, dose has been defined by air pollution vegetation researchers as ambient air quality concentration multiplied by time (O'Gara, 1922). The material in Chapter 5 discusses the uptake processes that are associated with affecting the ozone dose realized by vegetation. The application of the dosage concept for assessing the yield reduction of vegetation has been discussed by several authors. Munn (1970) was one of the first to draw attention to the need to distinguish among the concentration, dosage, and flux of a pollutant. Runeckles (1974) introduced the concept of "effective dose" as the amount or concentration of pollutant that was adsorbed by vegetation, in contrast to that present in the ambient air. Fowler and Cape (1982) developed this concept further and proposed that the "pollutant adsorbed dose" (PAD) be defined in units of $g\ m^{-2}$ (ground area or leaf area) and that it could be obtained as the product of concentration, time, and stomatal (or canopy) conductance for the gas in question. Taylor et al. (1982) suggested internal flux ($mg\ m^{-2}\ h^{-1}$) as a measure of the dose to which plants respond.

Although understanding the effects of specific doses is important when attempting to link exposures with the potential for air pollutant effects, little is known about the concentration, and in some cases, the form of the pollutant that enters the organism. Some suggestions concerning chemical and physical processes that affect vegetation have been provided by Tingey and Taylor (1982). The authors discussed a conceptual model with four sequential components: leaf conductance (gas and liquid phases), perturbation, homeostasis, and injury. Unfortunately, not enough is known to allow us to quantify the links between exposure and dosage. Hence, scientists have explored the relationships between various exposure characteristics and plant response, some of which may be explainable in terms of effects on the actual, internal dosage received.

The search for an exposure index that relates well with plant response has been the subject of intensive discussion in the research community (U.S. EPA, 1986; Lefohn and Runeckles, 1987; Hogsett et al., 1988; U.S. EPA, 1988a; Tingey et al., 1989; Lefohn et al., 1989). Both the magnitude of a pollutant concentration and the length of exposure are important considerations when attempting to develop a realistic exposure index. However, evidence exists in the literature to indicate that the magnitude of vegetation responses to air pollution is more an effect of the magnitude of the concentration than the length of the exposure (Stratmann, 1963; Heck et al., 1966; Heck and Tingey, 1971; Bennett, 1979; Heagle and Heck, 1980; Musselman et al., 1983, 1986; Amiro et al., 1984; Tonneijck, 1984; Hogsett et al., 1985a). For ozone, the short-term, high-concentration exposures were identified by many researchers as being more important than long-term, low-concentration exposures (Heck et al., 1966; Heck and Tingey, 1971; Bicak, 1978; Henderson and Reinert, 1979; Nouchi and Aoki, 1979; Reinert and Nelson, 1979; Bennett, 1979; Stan et al., 1981; Musselman et al., 1983, 1986; Ashmore, 1984; Amiro et al., 1984; Tonneijck, 1984; Hogsett et al., 1985a; Guderian et al., 1985; U.S. EPA, 1986).

Several different types of ozone exposure indices have been proposed as surrogates for dose. A 6-h long-term seasonal mean ozone exposure parameter was used by Heagle et al. (1974). Also, Heagle et al. (1979) reported the use of a 7-h experimental-period mean. The 7-h (0900 to 1559 h) mean, calculated over an experimental period, was adopted as the statistic of choice by the U.S. EPA National Crop Loss Assessment Network (NCLAN) program (Heck et al., 1982). The 7-h daily daylight period was selected by NCLAN because the parameter was believed to correspond to the period of greatest plant susceptibility to ozone pollution. In addition, the 7-h period of each day (0900 to 1559 h) was assumed to correspond to the time that the highest hourly ozone concentrations would occur. However, not all monitoring sites in the U.S. experience their highest ozone exposures within the 0900 to 1559 h 7-h time period (Lefohn and Jones, 1986; Lefohn and Irving, 1988; Logan, 1989). Toward the end of the program, NCLAN redesigned its experimental protocol and applied proportional additions of ozone to its crops for 12-h periods. The expanded 12-h window reflected NCLAN's desire to capture more of the daily ozone exposure. In the published literature, the majority of NCLAN's experiments were summarized using the 7-h experimental-period average.

Lefohn and Benedict (1982) introduced an exposure parameter based on the hypothesis that if the higher ozone concentrations were more important in eliciting adverse effects on agricultural crops than the lower values, then the higher hourly mean concentrations should be given more weight than the lower values. This integrated exposure parameter summed all hourly concentrations equal to and above a threshold level (i.e., 0.10 ppm). The exposure parameter was similar to that used by Oshima (1975), where the difference between the value above 0.10 ppm and 0.10 was summed.

Possibly one of the first studies that examined the importance of elevated ozone exposures on agricultural crop growth was published by Musselman et al. (1983). The authors demonstrated that when the shape of the ozone distribution followed an ambient-type exposure (i.e., increasing throughout the day to a peak and then decreasing), the result was significantly greater injury, less growth, and lower yield than an equivalent "concentration multiplied by time" square wave fumigation. Because peak concentrations of the two treatments were different and greater negative response occurred from the distribution with the greatest peak, the implication was that peaks were important in plant response to ozone. Musselman et al. (1986), in a subsequent experiment, applied equivalent exposures using a square wave and an ambient-type exposure regime with equal peak ozone concentrations. The experiment was conducted at two different peak concentration levels. Although plants responded differently to the exposures tested (i.e., two different peak concentrations), with more injury and reduced growth and yield at the treatment with the highest peak concentration, the shape of the distribution curve had no effect when the peak concentrations were equal. This experiment further suggested that peak ozone concentrations were important. There was no difference in response when the peaks were the same; however, dissimilarities were evident when the peaks were different.

Hogsett et al. (1985a) reported further evidence supporting weighting the peak concentrations more than the lower values. The authors, using field growth chambers, reported that over the period of three cuttings (133 days), alfalfa (*Medicago sativa* L.) growth was reduced more when exposed to an episodic ozone profile than to a regime of equal daily ozone peaks, with essentially identical long-term concentrations multiplied by time values.

Although little is known about the ozone distribution patterns that affect trees, some information does exist. Support for the hypothesis that peak concentrations are an important factor in determining the effects of ozone on trees comes from the work by (1) Hayes and Skelly (1977), who reported on injury to white pine in rural Virginia, and (2) Mann et al. (1980), who described oxidant injury in the Cumberland Plateau. Work by Hogsett et al. (1985b) with two varieties of slash pine (*Pinus elliottii* Engelm.) seedlings suggests that, over similar periods of time, the regimes containing peak ozone concentrations elicit a greater response than regimes containing mostly lower concentrations.

As additional evidence began to mount showing that higher concentrations of ozone should be given more weight than lower concentrations, concerns about the use of a long-term average to summarize exposures of ozone began appearing in the literature (Lefohn and Benedict, 1982; Tingey, 1984; Lefohn, 1984; Lefohn and Tingey, 1985). Specific concerns were focused on the fact that the use of a long-term average failed to consider the impact of peak concentrations. The 7-h seasonal mean contained all hourly concentrations between 0900 to 1559 h; this long-term average treated all concentrations within the fixed window in a similar manner. An infinite number of hourly distributions within the 0900 to 1559 h window could be used to generate the same 7-h seasonal mean, ranging from those containing many peaks to those containing none. Larsen and Heck (1984) pointed out that it was possible for two air sampling sites with the same daytime arithmetic mean ozone concentration to experience different estimated crop reductions. In an important experiment, Hogsett et al. (1985a) showed that plants could display a greater growth reduction with a lower 7-h seasonal mean concentration than with a higher 7-h seasonal mean, if the former was based on fluctuating episodic ozone exposures rather than on repetitive daily exposures. Upon reviewing the existing evidence and the peer-reviewed literature, the U.S. EPA (1988b) concluded that "the weight of the recent evidence seems to suggest that long-term averages, such as the 7-hour seasonal mean, may not be adequate indicators for relating ozone exposure and plant response."

In the late 1980s the focus of attention turned from the use of long-term seasonal means to cumulative indices (i.e., exposure parameters that sum the products of concentrations multiplied by time over an exposure period). As indicated previously, the cumulative index parameters proposed by Oshima (1975) and Lefohn and Benedict (1982) were similar. Both parameters gave equal weight to the higher hourly concentrations, but ignored the concentrations below a subjectively defined minimum threshold (e.g., 0.10 ppm). Besides the cumulative indices proposed by Oshima et al. (1976) and Lefohn and Benedict (1982), other cumulative indices, such as (1) the number of occurrences of daily maximum hourly averaged concentrations greater than a threshold level (Ashmore, 1984), and (2) the use of exponential functions (Nouchi and Aoki, 1979; Larsen and Heck, 1984) to assign unequal weighting to ozone concentrations, were suggested.

Early evidence for testing cumulative indices came from results reported by Oshima et al. (1976). Thompson et al. (1976) reported 33.3 and 42.2% yield reductions in the seasonal yield for El Dorado and Hayden alfalfa, respectively, when exposed to ambient pollution in Riverside, CA in 1974. Using an ozone exposure-response seasonal-yield conversion function for Moapa 69 alfalfa, Oshima et al. (1976) predicted a 32.4% reduction for the same period. Similarly, Lefohn and Benedict (1982), applying their cumulative integrated exposure index, reported fairly good agreement between exposures of ozone and predicted agricultural yield loss in California. The two exposure indices apparently performed well because of the frequent occurrence of high hourly mean ozone concentrations (e.g., ≥0.10 ppm) and, possibly, the short period between episodes. The high frequency of such concentrations was responsible for the large magnitude of the cumulative index as well as the impacts on agricultural crops, and thus, a favorable correlation existed between the index and the agricultural effect.

A possible disadvantage of applying an integrated exposure index, as defined by Oshima (1975) or Lefohn and Benedict (1982), is that the use of an artificial threshold concentration as a cutoff point eliminates any possible contribution of the lower concentrations to vegetation effects. Although this disadvantage may not be important when considering ozone exposures that occur in the South Coast Basin of California, where repeated high concentrations are experienced from day to day and where relatively short periods between episodes occur, it is important when assessing the typical exposures experienced in other parts of the U.S.

Recognizing the disadvantage, Lefohn and Runeckles (1987) suggested a modification to the Lefohn and Benedict (1982) exposure index by weighting individual hourly mean concentrations of ozone and summing over time. Lefohn and Runeckles (1987) proposed a sigmoidal weighting function that was used in developing a cumulative integrated exposure index. The sigmoid function was of the form:

$$w_i = \frac{1}{\left[1 + M \times \exp^{\left(-A \times c_i \right)} \right]} \tag{3.1}$$

where:
M and A = positive arbitrary constants
w_i = weighting factor for concentration
c_i = concentration i

The sigmoidal weighting function was multiplied by each of the hourly mean concentrations; thus, the lower, less biologically effective concentrations were included in the integrated exposure summation.

In addition to proposing the sigmoidally weighted integrated exposure index, Lefohn and Runeckles (1987) proposed an alternative cumulative exposure index. While recognizing that insufficient biological data were available, the authors suggested a potential means for incorporating the length of and time between episodic exposures. They adapted the results described by Mancini (1983), who described a method for calculating the effects of exposure to time-variable concentrations of toxicants on aquatic organisms. Lefohn and Runeckles (1987) proposed separating the integrated exposure parameter into two segments: a sequential exposure (the exposure realized during the episodic event) and a detoxification or respite period (the residual exposure remaining after the episodic event had occurred). The relationship was described as:

$$\text{Integrated Exposure} = \sum_{i=0}^{n} w_i \times c_i + \sum_{j=1}^{n} c_{tj} \exp\left(-k_r \Delta t_j\right) \tag{3.2}$$

where:

w_i = sigmoidal weighting function for hour i

c_i = ambient hourly average concentration at hour i

c_{tj} = last hourly concentration above threshold before episode j terminated

k_r = detoxification rate

Δt_j = time between episode j and j + 1

While this approach still failed to address the problem of the timing of the exposure in relation to the susceptibility of the target organism, it recognized the important role of recovery periods between exposures (Heck and Dunning, 1967; Tingey and Taylor, 1982).

Retrospective studies have been performed using NCLAN data to test the hypothesis that cumulative indices that describe ozone exposures may adequately serve as a dose surrogate for describing exposure/dose-response relationships for agricultural crops (Lefohn et al., 1988a; Lee et al., 1988, 1989, 1991). Lefohn et al. (1988a), using wheat and soybean data sets summarized by Kohut et al. (1986, 1987), compared the use of several exposure indices in describing the relationship between ozone and reduction in agricultural crop yield. NCLAN investigators normally fumigated for only a 7-h period each day. Recognizing that the experimental crops were exposed to ozone during the period that artificial fumigations were not applied, Lefohn et al. (1988a) estimated the 17-h period exposure outside the daily 7-h fumigation window by using air quality data recorded at the ambient plots. Thus, all cumulative exposure indices used in their analysis were determined for a daily 24-h period.

Two of the indices used by Lefohn et al. (1988a) were determined using a sigmoidally weighted function as proposed by Lefohn and Runeckles (1987). One of the two sigmoidally weighted functions used by Lefohn et al. (1988a) in their analysis was designed with an inflection point of approximately 0.065 ppm. Unlike the seasonal average index, the cumulative indices performed well when data were combined over a 2-year period. Lefohn et al. (1988a) reported that, while none of the exposure indices consistently provided a best fit with the models tested, their analysis indicated that exposure indices that weight peak concentrations of ozone differently than lower concentrations of an exposure regime can be used in the development of exposure-response functions.

During the course of their work, the authors demonstrated the advantages of using cumulative indices when comparing vegetation results obtained over different "time-of-exposure" periods. The results reported by Lefohn et al. (1988a) elicited comments from Runeckles (1988), Parry and Day (1988), and Ashmore (1988). In response, Lefohn et al. (1988b) and Lefohn (1988) stressed that the use of long-term seasonal or experimental means has serious flaws because the specific exposure duration component for each experiment is decoupled from the parameter. The use of long-term exposure-period means averages out the "time-of-exposure" element and, thus, makes it difficult to compare results obtained from different experiments operated under different air pollutant exposure durations. For example, Figure 3.1 (adapted from Lefohn et al., 1988a) illustrates the weakness of using a 7-h exposure-period mean when winter wheat data were combined for 1982 and 1983. Lefohn et al. (1988a) quantitatively showed that the cumulative index performed better than the 7-h average when the data were combined for the 2 years. The winter wheat experiment performed in 1982 was operated over a 71-d period, while the winter wheat experiment performed in 1983 lasted only 36 d. Although the cumulative indices used in the experiment appeared to take into consideration the differences in the length of ozone exposure over the 2-year period, the long-term mean was unable to differentiate between the 71- and 36-d "time-of-exposure" periods. Figure 3.1a shows the lack of fit at the highest treatment levels for the 7-h mean, while Figure 3.1b shows a closer fit to the data, using the sigmoidal, W126, cumulative index. The data were ordered properly using the cumulative index. The figure "C" distribution, observed when the 7-h average exposure data were used at the higher treatment region, indicated a lack of correlation between the exposure parameter and yield. Lefohn (1988) concluded that the analysis showed that a long-term average concentration cannot adequately describe the total exposure that a plant receives.

In a more extensive analysis of NCLAN data, Lee et al. (1988) fitted more than 600 exposure indices to response data from 7 crop studies. For most of the NCLAN experiments used in their analyses, they characterized the daily hourly mean ozone concentrations that were recorded over the 7-h period (0900 to 1559 h) by the original experimenters. The alfalfa experiments described by Hogsett et al. (1985a) collected exposure data over a 24-h period, and these data were included in the analysis of Lee et al. (1988). Using mostly the 7-h-windowed data provided by the NCLAN investigators, the "best" exposure indices were those that applied a general phenologically weighted, cumulative-impact (GPWCI) index with a sigmoid weighting on concentration and a gamma weighting function as a surrogate for changes in plant sensitivity over time. Cumulative indices with various threshold values performed as well as the GPWCIs. Lee et al. (1988) reported that mean indices (e.g., 7-h exposure-period means) did not perform well. The authors concluded that the top-performing indices were those whose form (1) accumulated the hourly ozone concentrations over time, (2) used

FIXED 7-H (0900-1559H)

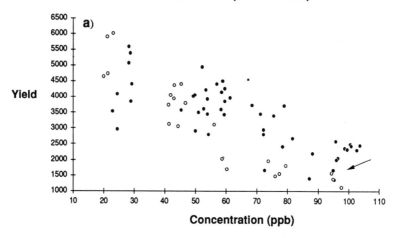

W126

INTEGRATED EXPOSURE INDEX

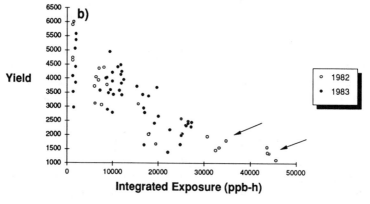

Figure 3.1.   Winter wheat ozone exposure-response relationships, 1982 to 1983 (adapted from Lefohn et al., 1988a).

a sigmoid weighting scheme that emphasized concentrations of 0.06 ppm and higher, and (3) phenologically weighted the exposure. The authors suggested that lower concentrations should be included, but given lesser weight, in the calculation of the exposure index. In a subsequent analysis using NCLAN data, Lee et al. (1989) reported that the phenologically weighted cumulative impact indices, as well as the sigmoidally weighted integrated index, centered at 0.062 ppm, and the cumulative censored indices that integrated hourly average concentrations of 0.06 and 0.07 ppm or higher performed at near optimal levels. The results reported by Lefohn et al. (1988a) and Lee et al. (1988, 1989) have demonstrated that cumulative indices can be used in relating ozone exposure to vegetation effects. Although some cumulative indices offer the advantage of focusing on the higher hourly average concentrations, not all cumulative indices (i.e., the summation of all hourly concentrations: SUM0) achieve this goal (Lefohn et al., 1989).

A consideration when testing for "top-performing indices" is the selection of the model to be used in comparing the performance of exposure indices. Lefohn et al. (1988a) noted that different results were obtained when attempting to identify the optimal exposure parameter. Runeckles and Wright (1989) also showed different results when different models were compared. When comparing indices, it is important that investigators clearly identify and defend the models of choice prior to initiating research to identify the most optimum exposure parameters. In addition, as suggested by Musselman et al. (1988), several ozone summary statistics should be examined, and where no clear statistical preference is indicated among several indices, those most biologically relevant should be selected.

It is important to note that for some experiments, it is difficult to use regression techniques to identify "top-performing" exposure indices similar to the techniques applied using data from the NCLAN program (e.g., Lee et al., 1988, 1989). Lefohn et al. (1992) summarized data obtained by exposing loblolly pine seedlings at an Auburn University site to ozone at several levels. The authors reported that the ozone exposures in the treatment chambers (i.e., charcoal-filtered, nonfiltered, nonfiltered times 1.7, and nonfiltered times 2.5) were distinctly different in their levels of application. This resulted in the observation that, when a given ozone exposure index was compared across the ozone treatments experienced in the chambers, the values of the surrogate maintained the order implied in the treatments. Figure 3.2 shows that the highest treatments were associated with the highest values of the four exposure indices tested (adapted from Lefohn et al., 1992). Thus, for this case, because of the nature of the ordering, the use of mathematical approaches to identify "top-performing" exposure indices would not provide a clear separation of performance for each of the indices.

Recognizing this problem, Lefohn et al. (1992) proposed an alternative approach. They investigated whether the absolute value associated with a specific exposure index, calculated from hourly averaged ozone data obtained from a specific experimental chamber, was similar to values calculated from ambient ozone data reported for monitoring sites. For a set of indices (SUM0, SUM06, W126, and SUM08), the exposures observed in the Auburn charcoal-filtered chambers were compared with those reported from two remote and isolated ozone monitoring sites. Because the ozone exposures in these chambers were so low, for comparison the authors selected ozone monitoring sites located at the South Pole (Antarctica) and Pt. Barrow (AK). Lefohn et al. (1992) reported that the values of the SUM0 index measured in the Auburn charcoal-filtered treatments were approximately 50% less than those experienced at either the

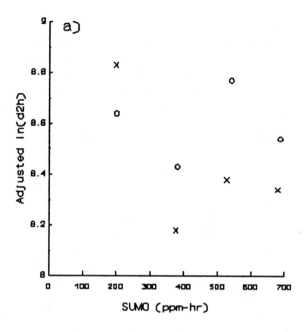

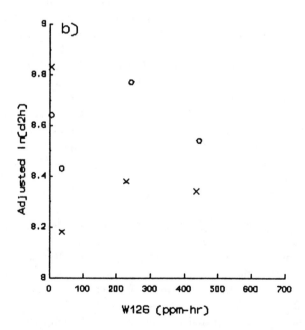

Figure 3.2A.   A comparison of the responses vs. the exposure indices for the (a) SUM0; (b) sigmoidally weighted cumulative index (W126) o = block 1, × block 2 (Lefohn et al., 1992)..

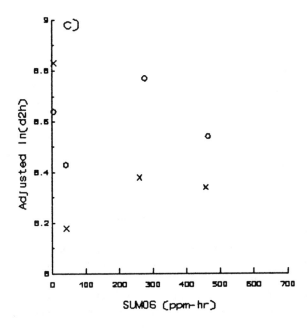

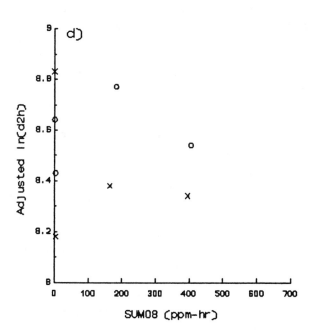

Figure 3.2B. A comparison of the responses vs. the exposure indices for the (c) SUM06; (d) SUM08 for family 91 at pH 4.3, o = block 1, x = block 2 (Lefohn et al., 1992).

South Pole or Pt. Barrow. The value of the SUM0 index was less in the Auburn charcoal-filtered chamber than at the other two sites because the charcoal-filtered chambers experienced approximately more than 90% of their hourly average concentrations below 0.03 ppm. Both the South Pole and Pt. Barrow sites experienced approximately 50% of their hourly average concentrations near 0.03 ppm (Lefohn et al., 1990b). However, there were more occurrences of hourly average concentrations greater than or equal to 0.06 ppm at the Auburn site than at the South Pole or Pt. Barrow sites. Thus, the SUM0 index failed to focus on the higher concentrations. On the other hand, Lefohn et al. (1992) reported that the values of the SUM06 and the W126 exposure indices, calculated using the charcoal-filtered ozone data, were higher than those calculated using the data from the South Pole and Pt. Barrow sites. The SUM06 and W126 indices focused more on the greater number of higher hourly average concentrations that occurred in the charcoal-filtered chambers.

In addition to using the charcoal-filtered data, Lefohn et al. (1992) used hourly ozone data measured at the ambient Auburn plots to calculate SUM0 values. The authors identified a set of "clean" ozone monitoring sites that experienced SUM0 values similar to the Auburn site. Although the SUM0 values were similar, the ambient Auburn plots experienced higher maximum hourly average concentrations, with more hourly average concentrations above 0.07 ppm, than did the "clean" sites. Accordingly, the values of the SUM06, W126, and SUM08 indices were greater at the ambient Auburn plots than at the "clean" ozone monitoring sites and ordered themselves properly. Thus, Lefohn et al. (1992) were able to eliminate the SUM0 exposure index from further consideration, using an alternative to the regression approach, which was unable to separate out any of the four indices.

There are weaknesses associated with using most exposure indices. Most are insensitive to the sensitivity of the plant. Unfortunately, the sensitivity of vegetation as a function of the time of day or period of growth has not been well defined. Because of this, it is premature to subjectively define weightings as a function of sensitivity. However, indications are that sensitivity may be an important factor for future considerations. Krupa and Nosal (1989) have reported that time-series, spectral-coherence analysis showed that at high ozone flux density, the cumulative integral of exposure appeared to perform well when predicting the effects of ozone on alfalfa. However, the authors pointed out that at lower flux density, the median hourly ozone concentration provided the highest coherence with alfalfa height growth. The authors recommended that, because the exposure parameter providing the best description of crop growth response varied with the ozone flux density, exposure terms that take into consideration such things as the sensitivity as a function of the specific growth stage will have to be identified and included in the final form of any crop growth equation. Showman (1991) reported further indications that sensitivity may be important. For field surveys in the midwestern U.S. in 1988, ozone levels were high, but injury to vegetation was low due to drought stress. In 1989 ozone exposures were much lower than in 1988, and optimum growing conditions resulted in greater foliar injury.

In addition, as described in the literature, the distribution patterns of the hourly average concentrations for some high-elevation and low-elevation sites are different. Most cumulative-type and other exposure indices cannot adequately describe the subtle differences in the two different types of exposure regimes. However, because there is little biological evidence published illustrating whether differences in effects exist when vegetation is exposed to these distinct distribution patterns of exposure, the use of cumulative-type indices should not be discouraged. These indices offer viable alternatives to the seasonal mean indices that have fallen out of favor with the U.S. EPA (1988a,b).

Besides sensitivity, the majority of dose surrogates used today do not address (1) the amount and chemical form of the pollutant that enters the target organism, (2) the length of the exposure within each episodic event, or (3) the time between exposures (i.e., the respite or recovery time). Taylor and Norby (1985) have discussed the possible importance of the time between exposures (i.e., respite) and its relevance to predicting the effects of ozone on trees. Only the work of Nouchi and Aoki (1979) addresses the question of repetitive exposures. It is unclear how important sensitivity and the amount and chemical form of the pollutant that enters the target organism are in an overall weighting scheme when predicting vegetation effects. If both the sensitivity of the target organism and the actual dose that enters the organism are as important as ambient air pollutant exposure, then a given pollutant exposure will elicit varying biological responses at different times for the same crop, as conjectured by Krupa and Teng (1982).

Techniques other than indices that accumulate exposures over time have been used to investigate varying exposure patterns. Investigators have utilized diagrams that illustrate composite diurnal patterns as a means to describe qualitatively the differences of ozone exposures between sites (Lefohn and Jones, 1986; Böhm et al., 1991). Although it might appear that composite diurnal pattern diagrams could be used to quantify the differences of ozone exposures between sites, Lefohn and Benkovitz (1990) caution their use for this purpose. The composite diurnal patterns are derived from long-term average calculations of the hourly concentrations, and the resulting diagram cannot adequately identify, at most sites, the presence of high hourly average concentrations and, thus, may not adequately be able to distinguish ozone exposure differences among sites.

Models similar to the one described by Benson et al. (1982) do address some of these considerations. Krupa and Kickert (1987) have discussed the possible importance of plant growth stage modifying the end response that is associated with biotic and abiotic stress and have discussed the concept of a biological time clock. As discussed by Lefohn and Runeckles (1987), should the sensitivity, amount, and chemical form of the pollutant that enters the target be less important than the pollutant exposure in determining biological response, then an exposure parameter that gives greater weight to higher, rather than lower, concentrations and takes into consideration the length of episodes, as well as the time between episodes, may adequately predict injury and yield effects. Clearly, additional research is required to describe the relationships between exposure and dose.

Several of the ozone exposure indices that have been utilized in exposure-response models are listed in Table 3.1. Descriptions of the above indices can be

Table 3.1. List of Ozone Exposure Indices[a]

One event	Mean	Cumulative	Concentration weighting	Multicomponent	Respite time
HDM2	M1	TOTDOSE	SUM06	PWCI	DAYBET08
P7	M7		SUM07		DAYBET10
P1	EFFMEAN		SUM08		
PER90			SUM10		
PER95			AOT08		
PER99			AOT10		
			TIMPACT		
			ALLOMETRIC		
			SIGMOID		
			(e.g., W126)		
			HRS08		
			HRS10		
			NUMEP08		
			NUMEP10		
			AVGEP08		
			AVGEP10		

[a] Adapted from Lee et al. (1989).

found in either Lee et al. (1989) or Lefohn et al. (1988a). Based on evidence published in the literature, as well as special analytical studies sponsored by the U.S. EPA (1988a,b), many in the research community have concluded that the use of cumulative indices to describe exposures of ozone for predicting agricultural crop effects appears to be a more rational approach than the use of long-term seasonal averages.

It may appear that one could use any index to characterize an ozone exposure. However, the choice of appropriate indices is important when attempting to describe ozone exposures on a regional basis (Lefohn and Benkovitz, 1990). The selection of an index can alter the appearance of region-wide ozone exposures. For example, Figure 3.3 shows the comparison of ozone exposures that occurred in 1985 in the northeastern U.S. with those that occurred in 1988 (reproduced from work reported by Lefohn et al., 1990a). Using the long-term 7-month average of the daily 7-h (0900 to 1559 h) average concentration index, although the ordering is correct, large differences are not readily apparent between the 2 years. However, when the number of hourly average concentrations greater than or equal to 0.08 ppm index is used, the figure shows that 1988 is the year in which much higher episodic exposures occurred. Thus, for assessing the potential effects of ozone on vegetation on a regional basis, it is important that careful attention be given to the selection of the exposure index.

Although it would be helpful if one could use actual ozone dose measurements to predict cause-and-effect relationships, the current state of science is limited mostly to relating ozone exposure measurements to vegetation effects. As will be described in Chapter 8, for establishing ozone standards that protect vegetation, ozone exposure information, and not dose, is the practical manner in which to proceed. Therefore, based on the state of knowledge concerning the best

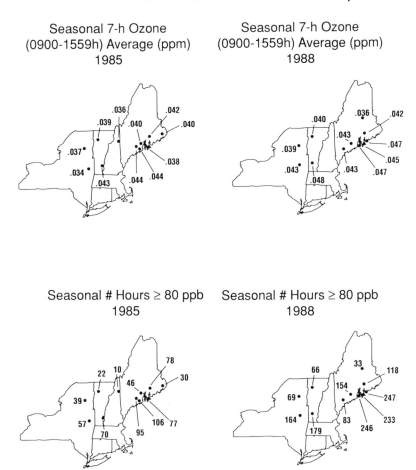

Figure 3.3.   Comparison of northern New England ozone exposures that occurred in 1985 and 1988 (Lefohn et al., 1991).

available exposure indices for relating exposure to vegetation effects, the emphasis in this chapter is on ozone exposure indices that give greater weight to the higher hourly average concentrations. Where appropriate, the long-term seasonal averages (i.e., 7- and 12-h averages) are included only for historical purposes.

3.2   SPATIAL ANALYSES BASED ON ACTUAL MEASUREMENTS FROM LONG-TERM AIR QUALITY MONITORING NETWORKS

Figures 3.4 and 3.5 show the nonattainment regions of the U.S. as of December 1989 and December 1990, respectively, where ozone concentrations

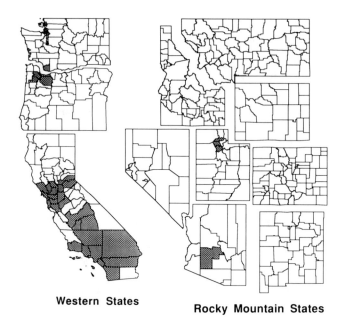

Western States

Rocky Mountain States

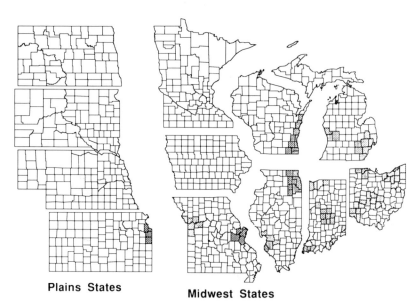

Plains States

Midwest States

Figure 3.4A. Nonattainment counties (identified in dark shading) for ozone for 1986–1988 (Lefohn et al., 1990a).

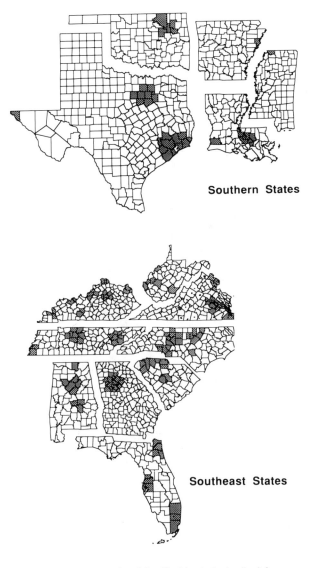

Figure 3.4B. Nonattainment counties (identified in dark shading) for ozone for 1986–1988 (Lefohn et al., 1990a).

exceed the current federal standard. The U.S. EPA (1990) has noted that ozone continues to be the most pervasive ambient air pollution problem in the U.S., with 101 and 96 areas failing to meet the ozone national ambient air quality standard for 1986 to 1988 and 1987 to 1989, respectively. Tables 3.2 and 3.3 summarize the total forested and agricultural acreage, respectively, located in ozone nonattainment counties (for the period 1986 to 1988) in the U.S., as well as the total forest and agricultural acres in each state.

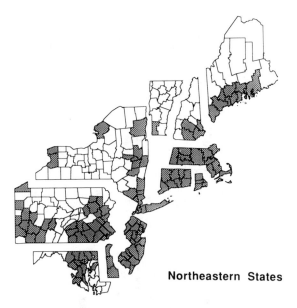

Northeastern States

Figure 3.4C. Nonattainment counties (identified in dark shading) for ozone for 1986–1988 (Lefohn et al., 1990a).

It is difficult to generalize about ozone exposures across the U.S. Ozone exposures experienced at a specific site in one year may not resemble those experienced at the same site the following year. Large year-to-year variations of ozone concentrations, measured at sites influenced by urban sources, do occur (U.S. EPA, 1989c; Lefohn et al., 1990a). For example, ozone exposures at some urban-influenced sites in the U.S. were higher in 1983 than in any other year for the period 1978 to 1985 (Lefohn and Pinkerton, 1988; U.S. EPA, 1989c). In 1988, ozone exposures experienced at many locations across the U.S. were higher than in previous years (U.S. EPA, 1989c, 1990, 1991). On the other hand, the EPA (1990) notes that monitoring information suggests that 1989 ozone levels were approximately 15% lower than 1988 levels; the 1989 ozone levels were comparable to those experienced in 1986. The downward trend is attributed to the cooler and wetter weather in 1989.

Sites located at higher elevations experience different ozone exposure profiles than those located near sea level (Berry, 1964; Stasiuk and Coffey, 1974; Mohnen et al., 1977; Miller et al., 1986; Lefohn and Jones, 1986; Lefohn and Mohnen, 1986). Many of the lowest hourly mean concentrations of ozone are near 0.04 ppm. The distribution of the highest concentrations at the higher elevation sites is not distinctly different from the distribution observed at the lower elevation sites. Thus, the difference in distribution is found mostly in the lower concentration region (Lefohn et al., 1990c; Winner et al., 1989).

It is possible to identify regions of the U.S. that are at higher risk from ozone exposures than other regions. For example, Lefohn and Pinkerton (1988)

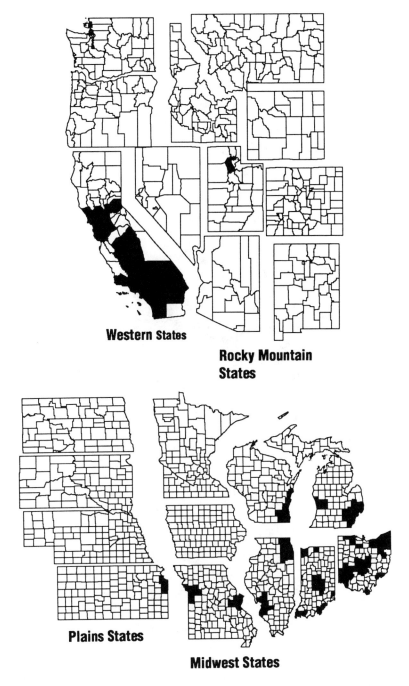

Figure 3.5A. Nonattainment counties (identified in dark shading) for ozone for 1987–1989.

Northeastern States

Southeast States

Southern States

Figure 3.5B.   Nonattainment counties (identified in dark shading) for ozone for 1987–1989.

Table 3.2.   The Total Forested Acreage that is Located in Ozone Nonattainment Counties in the U.S.[a]

Region/state	Total forest acres in the county (1,000s)	Total forest acres in the state (1,000s)
NORTHEAST		
Connecticut	1,860.8	1,860.8
Maine	3,295.0	17,607.4
Massachusetts	2,952.3	2,952.3
New Hampshire	1,434.2	4,987.2
New York	3,982.3	18,182.0
Rhode Island	404.2	404.2
Vermont	380.0	4,544.4
MID-ATLANTIC		
Delaware	380.6	380.6
Maryland	1,215.3	2,653.2
New Jersey	1,829.1	1,829.1
Pennsylvania	5,861.0	16,825.9
West Virginia	1,667.9	11,632.6
SOUTH		
Alabama	2,701.8	21,658.8
Arkansas	35.5	16,615.6
Florida	1,989.4	16,549.0
Georgia	54.4	24,242.4
Louisiana	1,301.3	13,882.7
Mississippi	93.6	16,990.1
North Carolina	2,953.4	18,952.9
South Carolina	2,216.8	12,257.0
Tennessee	1,841.8	12,879.0
Virginia	2,063.8	15,968.4
MIDWEST		
Illinois	420.7	4,265.5
Indiana	637.9	4,439.2
Kentucky	1,338.9	12,160.8
Michigan	928.8	18,368.8
Missouri	795.9	12,919.1
Ohio	2,292.6	7,120.1
Wisconsin	324.9	15,351.3
WEST		
California	12,190.0	38,947.0
PACIFIC NORTHWEST		
Oregon	2,822.0	29,473.0
Washington	232.0	21,520.0
ROCKY MOUNTAIN		
Arizona	879.1	21,730.4
Utah	371.5	18,535.7

[a]   Lefohn et al. (1990a).

Table 3.3. The Crop Sales and Agricultural Harvested Acreage that is Located in Ozone Nonattainment Counties in the U.S.[a]

Region/State	Crop sales county ($1,000s)	Total Crop sales state ($1,000s)	Harvested acres (county)	Total Harvested acres (state)
NORTHEAST				
Connecticut	102,057	102,057	171,229	171,229
Delaware	110,276	110,276	499,986	499,986
Maine	26,889	137,290	160,241	457,076
Maryland	168,419	339,429	756,997	1,528,994
Massachusetts	135,360	135,360	196,069	196,069
New Hampshire	20,286	26,207	49,756	116,613
New Jersey	315,821	315,821	570,031	570,031
New York	217,933	642,310	824,654	4,428,959
Pennsylvania	615,498	751,291	2,541,811	4,363,049
Rhode Island	18,139	18,139	21,252	21,252
Vermont	1,182	20,054	12,006	547,848
APPALACHIAN				
Kentucky	304,903	1,253,934	909,775	4,835,631
North Carolina	277,571	1,884,342	772,247	4,571,159
Tennessee	175,293	807,921	829,094	4,548,895
Virginia	97,174	626,544	439,810	2,779,282
West Virginia	4,978	48,414	57,072	576,392
SOUTHEAST				
Alabama	76,752	606,428	430,113	3,265,361
Florida	936,369	2,516,069	685,019	2,643,147
Georgia	20,532	1,175,065	141,336	4,760,285
South Carolina	40,843	601,021	304,431	2,474,025
DELTA STATES				
Arkansas	70,689	1,355,703	342,809	5,140,716
Louisiana	63,563	935,173	387,245	4,628,001
Mississippi	19,345	1,102,099	104,045	5,799,772
CORN BELT				
Illinois	530,479	5,020,577	2,233,856	23,008,246
Indiana	413,811	2,237,382	1,890,785	12,136,310
Missouri	159,483	1,523,053	1,223,918	12,725,381
Ohio	684,873	1,847,101	3,390,728	10,396,324
LAKE STATES				
Michigan	304,160	1,364,666	1,198,900	7,250,561
Wisconsin	180,251	937,848	1,357,974	10,050,692
PACIFIC				
California	5,760,474	8,158,495	4,746,894	7,203,825
Oregon	200,072	935,459	417,794	3,305,714
Washington	7,060	1,714,018	34,643	5,278,802
MOUNTAIN				
Arizona	317,880	806,847	370,978	1,033,130
Utah	24,449	129,948	72,313	1,118,486

[a] Lefohn et al. (1990a).

**Table 3.4. Summary of the High, Medium, and Low Years for
Ozone Exposures Occurring at Rural Sites Located
in Agricultural and Forestry Regions
of the U.S. (1979–1988)[a]**

Forestry Regions	High	Medium	Low
South	1988	1986	1982
Midwest	1988	1985	1986
West	1979	1984	1982
Pacific Northwest	1986	1984	1983
Plains	1988	1985	1982
Northeast	1988	1984	1986
Mid-Atlantic	1988	1984	1986
Rocky Mountains	1987	1986	1984
Agricultural Regions	**High**	**Medium**	**Low**
Pacific	1979	1984	1982
Mountain	1987	1986	1984
Northern Plains	1988	1985	1982
Southeastern Plains	1988	1985	1982
Lake States	1988	1985	1986
Corn Belt	1988	1985	1986
Northeast	1988	1984	1986
Southeast	1987	1986	1982
Appalachian	1988	1985	1981
Delta	1987	1986	1982

[a] Lefohn and Shadwick (1991).

characterized ozone levels across the U.S., using hourly averaged ambient ozone
monitoring data for an 8-year period, 1978 to 1985, for 8 forested areas. The
analysis focused on the annual number of occurrences of hourly averaged ozone
concentrations that were greater than or equal to 0.07, 0.08, and 0.10 ppm during
the growing season (April to October), as well as during the early (April to June)
and late (July to October) portions of the growing season. On the average the
authors reported that elevated ozone concentrations occurred more often in the
Piedmont/Mountain/Ridge-Valley and Ohio River Valley areas than in the
Pacific Northwest, upper Great Lakes, and northern New England/New York
areas. Air quality in the Southern California area was not investigated by Lefohn
and Pinkerton (1988). In the eastern U.S., 1978, 1980, and 1983 were generally
the years with the most occurrences of elevated ozone concentrations. In these
years the later part (July to October) of the growing season experienced more
elevated concentrations than the earlier part. The area-wide ozone statistics were
derived mainly from urban-oriented ozone monitoring stations, which the
authors believed would likely overestimate the number of elevated ozone
concentrations that would be observed in the majority of commercial forests.

Lefohn et al. (1991) have characterized ozone exposures across the U.S.
between 1979 and 1988 by agricultural and forest regions. High, medium, and
low years of exposure for each of the regions were summarized. Table 3.4 lists

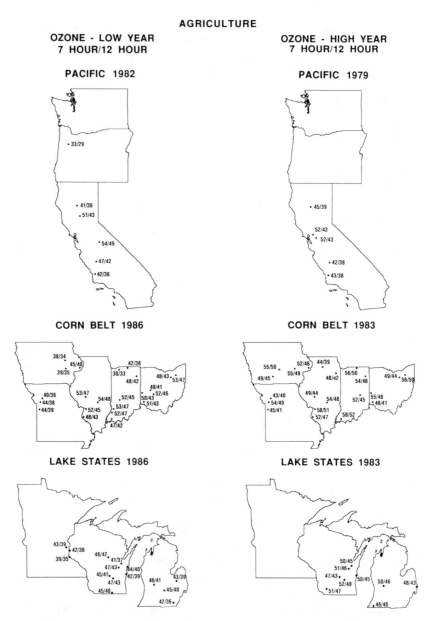

Figure 3.6.   Ozone characterization for the Pacific, Corn Belt, and Lake States agricultural regions showing the 7-/12-h values (units in ppb) for low- and high-exposure years (Lefohn et al., 1991).

the high, medium, and low years for ozone exposures in agriculture and forest regions, as reported by Lefohn and Shadwick (1991). Figures 3.6 to 3.9 summarize the highest and lowest ozone exposures by region (from Lefohn et al., 1991). Shadwick (1989) has estimated the standard errors for the above exposure indices. The standard errors for (1) the 7-month average of the daily maximum 7-h average concentration and (2) the 7-month average of the daily maximum

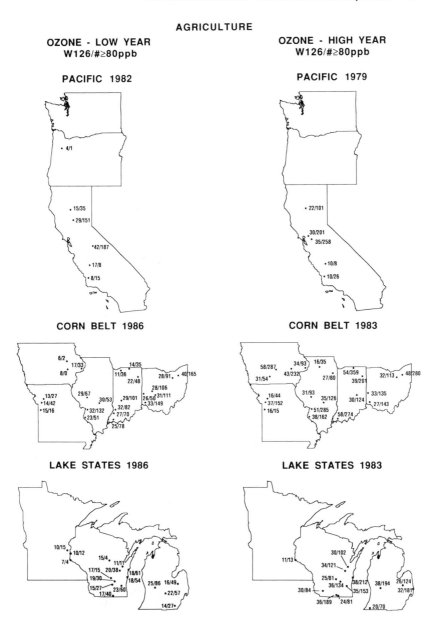

Figure 3.7. Ozone characterization for the Pacific, Corn Belt, and Lake States agricultural regions showing the W126/# ≥80 ppb values for low- and high-exposure years (Lefohn et al., 1991).

12-h average concentration are 0.12 ppb and 0.08 ppb, respectively. The standard error for the 7-month cumulative W126 integrated exposure parameter ranges from 6.5 to 22 ppb-h. The standard error for the 7-month accumulation of the number of hourly average concentrations equal to or greater than 0.08 ppm is near zero.

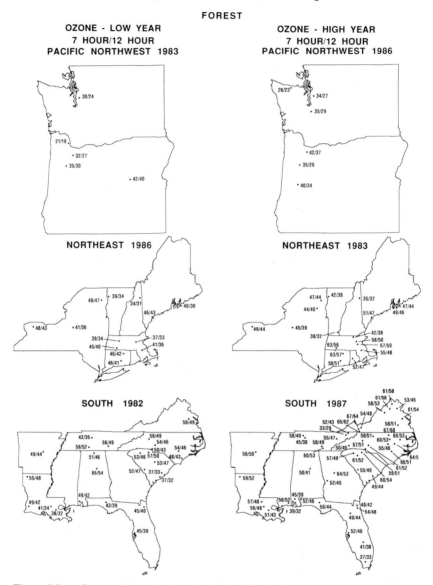

Figure 3.8.   Ozone characterization for the Pacific Northwest, Northeast, and South forest regions showing the 7-/12-h values (units in ppb) for low- and high-exposure years (Lefohn et al., 1991).

Although the highest ozone exposures from 1979 to 1987 occurred in 1983 in some forest regions of the U.S. (i.e., Northeast, Midwest, and Mid-Atlantic), the hot, dry summer of 1987 is associated with the highest ozone exposures for the South and Rocky Mountain regions. A similar pattern was observed for the agricultural regions. For many, 1983 was a high year (e.g., northern plains, lake states, corn belt, and the Northeast). In those areas where 1987 was hot and dry, 1987 ozone exposure was high (i.e., the mountain, Southeast, Appalachian, and

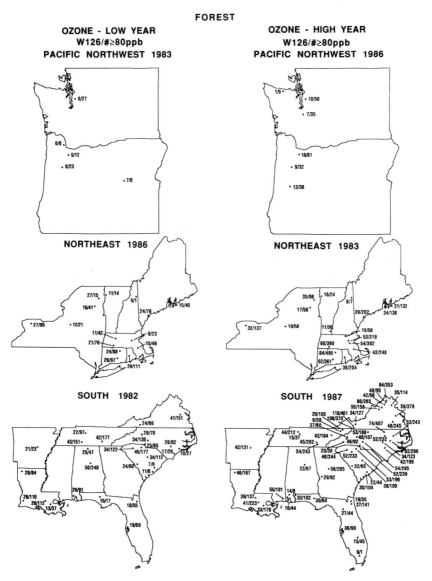

Figure 3.9. Ozone characterization for the Pacific Northwest, Northeast, and South forest regions showing the W126/# ≥80 ppb values for low- and high-exposure years (Lefohn et al., 1991).

delta regions). The year 1988 was also hot and dry, and several regions across the U.S. experienced high ozone exposures (U.S. EPA, 1989c, 1990, 1991; Aneja et al., 1990; Pagnotti, 1990; Lefohn and Benkovitz, 1990). For several regions of the country over the last 10 years, the highest ozone exposures occurred in 1980, 1983, and 1988 (U.S. EPA, 1991).

Areas downwind from large urban and industrial centers experience elevated ozone exposures. The highest ozone exposures in the U.S. occur in the California

South Coast Air Basin. The highest hourly average concentrations in this area of the country can exceed 0.30 ppm during the April to October period. In some locations, seasonal 7-h average concentrations routinely exceed 0.10 ppm (Lefohn and Benedict, 1985). Figures 3.10 and 3.11 show the ozone exposures that occur in the Southern California area for both a high- and a low-exposure year.

Extensive agricultural and forest damage resulting from ozone exposures have occurred in this area of the U.S. As a result of urban-generated ozone, injury symptoms to forests downwind in the San Bernardino Mountains were first observed in the early 1950s (Miller et al., 1989). Ozone hourly average concentrations for this area are among the highest experienced in rural locations across the U.S. Further north, Sequoia National Park, located in the Sierra Nevada, east of the San Joaquin Valley in California, has experienced hourly average concentrations exceeding the state of California 0.09-ppm ozone standard, and injury to local forests has been reported (Peterson et al., 1987). Although hourly average ozone concentrations do not appear to be exceeding the 0.09-ppm level, Pedersen and Cahill (1989) have indicated that part of the ozone exposure experienced at Emerald Lake, a remote, high-elevation site in Sequoia National Park, is associated with transport from nearby lower elevations.

The northeastern, midwestern, and southern regions of the U.S. occasionally experience ozone exposures where the hourly maximum concentrations exceed 0.20 ppm. In these regions of the country, seasonal 7-h (0900 to 1559 h) averages typically exceed 0.050 ppm (Reagan, 1984; Lefohn and Benedict, 1985; Pinkerton and Lefohn, 1986, 1987; Lefohn and Pinkerton, 1988). As indicated from the figures, these exposures are far below those in Southern California. However, because of differing growth conditions (e.g., humidity, temperature, and soil moisture), vegetation in these regions may be more sensitive to ozone exposure than in the West.

3.3   HIGH-ELEVATION OZONE EXPOSURES

3.3.1   General Observations

There appears to be no consistent conclusion concerning the relationship between ozone exposure and elevation. Wolff et al. (1987) have reported, for a short-term study at High Point Mountain in northwestern New Jersey, that both the daily maximum and midday ozone concentrations were similar at different altitudes, but that the ozone exposures increased with elevation. On the other hand, Winner et al. (1989) have reported that for three Shenandoah National Park sites (i.e., Big Meadows, Dickey Ridge, and Sawmill Run), the 24-h monthly mean ozone concentrations tended to increase with elevation, but that the number of elevated hourly occurrences equal to or above selected thresholds did not. The authors reported that the highest elevation site (Big Meadows) experienced a smaller number of concentrations at or below the minimum detectable level than did the other two sites. The larger number of hourly average concentrations that occurred at or below the minimum detectable level at both

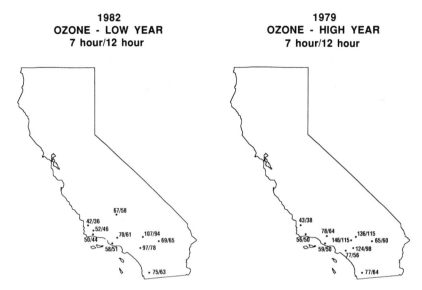

Figure 3.10. Ozone characterization for the southern California area showing the 7-/12-h values (units in ppb) for low- and high-exposure years (Lefohn et al., 1991).

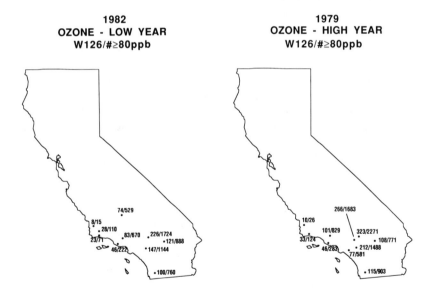

Figure 3.11. Ozone characterization for the southern California area showing the W126/# ≥80 ppb values for low- and high-exposure years (Lefohn et al., 1991).

Dickey Ridge and Sawmill Run resulted in lower 24-h averages at these sites.

Lefohn et al. (1990c) have summarized the characterization of gaseous exposures in 1986 and 1987 at several Mountain Cloud Chemistry Program (MCCP) high-elevation sites. The MCCP was the principal source of atmospheric chemical and physical information for addressing the hypothesis that acidic and other airborne chemicals may contribute to the observed decline in the high-elevation spruce-fir forests in the Adirondacks (NY), Green Mountains (VT), White Mountains (NH), and southern Appalachians (VA, NC, TN, WV). The authors have quantified exposures of trees and forests to ozone and investigated the regional (i.e., longitudinal) and elevational variability of exposures experienced in the spruce-fir and other forest ecosystems that cover high-elevation sites in the Appalachian Mountains of eastern North America.

Data from MCCP high-elevation sites, situated between 45°N and 35°N, were characterized. The sites are representative of the geographic and meteorological variability of the large northeastern region. In addition, a low-elevation site was established at Howland Forest, ME. The sites are listed in Table 3.5. Figure 3.12 shows the location of each of the sites.

For studying the elevation-gradient phenomenon, hourly mean concentrations of ozone data for 1987 for the three MCCP Shenandoah National Park sites (i.e., SH1, SH2, and SH3) and the three Whiteface Mountain sites (WF1, WF3, and WF4) were used. The Mt. Mitchell sites were not used in this analysis because only two locations were established. Aneja et al. (1991) have reported on ozone measurements made at the two high-elevation sites at Mt. Mitchell. The three Whiteface Mountain sites were close to one another and were mainly separated by elevation. The MCCP Shenandoah National Park sites were similarly located. To supplement the analysis, data from three additional Shenandoah National Park sites were used. Since 1983 the National Park Service has been monitoring ozone at Big Meadows (1067 m), Dickey Ridge (640 m), and Sawmill Run (457 m) in the Shenandoah National Park. Since these three sites are separated by both elevation and horizontal distances, differences in ozone exposure may be associated with both elevation and horizontal distance phenomena. For these three sites, ozone data for 1983 through 1987 were used.

Because of the interest in characterizing both high and low hourly mean ozone concentrations, cumulative indices were used. For comparative purposes the long-term seasonal averages were also included. The following exposure indices of ozone were used to characterize hourly mean concentrations over monthly and seasonal (May to September) periods:

- the sum of all hourly mean ozone concentrations using no threshold concentration. This is commonly referred to as "total dose" (SUM0)
- the sum of all hourly mean ozone concentrations (W126), where each hourly concentration is weighted by a sigmoidal weighting function (Lefohn and Runeckles, 1987; Lefohn et al., 1988a)
- the sum of all hourly mean ozone concentrations equal to or greater than 0.07 ppm (SUM07)
- the number of hourly mean ozone concentrations equal to or greater than 0.07 ppm
- the seasonal means of the average of the daily 7-h (0900 to 1559 h) concentrations
- the seasonal means of the average of the daily 12-h (0700 to 1859 h) concentrations

Table 3.5. Description of Sites Used in the MCCP Analyses[a]

Site	Elevation (m)	Latitude			Longitude		
Howland Forest (HF1), ME	65	45°	11′		68°	46′	
Mt. Moosilauke (MS1), NH	1000	43°	59′	18″	71°	48′	28″
Whiteface Mountain (WF1), NY	1483	44°	23′	26″	73°	51′	34″
Shenandoah Park (SH1), VA	1015	38°	37′	12″	78°	20′	48″
Shenandoah Park (SH2), VA	716	38°	37′	30″	78°	21′	13″
Shenandoah Park (SH3), VA	524	38°	37′	45″	78°	21′	28″
Whitetop Mountain (WT1), VA	1689	36°	38′	20″	81°	36′	21″
Mt. Mitchell (MM1), NC	2006	35°	44′		82°	16′	
Mt. Mitchell (MM2), NC	1760	35°	45′		82°	15′	

[a] Lefohn et al. (1990c).

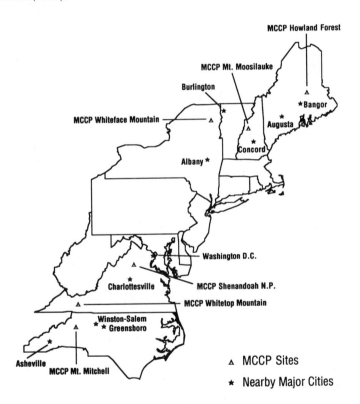

Figure 3.12.   The location of Mountain Cloud Chemistry Program monitoring sites (Lefohn et al., 1990c).

The characterized information was compared among MCCP sites. Those sites that experienced elevated levels of ozone were identified, and their exposure levels were compared with other MCCP sites. No attempt was made to correlate meteorological conditions with the air pollutant episodic exposures. However, Mohnen (1988a) has provided a summary of the relationship between mean ozone concentrations at the MCCP sites with respect to locally measured wind direction.

Lefohn et al. (1990c) reported that a fairly flat diurnal pattern for the Whiteface Mountain summit site (WF1) was observed. A similar pattern, but not as high, was observed for the WF3 site. The site at the base of Whiteface Mountain (WF4) showed the typical diurnal pattern expected from sites that experience some degree of ozone scavenging. More variation in the diurnal pattern for the highest Shenandoah National Park sites occurred than for the higher-elevation Whiteface Mountain sites, with the typical diurnal pattern at the lower-elevation Shenandoah National Park site.

The 7- and 12-h seasonal means were insensitive to the differences in the exposures experienced at the three Whiteface Mountain sites. In some cases the 12-h mean was slightly higher than the 7-h mean value. This occurred when the 7-h mean period (0900 to 1559 h) did not capture the period of the day when the highest hourly mean ozone concentrations were experienced. A similar observation was made for the MCCP Shenandoah National Park sites when the data were reviewed. The exposure differences for ozone between the SH1 and SH2 sites were minimal when the 7- and 12-h parameters were used.

The ordering of sites from highest to lowest did not appear to depend on the surrogate used (Lefohn et al., 1990c). The means and cumulative exposure indices showed the same qualitative patterns. However, the patterns appeared more pronounced when using cumulative indices. Based on cumulative indices, the Whiteface Mountain summit site (WF1) experienced a slightly higher exposure than did the WF3 site. The site at the base of the mountain (WF4) experienced the lowest exposure of the three ozone sites. Among the MCCP Shenandoah National Park sites, the SH2 site experienced higher ozone exposures than the high-elevation site (SH1). The total-dose and sigmoidal (W126) indices were slightly higher at the SH2 site than at the SH1 site. The data capture at the two sites for the 5-month period was similar. However, the sum of the concentrations greater than or equal to 0.07 ppm and the number of hourly concentrations greater than or equal to 0.07 ppm were slightly higher at the SH1 site than the SH2 site. For the Whiteface Mountain sites, both the sum of the concentrations greater than or equal to 0.07 ppm (SUM07) and the number of hourly concentrations greater than or equal to 0.07 ppm were higher at the WF1 site than at the WF3 site.

When the Big Meadows, Dickey Ridge, and Sawmill Run Shenandoah National Park data for 1983 to 1987 were compared, it was again found that the 7- and 12-h seasonal means were insensitive to the different ozone exposure regimes. A higher resolution of the differences among the regimes was observed when the cumulative indices were used. No specific trend could be identified that showed the higher-elevation, Big Meadows site had consistently experienced higher ozone exposures than had the lower-elevation sites. In 2 of the 5 years, the higher-elevation site experienced lower exposures than the Dickey Ridge and Sawmill Run sites, based on total-dose or sigmoidal indices. For 4 of the 5 years the SUM07 index yielded the same result.

Based on ozone data for ozone monitoring sites located near the MCCP sites, in most cases the high-elevation sites appeared to receive greater ozone exposure than the nearby, lower-elevation sites. Six sites in the AIRS database were located near the Howland Forest site in Maine. Of the seven sites, Howland Forest appeared to experience the lowest ozone exposure, based on five of the six indices. The data capture at the Howland Forest site was greater than at most of the sites. Three monitoring sites were near Mt. Moosilauke, NH. The New Hampshire MCCP site experienced higher exposures than the sites to the north or south for all indices. Several monitoring sites were located near the three Whiteface Mountain sites. In 1986 the WF1 Whiteface Mountain site experienced higher exposures of ozone than the Cornwall (Ontario, Canada), Burlington (VT) or Essex County (NY) site at Huntington Forest. A similar observation was made when the 1987 data were characterized.

The 1987 MCCP Shenandoah National Park data were compared with the three National Park sites (i.e., Dickey Ridge, Big Meadows, and Sawmill Run), as well as the nearby Fauquier County site. In 1987 it appeared that the three MCCP sites experienced fewer ozone exposures during the 5-month period than did the three National Park sites. The SH1 and SH2 sites experienced higher exposures of ozone than the Fauquier County site for all indices except the 7-h seasonal mean.

In 1987 four ozone monitoring sites were located near the Whitetop Mountain MCCP site. The ozone exposures at the Whitetop Mountain site were the highest characterized for all indices except the seasonal 7-h mean, which equalled that for Vinton, VA. In 1987 only two monitoring sites were located near the Mt. Mitchell sites. The Greenville County site experienced higher ozone seasonal means and numbers of occurrences than did the Mt. Mitchell sites, although the integrated exposure index values at the Greenville County site were, in some cases, lower. Greenville County is a fairly large metropolitan area with significant automotive traffic. This might explain the higher seasonal means and number of occurrences relative to Mt. Mitchell. The highest-elevation Mt. Mitchell site apparently experienced a greater ozone exposure during the 5-month period than did the lower-elevation site.

Lefohn et al. (1990c) reported that, based on the W126 sigmoidal cumulative index for the period May through September 1987, the MCCP sites in the south appeared to experience higher cumulative ozone exposures than the northern sites, although the data on the number of occurrences show little consistent geographic trend if one excludes the Whitetop Mountain site. The W126 sigmoidal cumulative index (in units of ppm-h) and the maximum hourly mean ozone concentration (arranged in descending order of W126 index values) are shown in Table 3.6.

A preliminary review of data collected in 1988 at the MCCP sites indicates that several locations experienced higher ozone levels in 1988 than in 1986 or 1987. For example, data from 1988 indicate that the sites at Mt. Mitchell experienced higher ozone exposures than in the previous 2 years (Aneja et al., 1989, 1990).

Table 3.6. The W126 Sigmoidal Cumulative Index and the Maximum Hourly Average Ozone Concentration Experienced in 1987 (Arranged in Descending Order of W126 Index Values) Calculated for 1987 MCCP Sites[a]

MCCP sites	W126 value (ppm-h)	Max. hourly concentration (ppm)
Whitetop	106.7	0.124
Mt. Mitchell 1	43.1	0.105
Shenandoah 2	41.8	0.145
Shenandoah 1	40.2	0.135
Whiteface Mountain 1	40.0	0.104
Mt. Mitchell 2	37.4	0.095
Whiteface Mountain 3	36.7	0.117
Mt. Moosilauke	34.9	0.102
Shenandoah 3	18.2	0.108
Whiteface Mountain 4	17.9	0.117
Howland Forest	6.3	0.076

[a] Lefohn et al. (1990c).

3.3.2 Absolute Concentrations vs. Mixing Ratios

In most cases, mixing ratios (e.g., ppm) or mole fractions are used to describe ozone concentrations. Lefohn et al. (1990c) have pointed out that the manner in which the concentration is reported may be important when assessing the potential impacts of air pollution on high-elevation forests. Concentration (in units of micrograms per cubic meter) varies as a function of altitude. Although the change in concentration is small when the elevational difference between sea level and the monitoring site is small, it becomes substantial at high-elevation sites. Given the same part-per-million value experienced at both a high- and low-elevation site, the absolute concentrations (i.e., micrograms per cubic meter) at the two elevations will be different. Since both pollutants and ambient air are gases, changes in pressure directly affect their volume. According to Boyle's law, if the temperature of a gas is held constant, the volume occupied by the gas varies inversely with the pressure (i.e., as pressure decreases, volume increases). This pressure effect must be considered when measuring absolute pollutant concentrations. At any given sampling location, normal atmospheric pressure variations have very little effect on air pollutant measurements. However, when mass/volume units of concentration are used and pollutant concentrations measured at significantly different altitudes are compared, pressure (and hence volume) adjustments are necessary.

Moles of gaseous pollutant per liter of air is the most useful parameter when considering health effects caused by exposure to air pollution (U.S. EPA, 1978). The same should be true when considering effects of air pollution on vegetation. Thus, as indicated above, exposure to identical part-per-million concentrations at differing pressures does not expose the receptor to the same number of moles of pollutant per unit volume of air.

These exposure considerations are trivial at low-elevation sites. However, when one compares exposure-effects results obtained at high-elevation sites with those from low-elevation sites, the differences may become significant. In particular, assuming that the sensitivity of the biological target is identical at both low and high elevations, some adjustment will be necessary when attempting to link experimental data obtained at low-elevation sites with air quality data monitored at the high-elevation stations.

These adjustments become particularly important when computing threshold-based exposure indices. Using Boyle's Law, Lefohn et al. (1990c) provided an example of the magnitude of elevation-related adjustments for the SUM07 exposure parameter (Table 3.7.) Each hourly averaged concentration, in parts-per-million units, was converted to micrograms per cubic meter at the reference of 298° K and 1.0133×10^5 pascal (Pa) (standard temperature and pressure). Using the elevation information for each of the MCCP sites, an adjusted concentration, in units of micrograms per cubic meter, was determined at both 298° K and 273° K. The pressure corresponding to the elevation was calculated from an equation published by the U.S. EPA (1978). Assuming the same temperature (i.e., 298° K) at sea level and at the MCCP site, the adjusted hourly average concentrations equal to and above $137 \, \mu g \, m^{-3}$ (0.07 ppm) were summed. The percent difference between the unadjusted and adjusted SUM07 values was from 11.8% (Howland Forest) to 98.7% (Whiteface Mountain). Although the magnitude of adjustment to each hourly average concentration was less than 20%, the cumulative effect of summing the hourly average concentrations equal to and above a threshold (expressed in equivalent absolute units of concentration) was substantial. For purposes of discussion, a temperature of 273° K was also used in the same analysis. As Table 3.7 shows, the reduction in temperature reduces the effect of elevation and, at low elevations, may result in significant adjustments in the opposite direction. However, for practical purposes, the average temperature during the ozone season (e.g., May through September) would not be 273° K, and the temperature factor would not compensate for the pressure effect. To calculate the changes presented in Table 3.7, the average pressure and temperature at each site were used to estimate changes in concentration as a function of elevation. For those researchers interested in absolute hourly average concentrations, actual pressure and temperature measurements for each hour will have to be used to compute absolute hourly average concentrations from ppm readings.

As indicated, the results obtained when summing adjusted hourly average concentrations equal to and above a threshold are, in some cases, substantially different from the results obtained using unadjusted values. The difference occurs because of the cumulative effect of applying a number of small corrections that were *biased* in the same direction.

A substantial adjustment might be required when using a deposition parameter that was cumulative and exhibited a threshold. The change in deposition as a function of elevation, based on models, has been estimated to be in the range of 4% to approximately 8% (Larson and Vong, 1990). *At this time, there is no parameter that links deposition with direct vegetation effects.* If an index were

Table 3.7.   Comparison of Unadjusted and Adjusted Seasonal SUM07[a] at 298 K, 1.0133 × 10⁵ Pascal, and Corrected for Altitude at 298 K and 273 K[b]

Site	Year	Pressure (Pascal) $\times 10^5$	Unadjusted SUM07	Adjusted SUM07	(T = 298 K) % Diff.	Adjusted SUM07	(T = 273 K) % Diff.
HF1 (65 m)	1987	1.00497	1,745	1,540	−11.8	5,307	204.1
MS1 (1000 m)	1986	0.89311	1,097	576	−47.5	975	−11.1
	1987		40,488	16,407	−59.4	33,285	−17.7
WF1 (1483 m)	1986	0.84030	11,203	142	−98.7	2,883	−74.3
	1987		52,322	9,913	−81.1	25,764	−50.8
Essex Co. (WFM) 360310002 (1494 m)	1986	0.83914	15,011	688	−95.4	3,800	−74.7
	1987		53,833	10,323	−80.8	26,594	−50.6
WF3 (1026 m)	1987	0.89019	50,870	24,683	−51.5	42,208	−17.0
WF4 (604 m)	1987	0.93889	21,786	12,738	−41.5	22,035	1.1
Essex Co. (HUNT) 360310005 (500 m)	1986	0.95129	12,169	7,975	−34.5	13,359	9.8
	1987		19,646	10,748	−45.3	21,522	9.5
SH1 (1015 m)	1987	0.89143	52,258	14,699	−71.9	38,058	−27.2
SH2 (716 m)	1987	0.92571	50,244	21,530	−57.1	45,968	−8.5
SH3 (524 m)	1987	0.94841	17,363	9,373	−46.0	20,037	15.4

Site	Year					
SH Big Meadows 48289003N05 (1067 m)	1983	0.88559	60,365	30,879	−48.8	33,706	−44.2
	1984		79,196	23,548	−70.3	47,268	−40.3
	1985		24,251	6,522	−73.1	11,260	−53.6
SH Big Meadows 511130003 (1067 m)	1986		36,829	8,519	−76.9	18,724	−49.2
	1987		106,390	34,069	−68.0	73,542	−30.9
SH Dickey Ridge 482890002N05 (640 m)	1983	0.93463	100,740	63,508	−37.0	101,433	0.7
	1984		86,756	56,157	−35.3	87,352	0.7
	1985		64,871	43,452	−33.0	65,317	0.7
SH DickeyRidge 511870002	1986		28,076	12,480	−55.5	28,269	0.7
	1987		117,772	74,829	−36.5	118,582	0.7
SH Sawmill Run 482890004N05 (457 m)	1983	0.95646	60,567	41,593	−31.3	62,408	3.0
	1984		60,206	38,876	−35.4	62,036	3.0
	1985		28,083	16,080	−42.7	28,937	3.0
SH Sawmill Run 510150004 (457 m)	1986		42,087	22,829	−45.8	43,367	3.0
	1987		66,195	40,205	−39.3	76,964	16.3
MM1 (2006 m)	1986	0.79221	43,799	4,099	−90.6	12,396	−71.7
	1987		39,877	1,740	−95.6	7,993	−80.0
MM2 (1760 m)	1987	0.80990	33,803	1,685	−95.0	7,534	−77.7
WT1 (1689 m)	1987	0.81905	174,238	30,057	−82.7	68,538	−60.7

[a] In units of $\mu g\ m^{-3}\ h$.
[b] Lefohn et al. (1990c).

developed that exhibited cumulative properties with a threshold deposition value, an adjustment would be required.

The preceding comments concerning elevational gradients and spatial variability were based on the recorded parts-per-million units. However, as indicated above, an adjustment should be made for elevational pressure differences. Using Boyle's Law, adjusted W126 sigmoidal cumulative index values (in units of micrograms per cubic meter-h) were compared with the unadjusted values (in units of ppm-h). When the data are arranged in descending order of adjusted W126 index values, the ordering is shown in Table 3.8.

With adjusted values determined using Boyle's Law for the period May through September 1987, the elevational gradient reported for Whiteface Mountain is no longer observed. Furthermore, the case for the existence of a geographic gradient decreasing from south to north becomes weak. Although the Whitetop Mountain (WT1) and the Shenandoah sites (SH2 and SH1) still top the list, with Howland Forest (HF1) at the bottom, the north-south distribution has become much less pronounced. In addition, as described previously, in most cases the unadjusted exposures indicated that the MCCP sites appeared to receive greater ozone exposure than the non-MCCP lower-elevation sites. When the pressure adjustments were made to the SUM07 index, Mt. Moosilauke experienced higher exposures than the sites nearby, but the Whiteface Mountain (WF1), the Shenandoah (SH1 and SH2), and Mt. Mitchell (MM1 and MM2) MCCP sites experienced lower SUM07 values than the surrounding lower-elevation sites. Except for the Vinton site, the Whitetop Mountain (WT1) site experienced a higher SUM07 value.

Neither target sensitivity nor temperature considerations were integrated into the adjusted sigmoidal cumulative exposure values described above. However, temperature is not considered an important ameliorating factor when actual ambient temperatures are used. Nevertheless, the sensitivity of the target organism may be an important consideration. Taylor and Norby (1985) have raised additional considerations relating to elevation. Unfortunately, the relationship of target organism sensitivity to ozone and to elevation and temperature has not been evaluated. In the absence of such information, the biological consequences of high-elevation exposures to the reduced absolute concentration of ozone, which are disguised by the use of mole fraction units of concentration, are a matter of conjecture. However, intuitively one would predict that the biological consequences would be reduced, since ultimately these effects are caused by the numbers of ozone molecules that reach the reactive sites within the target organism. Additional research is required to quantify the level of error one might introduce by using mixing ratios instead of absolute concentrations in describing exposure.

Based on their analysis of the 1986 to 1987 MCCP data, Lefohn et al. (1990c) concluded:

- Assuming that target sensitivity remains nearly constant as elevation changes, adjusted concentrations are required when evaluating the rela-

Table 3.8. Comparison of the Unadjusted W126 Cumulative Index Values With the Adjusted W126 Values[a]

MCCP Sites	Unadjusted (ppm-h)	Adjusted (μg m^{-3}-h)
Whitetop	106.7	80,721
Shenandoah 2	41.8	51,948
Shenandoah 1	40.2	41,698
Whiteface Mountain 3	36.7	36,807
Mt. Moosilauke	34.9	36,775
Whiteface Mountain 1	40.0	32,167
Shenandoah 3	18.2	24,973
Whiteface Mountain 4	17.9	21,615
Mt. Mitchell 1	43.1	21,467
Mt. Mitchell 2	37.4	20,629
Howland Forest	6.3	11,364

[a] Lefohn et al. (1990c).

tionship between ozone exposures at high-elevation sites and biological effects.
- When unadjusted concentrations are used, the Whiteface Mountain sites show what appears to be an ozone elevational gradient. However, the MCCP Shenandoah sites do not show a strong elevational-gradient trend; the exposure profile at the lower-elevation site (SH2) was similar to the profile characterized at the higher-elevation location (SH1).
- Using Boyle's Law, when adjusted ozone values are used for the period May through September 1987, the elevational gradient reported for Whiteface Mountain was no longer observed.
- When unadjusted concentrations were used, in most cases the high-elevation sites appeared to be receiving greater ozone exposure than the nearby lower-elevation sites. Using Boyle's Law, when adjusted ozone values were used, a consistent conclusion was not evident.
- When unadjusted concentrations are used for the period May through September 1987, the MCCP sites in the south appeared to experience higher cumulative ozone exposures than sites in the north.
- Using Boyle's Law, when adjusted ozone values are used, the geographic difference previously observed was not strong.

3.4   CO-OCCURRENCE OF POLLUTANTS AT SPECIFIC SITES

3.4.1   Gaseous Mixtures

Increasing attention in the plant sciences is focusing on multiple interacting stresses of anthropogenic origin. Although it is recognized that most ecosystems experience prolonged exposure to ozone, they also are exposed to pollutants associated with nitrogen, sulfur, hydrogen ion, and hydrogen peroxide (Mohnen, 1988a). Unfortunately, extensive monitoring networks do not exist for many of these pollutants. Therefore, this section focuses its attention on the co-occurrence patterns of ozone and sulfur dioxide, and ozone and nitrogen dioxide. Until extensive monitoring activities are implemented, our knowledge of co-occurrence patterns will be limited to the few pollutants for which extensive monitoring data are available.

There have been increasing efforts to assess the effects of air pollutant mixtures on vegetation (e.g., Tingey et al., 1971; Oshima, 1978; Ashenden, 1979; Ashenden and Williams, 1980; Heggestad and Bennett, 1981; Foster et al., 1983; Heagle et al., 1983; Thompson et al., 1984). In most studies, pollutant combinations have been applied for several hours a day over an extended period, although such exposure regimes are not typical of ambient air quality in many agricultural areas in the U.S. (Lefohn and Tingey, 1984; Lefohn et al., 1987).

Pollutant combinations can occur at or above a threshold concentration either together or temporally separated from one another. For characterizing the different types of co-occurrence patterns, Lefohn et al. (1987) grouped air quality data within a 24-h period starting at 0000 h and ending at 2359 h. Patterns that showed air pollutant pairs appearing at the same hour of the day at concentrations equal to or greater than a minimum hourly mean value were defined as *simultaneous-only* daily co-occurrences. When pollutant pairs occurred at or above a minimum concentration during the 24-h period without occurring during the same hour, a *sequential-only* co-occurrence was defined. During a 24-h period, if the pollutant pair occurred at or above the minimum level at the same hour of the day *and* at different hours during the period, the co-occurrence pattern was defined as *complex-sequential*. A co-occurrence was not indicated if one pollutant exceeded the minimum concentration just before midnight and the other pollutant exceeded the minimum concentration just after midnight. The authors used only comonitored sites in the SAROAD (1978 to 1982), EPRI-SURE and -ERAQS (1978 to 1979), and TVA (1978 to 1982) databases that were designated as rural and remote. Many of the sites selected were associated with point sources such as power plants.

To put into perspective the potential number of co-occurrences that might be experienced, Lefohn et al. (1987) characterized the number of days with hourly occurrences equal to or greater than 0.03 ppm and 0.05 ppm for SO_2, NO_2, and O_3. The authors reported that daily occurrences of O_3 with hourly concentrations equal to or greater than 0.03 ppm were frequent. On the other hand, at any specific site the number of days with SO_2 and NO_2 hourly concentrations equal to or greater than 0.03 ppm was small. Because of the limiting nature of the occurrence of SO_2 and NO_2, the co-occurrence patterns of pollutant pairs, such as O_3/NO_2, SO_2/O_3, and NO_2/SO_2, tended to be infrequent at most monitoring sites in the U.S. (Lefohn and Tingey, 1984). Lefohn and Tingey (1984), using a minimum concentration of 0.05 ppm, reported that hourly SO_2/NO_2 co-occurrences were rare. In their study, Lefohn and Tingey (1984) used 1981 hourly averaged ozone data for all comonitored sites in the SAROAD database and data from the EPRI-SURE and -ERAQS (1978 to 1979) and TVA (1978 to 1982) databases. Jacobson and McManus (1985), studying air quality data from an area of the Ohio River Valley that contained four coal-fired power plants, also reported that the hourly simultaneous co-occurrence of SO_2 and NO_x was infrequent. Using minimum concentrations of 0.05 ppm and 58μg m^{-3} for SO_2 and NO_x, respectively, the authors reported that the two gases simultaneously co-occurred for less than 1% of the total hours monitored. Lane and Bell (1984), using air monitoring data from central London, England, reviewed 3 months (January through March) of hourly concentration data for SO_2, NO_2, NO, and NO_x and found that the co-occurrence of SO_2 and NO_2 hourly concentrations was infrequent. When a minimum concentration of 0.05 ppm was applied, the simultaneous occurrence of both pollutants was less than 1% of the monitoring period.

Meagher et al. (1987) reported that several documented ozone episodes at specific rural locations appeared to be associated with elevated sulfur dioxide levels. The investigators defined the co-occurrence of ozone and sulfur dioxide to be when hourly mean concentrations were equal to or greater than 0.10 ppm and 0.01 ppm, respectively. Upon reviewing the hourly mean ozone and sulfur dioxide data used by Lefohn et al. (1987), in 1980 (using a threshold of 0.05 ppm for both pollutants) the Paradise No. 23 (KY), Giles County (TN), Murphy Hill (reported as Marshall Co. by Meagher et al., 1987) (AL), and Saltillo (reported as Hardin Co. by Meagher et al., 1987) (TN) sites experienced fewer than 7 d over a 153-d period for a co-occurrence of any form (i.e., simultaneous only, sequential, and complex co-occurrence). Thus, as reported by Lefohn et al. (1987), the co-occurrence pattern of ozone and sulfur dioxide was infrequent.

Ozone occurs frequently at concentrations equal to or greater than 0.03 ppm at many rural and remote monitoring sites in the U.S. (Evans et al., 1983; Lefohn, 1984; Lefohn and Jones, 1986; Lefohn et al., 1990b). Lefohn et al. (1987) reported that the pollutant pair O_3/NO_2 co-occurred more often than the SO_2/O_3 and NO_2/SO_2 pollutant pairs. Lefohn et al. (1987) cautioned that if their 5-month observation period had been extended to 12 months, the co-occurrence patterns might have been different. Because O_3 concentrations are at their highest value during April through October, they predicted that the co-occurrence patterns for SO_2/O_3 and O_3/NO_2 would be similar to the ones reported. However, because both SO_2 and NO_2 tend to have their highest concentrations in the fall and winter months (Lefohn et al., 1987), the number of SO_2/NO_2 co-occurrences might change.

The work by Lefohn and Tingey (1984) concluded that (1) the occurrence of pollutant pairs together (SO_2/NO_2, SO_2/O_3, or O_3/NO_2) at hourly mean concentrations equal to or greater than 0.05 ppm lasted only a few hours per episode, and (2) there were long intervals between episodes (weeks, sometimes months). Subsequently, Lefohn et al. (1987) reported similar results for rural sites located in the U.S. In addition, the authors stated that, because of the rarity of these co-occurrences, the number of co-occurrences involving the three-pollutant mixture, $O_3/SO_2/NO_2$, would also be small. The authors recommended that, because there is a lack of research in which pollutant mixtures (1) mimic ambient concentrations and (2) reproduce realistic frequency of occurrences for *sequential* and *complex-sequential* co-occurrences, researchers attempting to assess the potential effects of SO_2/NO_2, SO_2/O_3, and O_3/NO_2 on vegetation in the U.S. should construct simulated exposure regimes so that (1) hourly simultaneous and daily simultaneous-only co-occurrences are fairly rare and (2) when co-occurrences are present, complex-sequential and sequential-only co-occurrence patterns predominate.

3.4.2 Ozone and Wet-Deposited Hydrogen Ion

Concern has been expressed about the possible effects on vegetation from co-occurring exposures of ozone and acidic precipitation (Prinz et al., 1985; National Acid Precipitation Assessment Program, 1987; Prinz and Krause, 1988). Little information has been published concerning the co-occurrence patterns associated with the joint distribution of ozone and acidic deposition (i.e., H^+). Lefohn and Benedict (1983) reviewed the Environmental Protection Agency's (EPA) Storage and Retrieval of Aerometric Data (SAROAD) monitoring data and the National Atmospheric Deposition Program (NADP) and Electric Power Research Institute (EPRI) wet-deposition databases for 1977 through 1980 and concluded that little data were available for characterizing co-occurrence patterns at comonitored locations.

As a result, Lefohn and Benedict (1983) focused their attention on ozone and acidic-deposition monitoring sites that were closest to one another. In some cases, the sites were as far apart as 144 km. Using hourly ozone monitoring data, and weekly and event acidic-deposition data from the NADP and EPRI databases, the authors evaluated the frequency distribution of pH events for 34 NADP and 8 EPRI chemistry monitoring sites located across the U.S. The authors identified specific locations where the hourly mean ozone concentrations were greater than or equal to 0.10 ppm, and 20% of the wetfall daily or weekly samples were below pH 4.0. Based on their analysis using data collected between 1977 and 1980, Lefohn and Benedict (1983) reported five sites where agricultural crops may have the potential for experiencing additive, less than additive, or synergistic (i.e., greater than additive) effects from elevated ozone and hydrogen ion exposures. The authors concluded, based on the available data, the greatest potential for interaction between acidic precipitation and ozone exposures in the U.S., with possible effects on crop yields, may be in the most industrial areas (e.g., Ohio and Pennsylvania). However, they cautioned that, because no documented evidence existed to show that pollutant interaction had occurred under field growth conditions and ambient exposures, their conclusions should only be used as a guide for further research.

In the late 1970s and throughout the 1980s, both the private sector and governments funded research efforts to better characterize gaseous air pollutant concentrations and wet deposition. Building upon the preliminary results reported by Lefohn and Benedict (1983), Smith and Lefohn (1991) examined data from two of the largest U.S. wet-deposition networks, the weekly NADP and the event-oriented Utility Acid Precipitation Study Program (UAPSP), operated by EPRI, to investigate the co-occurrence patterns of ozone and acidic deposition.

The authors obtained ozone data from the U.S. EPA Aerometric Information Retrieval System (AIRS), the EPA Mountain Cloud Chemistry Program (MCCP), the EPA National Dry Deposition Network (NDDN), the EPRI Sulfate Regional Experiment (SURE), and a field study that investigated the ozone effects on crops.

Based on vegetation effects concerns, the following cumulative ozone exposure indices were considered for use:

- the summation of all hourly mean concentrations over the specified period greater than or equal to 0.06 ppm (SUM06)
- the summation of all hourly mean concentrations over the specified period greater than or equal to 0.07 ppm (SUM07)
- the summation of all hourly mean concentrations over the specified period greater than or equal to 0.08 ppm (SUM08)
- the W126 exposure index as per Lefohn and Runeckles (1987)

The analysis involved the use of both weekly and event-based precipitation data where ozone was also monitored in the U.S. Collocation of ozone and wet-deposition monitoring was defined as occurring simultaneously and within 1 min of latitude and longitude of each other. Based on a review of the annual data capture for the ozone monitoring sites, specific sites were identified that had sufficient data for a co-occurrence analysis. Figure 3.13 shows a map of site locations used by Smith and Lefohn (1991).

Two different aspects of the investigation were addressed: short-term and annual exposures. Similar analyses were performed utilizing the weekly NADP and daily UAPSP wet-deposition data; however, the UAPSP data allowed a finer time scale for examining the short-term question. For data to be included in the analysis, both short-term analyses required (1) a valid precipitation sample and (2) ozone data capture of at least 75% (over the time period involved).

Smith and Lefohn (1991) reported that the annual exposure and short-term exposure investigations, based on weekly precipitation collection, provided similar results in identifying sites possibly at risk from high levels of both ozone and acidic deposition. Relatively high-level acidic deposition and ozone co-occurrences happened most frequently at Clingman's Peak, Whitetop Mountain, Shenandoah National Park, Penn State, Walker Branch Watershed, Finley Farm, and West Point. The first four sites are located in mountainous areas, three at high elevations. In addition, if one divides the group of sites into eastern and western parts, these sites all lie in the eastern part. The examination of the collocated ozone and daily wet-deposition data found that the co-occurrence of relatively

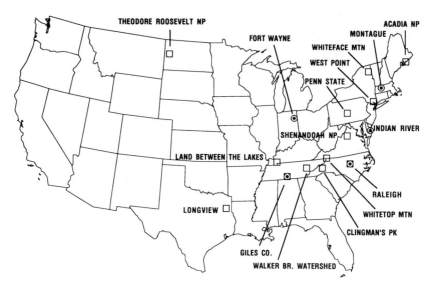

Figure 3.13. Locations for comonitored ozone and wet deposition sites used in the analysis. The Raleigh and Giles County samplers were located at the same spot for the NADP and UAPSP networks (◉ = daily, □= weekly). (Smith and Lefohn, 1991).

high levels of ozone and acidic deposition was more frequent at Indian River, Fort Wayne, Montague, and Raleigh than at Giles County. An east-west division was not as clearly drawn as for the weekly NADP data. Each "eastern" site showed relatively more frequent high-level co-occurrences, but the "western" sites were split. None of the UAPSP sites examined were located in mountainous areas.

Smith and Lefohn (1991) pointed out that one might have expected sites in the eastern portion of the country to have exhibited more frequent co-occurrences of high levels of ozone and acidic deposition, due to the greater influence of industrialized areas on such sites. The results of their analysis were generally in agreement with this, though it appeared that mountainous areas particularly experienced relatively high levels of ozone and acidic deposition.

Smith and Lefohn (1991) pointed out that no attempt was made to include hydrogen ion cloud deposition information in their analysis. In some cases, for mountaintop locations (e.g., Clingman's Peak, Shenandoah National Park, Whiteface Mountain, and Whitetop Mountain), the hydrogen ion cloud water deposition was greater than the hydrogen ion deposition in precipitation (Mohnen, 1988b), and the co-occurrence patterns associated with ozone and cloud deposition may be different than those patterns associated with ozone and deposition in precipitation.

Smith and Lefohn (1991) reported that individual sites experienced years in which both hydrogen ion deposition and total ozone exposure were at least moderately high (i.e., annual H^+ deposition ≥ 0.5 kg ha^{-1}, and an annual ozone cumulative sigmoidally weighted exposure (W126) value ≥ 50 ppm-h). With data compiled from all sites, it was found that relatively acidic precipitation (pH ≤ 4.31 on a weekly basis or pH ≤ 4.23 on a daily basis) occurred together with relatively high ozone levels (i.e., W126 values ≥ 0.66 ppm-h for the same week or W126 values ≥ 0.18 ppm-h immediately before or after the rainfall event) approximately 20% of the time, and highly acidic precipitation (i.e, pH ≤ 4.10 on a weekly basis or pH ≤ 4.01 on a daily basis) occurred together with a high ozone level (i.e., W126 values ≥ 1.46 ppm-h for the same week, or W126 values ≥ 0.90 ppm-h immediately before or after the rainfall event) approximately 6% of the time. Whether during the same week or before, during, or after a precipitation event, correlations between ozone level and pH (or H^+ deposition) were weak to nonexistent. Sites most subject to relatively high levels of both hydrogen ion and ozone were located in the eastern portion of the U.S., often in mountainous areas.

3.5 TRENDS

The U.S. EPA has established national ambient air quality standards to protect human health and public welfare (i.e., vegetation, visibility, and materials). To assess the efficacy of emission control measures, investigators have performed trend analyses (e.g., St. John et al., 1981; Kumar and Chock, 1984; Walker, 1985; Hemphill et al., 1988; Yarbrough, 1988; Allande et al., 1988; Hunt et al., 1988; U.S. EPA, 1989c; Wakim, 1989; U.S. EPA, 1990, 1991). Most of the trend analyses have used both urban and rural information; no attempt has been made to identify whether emission control measures have been effective in protecting vegetation from air pollution exposure in rural areas of the U.S..

The majority of (1) agricultural lands and (2) commercial and public forest lands are located in areas that are more than 50 km from major metropolitan centers (Lefohn et al., 1990a). On average, 40% of the agricultural lands in the U.S. are located within 50 km of a population center of 50,000 or more; 48% of the agricultural lands in the eastern 31 states are located within 50 km of a population center of 50,000 or more. On average, 27% of the forest lands are located within 50 km of a population center of 50,000 or more; 41% of the forest lands in the eastern 31 states are located within 50 km of a population center of 50,000 or more.

Although not extensive, air quality data exist in the U.S. to characterize ozone hourly average concentrations in some areas where trees and agricultural crops are grown. Using the W126 exposure index that was integrated over a seasonal (April to October) period, Lefohn and Shadwick (1991) investigated whether statistically significant linear ozone trends were present. In addition, they identified by region those years in which the highest ozone exposures occurred. In 1988 the ozone exposures experienced at many locations across the U.S. were higher than in previous years (U.S. EPA, 1990; Aneja et al., 1990; Pagnotti, 1990; Lefohn et al., 1990a; Lefohn et al., 1991b).

A list of ozone monitoring sites used in the U.S. Environmental Protection

Agency long- and short-term trend analyses (U.S. EPA, 1990) was obtained from
the agency. From these lists, sites designated in the database by the U.S. EPA as
"rural" were selected. To this rural site list, Tennessee Valley Authority (TVA),
EPRI, and National Park Service (NPS) sites were added. In addition, sites in the
southern U.S. were included if they had been identified, through photo interpretation by the U.S. Forest Service, as located in agricultural or forested areas,
independent of land-use classification. After the identification of the sites, only
those sites meeting a 75% data capture criterion were included in the analysis.
Figure 3.14 illustrates the locations of the ozone monitoring sites. The sites were
segregated into specific agricultural and forestry regions as described by Lefohn
et al. (1991). Figures 3.15 through 3.20 summarize the agricultural and forested
resources within these regions.

   To compare the exposure indices across years, Lefohn and Shadwick (1991)
applied a correction for missing data. The corrections were determined for each
site on a monthly basis using the following five steps: (1) an initial estimate of
the monthly values of an exposure index was computed as corrected exposure
index = (uncorrected exposure index)/(data capture as a fraction); (2) the
monthly values computed in step (1) were summed over the time period of
interest to provide an estimate of the time period's corrected exposure index; (3)
the proportion of the corrected index attributed to each month was calculated as
a proportion of the corrected index = (corrected monthly index value)/(time-
period-corrected exposure index) for each year within a site; (4) the arithmetic
average of the proportions calculated in step (3) over all years of record was
computed; and (5) for months with at least 75% data capture, the monthly
estimate from step (1) was used to compute the final time-period-corrected
exposure index. For months with less than 75% data capture, the estimate
(arithmetic average of monthly proportions from step [4] multiplied by the time-
period-corrected exposure index from step [2]) was used.

   The Kendall's K statistic (Mann-Kendall test) was used to identify linear
trends (H_0: $\beta = 0$, where β is the slope parameter in the regression equation $Y_i =
\alpha + \beta X_i + \varepsilon_i$, $i = 1, \ldots, N$; Y_i is the index; X_i is the year; and ε_i is the error term).
Estimates of the rate of change (slope) for the index were calculated. The analysis
was limited to the case of linear trends. Individual 5- and 10-year ozone trends
(1984 to 1988 and 1979 to 1988) for each monitor were examined using the K
statistic. A test of significance (H_0: $\beta = 0$, H_A: $\beta \neq 0$) was performed. The Thiel
(median) slope estimate was also calculated to give an indication of the
magnitude of the linear trend. The data were not aggregated from different sites
to form an overall or "regional" estimate of trend. Trends were examined on a
per site basis. Regional statements of trend were made on specific patterns
observed, using each individual site analysis. Given the problems associated
with much of the ambient gas monitoring data (e.g. missing data and unknown
precision and accuracy of particular monitoring instruments), ozone trends had
to be very pronounced before any final conclusions could be drawn.

   The percent difference per year for each pollutant was calculated as follows:
([year$_2$ value - year$_1$ value]/[(year$_2$ - year$_1$) × (year$_1$ value)]) ×100% (with year$_2$
greater than year$_1$) and was made for all possible pairs of yearly values. Note that

5-Year

10-Year

Figure 3.14.   Location of ozone trend sites in rural locations for the 5- and 10-year analyses (Lefohn and Shadwick, 1991).

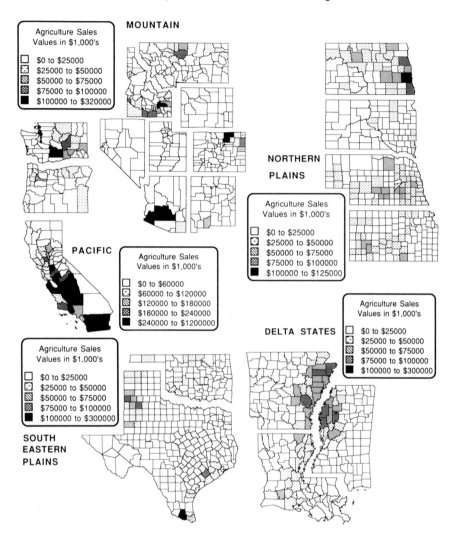

Figure 3.15.   Agricultural sales ($1000s) by county for 1982 (regions: Mountain, Northern Plains, Pacific, Southeastern Plains, and Delta States). (Lefohn et al., 1991).

the estimates are in the units "percent increase (or decrease) relative to the earlier year per year."

Lefohn and Shadwick (1991) reported that indications of trends for both the 10- and 5-year analyses were weak. At 54 of the 77 sites (70%) for the 10-year analysis (1979 to 1988) and at 118 of 147 sites (80%) with at least 4 years of data for the 5-year analysis (1984 to 1988), there was no indication of trends. Most of the significant trends occurred in the southern and midwestern forestry regions and in the Appalachian agricultural region. For the two forestry regions, assessing only significant trends, there were more positive than negative slope

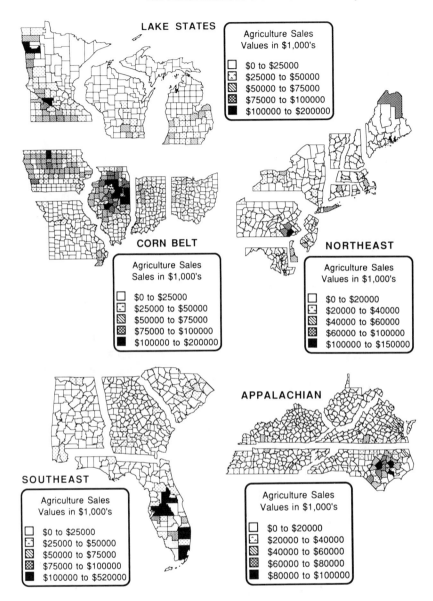

Figure 3.16. Agricultural sales ($1000s) by county for 1982 (regions: Lake States, Corn Belt, Northeast, Appalachian, and Southeast). (Lefohn et al., 1991).

estimates for the 10-year analysis of ozone sites. For the 5-year analysis, the above was true, except that the Mid-Atlantic forestry region also experienced more positive than negative slope estimates. Few sites showed a significant trend for sites in the northeastern region. For sites in agricultural regions, those in the Appalachian region showed a predominance of positive significant trends for both the 5-year and 10-year analyses. In the other regions, few significant trends

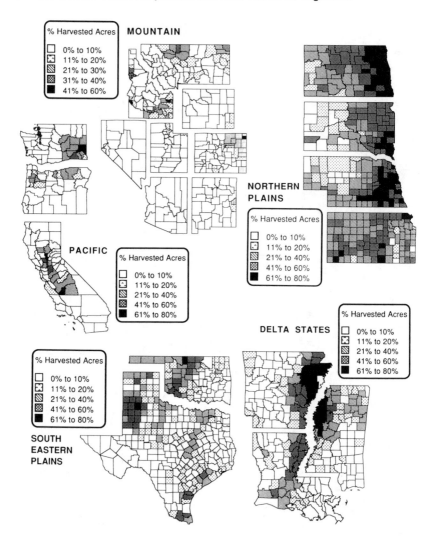

Figure 3.17.   Acres harvested for agriculture as a percent of the county for 1982 (regions: Mountain, Northern Plains, Pacific, Southeastern Plains, and Delta States). (Lefohn et al., 1991).

were detected. Significant positive 10- and 5-year trends were at least as numerous as negative significant trends.

The ozone results produced patterns that were not pronounced enough to draw more than tentative conclusions. However, when the 5-year trends were significant, they were almost always increasing. The 5-year results appear to reflect, in a disproportionate manner, the effect of the high exposures that occurred in some regions in 1988.

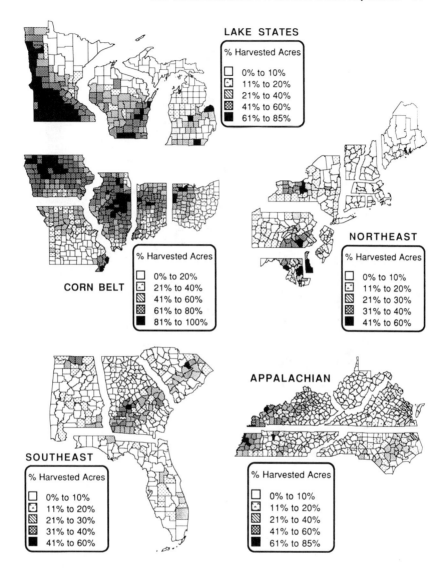

Figure 3.18.   Acres harvested for agriculture as a percent of the county for 1982 (regions: Lake States, Corn Belt, Northeast, Appalachian, and Southeast). (Lefohn et al., 1991).

Prior to 1988, the highest ozone exposures from 1979 to 1987 occurred in 1983 in some forest regions of the U.S. (i.e., Northeast, Midwest, and Mid-Atlantic). The hot, dry summer of 1988 was associated with the highest ozone exposures for the South, Midwest, Plains, Northeast, and Mid-Atlantic regions. For many agricultural regions, 1988 was a high-ozone-exposure year (e.g.,

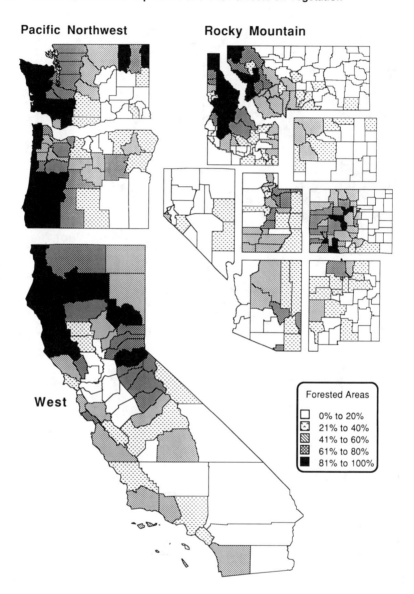

Figure 3.19.   Percent forest cover by county (regions: Pacific Northwest, Rocky Mountain, and West). (Lefohn et al., 1991).

northern plains, southeastern plains, lake states, corn belt, Northeast, and Appalachian).

A comparison of the EPA's results (EPA, 1990) and those reported by Lefohn and Shadwick (1991) is important. The statistical methods used in the EPA trends report (EPA, 1990) and those used by Lefohn and Shadwick (1991) were different. The major difference between the two approaches is that for the EPA

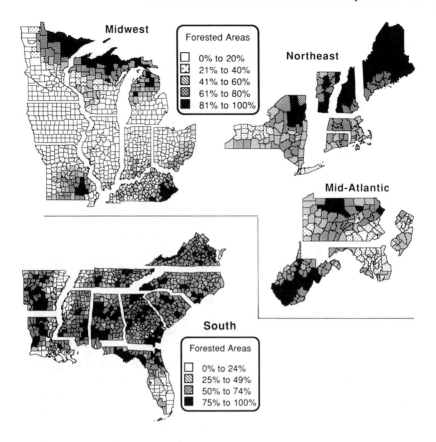

Figure 3.20.   Percent forest cover by county (regions: Midwest, Northeast, South, and Mid-Atlantic). (Lefohn et al., 1991).

trends reports, statistics are aggregated over sites within years and inference is drawn on trends across years. In the analysis by Lefohn and Shadwick (1991), using the Mann-Kendall approach, estimates of the trend were done first within sites across years, and then regional or overall statements about trends were drawn from the individual site analyses. Also, the rural sites used in the Lefohn and Shadwick (1991) analysis were mostly a subset of the trend sites (both urban and rural) employed in the EPA analyses.

Using data from both rural and urban monitoring sites, the EPA (1990) reported that the 1988 national composite ozone average, using the second highest daily maximum 1 h, was 2% higher than the 1979 average. For the 5-year period, the national composite average increased 9% between 1984 and 1988. The composite average increased 8% from 1987 to 1988. The hot, dry summer of 1988 was associated with the highest ozone exposures over a 10-year period for several regions in the U.S.

Meteorological factors influence ozone exposures, and therefore, it is difficult to quantify the efficacy of emission-control strategies that focus on the reduction of ozone exposures. For example, although 1988 was one of the highest exposure years over the 1979 to 1988 period, EPA reported that 1989 ozone levels were 15% lower than those experienced in 1988 (EPA, 1991).

3.6   FUTURE RESEARCH DIRECTIONS

There is some concern that exposure may not be an adequate substitute for dose. However, the data obtained from gaseous air pollution monitoring sites are usually provided in the form of hourly average concentrations. This information represents the potential exposure that biological targets receive. Therefore, any assessment of nonattainment status for a particular region by regulatory agencies is based on exposure, not dose, information. Additional research should be undertaken to identify sets of exposure indices that are biologically relevant and can possibly serve as substitutes for dose. Both retrospective studies as well as prospective analyses should continue with efforts to carefully identify those indices that order themselves properly in exposure-response relationships.

To better link exposure with dose, additional research is required to learn more about the biological mechanisms that are responsible for the cause-and-effect relationships between air pollution exposure and vegetation effects. Although exposure-response equations have been developed to predict vegetation effects caused by air pollution exposure, the uncertainties in these predictions should be refined by better characterizing the exposure/dose linkage.

Additional efforts are required to quantify the relative importance of absolute concentration, length of exposure, time between episodic exposures, and vegetation sensitivity to air pollution exposures. Much has been discussed in the literature recognizing the importance of these components for developing adequate predictive powers. However, little research has been performed developing the weighting functions that are required to identify their relative importance to one another. Results reported by Showman (1991) emphasize the importance of target sensitivity at the time of exposure.

Lefohn et al. (1990c) have emphasized the possible importance of elevation-related changes in concentration. On the other hand, Larson and Vong (1990), using mathematical models, have predicted that only small elevation-related adjustments are required. It is important that additional research be undertaken to clarify the level of adjustment required when laboratory and field research results obtained at low-elevation locations are applied to predict ozone effects on vegetation located at high-elevation sites.

REFERENCES

Allande, G., DiMarcello, D. B., and Kapichak, R. K. (1988) A regional perspective on ozone trends from the New York Metropolitan area. In *Proc. 81st Annual Meeting of the Air Pollution Control Association*. Air Pollution Control Association, Pittsburgh.

Amiro, B. D., Gillespie, T. J., and Thurtell, G. W. (1984) Injury response of Phaseolus vulgaris to ozone flux density. *Atmos. Environ.* **18**, 1207–1215.

Aneja, V. P., Claiborn, C. S., Li, Z., and Murthy, A. (1990) Exceedances of the national ambient air quality standard for ozone occurring at a "pristine" area site. *J. Air Waste Manage. Assoc.* **40**, 217–220.

Aneja, V. P., Businger, S., Li, Z., Claiborn, C. S., and Murthy, A. (1991) Ozone climatoloy at high elevations in the southern Appalachians. *J. Geophys. Res.* **96**, 1007–1021.

Aneja, V. P., Businger, S., Li Z., Claiborn, C., and Murthy, A. (1989) Ozone climate at Mt. Mitchell, North Carolina, and its association with synoptic episodes. In *Proc. 82nd Annual Meeting of the Air and Waste Management Association*, June 25–30, 1989, Anaheim, CA. Air and Waste Management Association, Pittsburgh.

Ashenden, T. W. (1979) The effects of long-term exposures to SO_2 and NO_2 pollution on the growth of *Dactylis glomerata* L. and *Poa pratensis* L. *Environ. Pollut.* **18**, 249–258.

Ashenden, T. W. and Williams, I. A. D. (1980) Growth reductions in *Lolium multiflorum* Lam. and *Phleum pratense* L. as a result of SO_2 and NO_2 pollution. *Environ. Pollut.* **21**, 131–139.

Ashmore, M. R. (1984) Effects of ozone on vegetation in the United Kingdom. In *Proc. International Workshop on the Evaluation of the Effects of Photochemical Oxidants on Human Health, Agricultural Crops, Forestry, Materials and Visibility* (ed., Grennfelt, P.). Swedish Environmental Research Institute, Göteborg, Sweden.

Ashmore, M. R. (1988) A comparison of indices that describe the relationship between exposure to ozone and reduction in the yield of agricultural crops [Comments on article by Lefohn et al. (1988a)]. *Atmos. Environ.* **22**, 2060–2061.

Bennett, J. H. (1979) Foliar exchange of gases. In *Methodology for the Assessment of Air Pollution Effects on Vegetation: A Handbook* (eds., Heck, W. W., Krupa, S. V., and Linzon, S. N.). Air Pollution Control Association, Pittsburgh, 10.1–10.29.

Benson, F. J., Krupa, S. V., Teng, P. S., Welsch, D. W., Chen, C., and Kromroy, K. (1982) *Economic Assessment of Air Pollution Damage to Agricultural and Silvicultural Crops in Minnesota*. University of Minnesota, St. Paul.

Berry, C. R. (1964) Differences in concentrations of surface oxidant between valley and mountaintop conditions in the southern Appalachians. *JAPCA* **14**, 238–239.

Bicak, C. (1978) Plant response to variable ozone regimes of constant dosage. M.Sc. Thesis, Department of Plant Science, University of British Columbia, Vancouver, B.C., p. 115.

Böhm, M., McCune, B., and Vandetta, T. (1991) Diurnal curves of tropospheric ozone in the western United States. *Atmos. Environ.* **25A**, 1577-1590.

Evans, G. F., Finkelstein, P., Martin, B., Possiel, N., and Graves, M. (1983) Ozone measurements from a network of remote sites. *JAPCA* **33**, 291–296.

Foster, K. W., Timm, H., Lawrence, C. K., and Oshima, R. J. (1983) Effects of ozone and sulfur dioxide on tuber yield and quality of potatoes. *J. Environ. Qual.* **12**, 75–80.

Fowler, D. and Cape, J. N. (1982) Air pollutants in agriculture and horticulture. In *Effects of Gaseous Air Pollution in Agriculture and Horticulture* (eds., Ormrod, D. P. and Unsworth, M. H.). Butterworth Scientific, London, 3–26.

Guderian R., Tingey, D. T., and Rabe, R. (1985) Effects of photochemical oxidants on plants. In *Air Pollution by Photochemical Oxidants* (ed., Guderian, R.). Springer-Verlag, Berlin, 129–333.

Hayes, E. M. and Skelly, J. M. (1977) Transport of ozone from the northeast U.S. into Virginia and its effect on eastern white pines. *Plant Dispos. Rep.* **61**, 778–782.

Heagle, A. S. and Heck, W. W. (1980) Field methods to assess crop losses due to oxidant air pollutants. In *Crop Loss Assessment: Proceedings of E.C. Stakman Commemorative Symposium* (edited by Krupa S. V. and Teng P. S.), Misc. Publication No. 7. University of Minnesota, St. Paul, 296–305.

Heagle, A. S., Body, D. E., and Neely, G. E. (1974) Injury and yield response of soybean to chronic doses of ozone and sulfur dioxide in the field. *Phytopathology* **64,** 132–136.

Heagle, A. S., Spencer, S., and Letchworth, M. B. (1979) Yield response of winter wheat to chronic doses of ozone. *Can. J. Bot.* **57,** 1999–2005.

Heagle, A. S., Heck,W. W., Rawlings, J. O., and Philbeck, R. B. (1983) Effects of chronic doses of ozone and sulfur dioxide on injury and yield of soybeans in open-top chambers. *Crop Sci.* **26,** 1184–1191.

Heck, W. W. and Dunning, J. A. (1967) The effects of ozone on tobacco and pinto bean as conditioned by several ecological factors. *JAPCA* **17,** 112–114.

Heck, W. W. and Tingey, D. T. (1971) Ozone time-concentration model to predict acute foliar injury. In *Proc. 2nd International Clear Air Congress* (eds., Englund, H. M. and Beery, W. T.). Acdemic Press, New York, 249–255.

Heck, W. W., Dunning, J. A., and Hindawi, I. J. (1966) Ozone: nonlinear relation of dose and injury to plants. *Science* **151,** 511–515.

Heck, W. W., Taylor, O. C., Adams, R. M., Bingham, G. E., Miller, J. E., Preston, E. M., and Weinstein, L. H. (1982) Assessment of crop loss from ozone. *JAPCA* **32,** 353–361.

Heggestad, H. E. and Bennett, J. H. (1981) Photochemical oxidants potentiate yield losses in snap beans attributable to sulfur dioxide. *Science* **213,** 1008–1010.

Hemphill, M. W., Gise, J. P., Broberg B. A., and Sager, T. W. (1988) Statewide ozone trends — Texas. In *Proc. 81st Annual Meeting of the Air Pollution Control Association*. Air Pollution Control Association, Pittsburgh.

Henderson, W. R. and Reinert, R. A. (1979) Yield response of four fresh market tomato cultivars after acute ozone exposure in the seedling stage. *J. Am. Soc. Hort. Sci.* **104,** 754–759.

Hogsett, W. E., Tingey, D. T., and Holman, S. R. (1985a) A programmable exposure control system for determination of the effects of pollutant exposure regimes on plant growth. *Atmos. Environ.* **19,** 1135–1145.

Hogsett, W. E., Tingey, D. T., and Lee, E. H. (1988) Exposure indices: concepts for development and evaluation of their use. In *Assessment of Crop Loss from Air Pollutants* (eds., Heck, W. W., Taylor, O. C., and Tingey, D. T.). Elsevier, London, 107–138.

Hogsett, W. E., Plocher, M., Wildman, V., Tingey, D. T., and Bennett, J. P. (1985b) Growth response of two varieties of slash pine seedlings to chronic ozone exposures. *Can. J. Bot.* **63,** 2369–2376.

Hunt, W. F., Jr., Curran, T. C., and Freas, W. P. (1988) National and regional trends in ambient ozone measurements, 1977-1986. In *Proc. 81st Annual Meeting of the Air Pollution Control Association*. Air Pollution Control Association, Pittsburgh.

Jacobson, J. S. and McManus, J. M. (1985) Pattern of atmospheric sulfur dioxide occurrence: an important criterion in vegetation effects assessment. *Atmos. Environ.* **19,** 501–506.

Kohut, R. J., Amundson, R. G., and Laurence, J. A. (1986) Evaluation of growth and yield of soybean exposed to ozone in the field. *Environ. Pollut.* (Series A) **41,** 219–234.

Kohut, R. J., Amundson, R. G., Laurence, J. A., Colavito, L., Van Leuken, P., and King, P. (1987) Effects of ozone and sulfur dioxide on yield of winter wheat. *Phytopathology* **77,** 71–74.

Krupa, S. V. and Kickert, R. N. (1987) An analysis of numerical models of air pollutant exposure and vegetation response. *Environ. Pollut.* **44,** 127–158.

Krupa, S. V. and Nosal, M. (1989) Effects of ozone on agricultural crops. In *Atmospheric Ozone Research and its Policy Implications* (eds., Schneider, T., Lee, S. D., Wolters, G. J. R., and Grant, L. D.). Elsevier, Amsterdam, 229–238.

Krupa, S. V. and Teng, P. S. (1982) Uncertainties in estimating ecological effects of air pollutants. In *Proc. 75th Annual Meeting of the Air Pollution Control Association,* New Orleans. Air Pollution Control Association, Pittsburgh.

Kumar, S. and Chock, D. P. (1984) An update on oxidant trends in the South Coast Air Basin of California. *Atmos. Environ.* **18**, 2131–2134.

Lane, P. I. and Bell, J. N. B. (1984) The effects of simulated urban air pollution on grass yield: Part 1-description and simulation of ambient pollution. *Environ. Pollut.* (Series B) **8**, 245–263.

Larsen, R. I. and Heck, W. W. (1984) An air quality data analysis system for interrelating effects, standards, and needed source reductions: Part 8. An effective mean ozone crop reduction mathematical model. *JAPCA* **34**, 1023–1034.

Larson, T. V. and Vong, R. J. (1990) A theoretical investigation of the pressure and temperature dependence of atmospheric ozone deposition to trees. *Environ. Pollut.* **67**, 179–189.

Lee, E. H., Hogsett, W. E., and Tingey, D. T. (1991) Efficacy of ozone exposure indices in the standard setting process. In *Trans. Atmospheric Ozone and the Environment Specialty Conference* (ed., Berglund, R., Lawson, D., and McKee, D.). Air and Waste Management Association, Pittsburgh, 255–271.

Lee, E. H., Tingey, D. T., and Hogsett, W. E. (1988) Evaluation of ozone exposure indices in exposure-response modeling. *Environ. Pollut.* 53, 43–62.

Lee, E. H., Tingey D. T., and Hogsett, W. E. (1989) Interrelation of experimental exposure and ambient air quality data for comparison of ozone exposure indices and estimating agricultural losses. Contract No. 68-C8-0006, U.S. Environmental Protection Agency, Corvallis Environmental Research Laboratory, Corvallis, OR.

Lefohn, A. S. (1984) A comparison of ambient ozone exposures for selected nonurban sites. In *Proc. 77th Annual Meeting of the Air Pollution Control Association*. Air Pollution Control Association, Pittsburgh, 84-104.1.

Lefohn, A. S. (1988) A comparison of indices that describe the relationship between exposure to ozone and reduction in the yield of agricultural crops (a response to comments by Parry, M. A. J., Day, W., and Ashmore, M. R.). *Atmos. Environ.* **22**, 2058–2060.

Lefohn, A. S. and Benedict, H. M. (1982) Development of a mathematical index that describes ozone concentration, frequency, and duration. *Atmos. Environ.* **16**, 2529-2532.

Lefohn, A. S. and Benedict, H. M. (1983) The potential for the interaction of acidic precipitation and ozone pollutant doses affecting agricultural crops. In *Proc. 76th Annual Meeting of the Air Pollution Control Association*. Air Pollution Control Association, Pittsburgh.

Lefohn, A. S. and Benedict, H. M. (1985) Exposure considerations associated with characterizing ozone ambient air quality monitoring data. In *Evaluation of the Scientific Basis for Ozone/Oxidants Standards* (ed., Lee, S. D.). Air Pollution Control Association, Pittsburgh, 17–31.

Lefohn, A. S. and Benkovitz, C. M. (1990) Air quality measurements and characterizations for vegetation effects research. In *Proc. 83rd Annual Meeting of the Air and Waste Management Association*, Pittsburgh, 90-98.1.

Lefohn, A. S. and Irving, P. M. (1988) Characterizing ambient ozone exposure regimes in agricultural areas. In *Proc. 81st Annual Meeting of the Air Pollution Control Association*, June 21–24, 1988, Dallas, TX. Air Pollution Control Association, Pittsburgh, 88-69.1

Lefohn, A. S. and Jones, C. K. (1986) The characterization of ozone and sulfur dioxide air quality data for assessing possible vegetation effects. *JAPCA* **36**, 1123–1129.

Lefohn, A. S. and Mohnen, V. A. (1986) The characterization of ozone, sulfur dioxide, and nitrogen dioxide for selected monitoring sites in the Federal Republic of Germany. *JAPCA* **36**, 1329–1337.

Lefohn, A. S. and Pinkerton, J. E. (1988) High resolution characterization of ozone data for sites located in forested areas of the United States. *JAPCA* **38**, 1504–1511.

Lefohn, A. S. and Runeckles, V. C. (1987) Establishing a standard to protect vegetation — ozone exposure/dose considerations. *Atmos. Environ.* **21**, 561–568.

Lefohn, A. S. and Shadwick, D. S. (1991) Ozone, sulfur dioxide, and nitrogen dioxide trends at rural sites located in the United States. *Atmos. Environ.* **25A,** 491–501.

Lefohn, A. S. and Tingey, D. T. (1984) The co-occurrence of potentially phytotoxic concentrations of various gaseous air pollutants. *Atmos. Environ.* **18,** 2521–2526.

Lefohn, A. S. and Tingey, D. T. (1985) Comments on atmospheric environment paper: injury response of *Phaseolus vulgaris* to ozone flux density. *Atmos. Environ.* **19,** 206–208.

Lefohn, A. S., Krupa, S. V., and Winstanley, D. (1990b) Surface ozone exposures measured at clean locations around the world. *Environ. Pollut.* **63,** 189–224.

Lefohn, A. S., Laurence, J. A., and Kohut, R. J. (1988a) A comparison of indices that describe the relationship between exposure to ozone and reduction in the yield of agricultural crops. *Atmos. Environ.* **22,** 1229–1240.

Lefohn, A. S., Laurence, J. A., and Kohut, R. J. (1988b) A comparison of indices that describe the relationship between exposure to ozone and reduction in the yield of agricultural crops (response to comments by V. C. Runeckles). *Atmos. Environ.* **22,** 1242–1243.

Lefohn, A. S., Shadwick, D. S., and Foley J. K., (1991) The quantification of surface level ozone exposures across the United States. In *Trans. Tropospheric Ozone and the Environment Specialty Conference.* (ed., Berglund, R., Lawson, D., McKee, D.) Air and Waste Management Association, Pittsburgh, 197–224.

Lefohn, A. S., Shadwick, D. S., and Mohnen, V. A. (1990c) The characterization of ozone concentrations at a select set of high-elevation sites in the eastern United States. *Environ. Pollut.* **67,** 147–178.

Lefohn, A. S., Runeckles, V. C., Krupa, S. V., and Shadwick, D. S. (1989) Important considerations for establishing a secondary ozone standard to protect vegetation. *JAPCA* **39,** 1039–1045.

Lefohn,, A. S., Davis, C. E., Jones, C. K., Tingey, D. T., and Hogsett, W. E. (1987) Co-occurrence patterns of gaseous air pollutant pairs at different minimum concentrations in the United States. *Atmos. Environ.* **21,** 2435–2444.

Lefohn, A. S., Benkovitz, C. M., Tanner, R. L., Shadwick, D. S., and Smith, L. A. (1990a) Air quality measurements and characterization for terrestrial effects research. In *NAPAP State of Science and State of Technology,* Report 7, National Acid Precipitation Assessment Program, Washington, DC. September.

Lefohn, A. S., Shadwick, D. S., Somerville, M. C., Chappelka, A. H., Lockaby, B. G., and Meldahl, R. S. (1992) The characterization and comparison of ozone exposure indices used in assessing the response of loblolly pine to ozone. *Atmos. Environ.* **26A,** 287–298.

Logan ,J. A. (1989) Ozone in rural areas of the United States. *J. Geophys. Res.* **94,** 8511–8532.

Mancini, J. L. (1983) A method for calculating effects, on aquatic organisms, of time varying concentrations. *Water Res.* **10,** 1355–1362.

Mann, L. K., McLaughlin, S. B., and Shriner, D. S. (1980) Seasonal physiological responses of white pine under chronic air pollution stress. *Environ. Exp. Bot.* **20,** 99–105.

Meagher, J. F., Lee, N. T., Valente, R. J., and Parkhurst, W. J. (1987) Rural ozone in the southeastern United States. *Atmos. Environ.* **21,** 605–615.

Miller, P. R., Taylor, O. C., and Poe, M. (1986) Spatial variation of summer ozone concentrations in the San Bernardino Mountains. In *Proc. 79th Annual Air Pollution Control Association Meeting, Minneapolis, MN.* Air Pollution Control Association, Pittsburgh.

Miller, P. R., McBride, J. R., Schilling, S. L., and Gomez, A. P. (1989) Trend of ozone damage to conifer forests between 1974 and 1988 in the San Bernardino mountains of Southern California. In *Effects of Air Pollution on Western Forests* (eds., Olson, R. K. and Lefohn, A. S.). Air and Waste Management Association, Pittsburgh, 309–323.

Mohnen, V. A. (1988a) Exposure of forests to air pollutants, clouds, precipitation and climatic variables. Contract Number CR 813934-01-02, U.S. Environmental Protection Agency, Office of Research and Development, Research Triangle Park, NC.

Mohnen, V. A. (1988b) Mountain cloud chemistry project — wet, dry and cloud water deposition. Contract Number CR 813934-01-02, U.S. Environmental Protection Agency, Office of Research and Development, Research Triangle Park, NC.

Mohnen, V. A., Hogan, A., and Coffey, P. (1977) Ozone measurements in rural areas. *J. Geophys. Res.* **82,** 5889–5895.

Munn, R. E. (1970) *Biometeorological Methods.* Academic Press, New York, 336.

Musselman, R. C., McCool, P. M., and Younglove, T. (1988) Selecting ozone exposure statistics for determining crop yield loss from air pollutants. *Environ. Pollut.* **53,** 63–78.

Musselman, R. C., Oshima, R. J., and Gallavan, R. E. (1983) Significance of pollutant concentration distribution in the response of "red kidney" beans to ozone. *J. Am. Soc. Hort. Sci.* **108,** 347–351.

Musselman, R. C., Huerta, A. J., McCool, P. M., and Oshima, R. J. (1986) Response of beans to simulated ambient and uniform ozone distribution with equal peak concentrations. *J. Am Soc. Hort. Sci.* **111,** 470–473.

National Acid Precipitation Assessment Program (1987) *Interim Assessment — The Causes and Effects of Acid Deposition of Vol. I — Executive Summary.* U.S. Government Printing Office, Washington, D.C.

Nouchi, I. and Aoki, K. (1979) Morning glory as a photochemical oxidant indicator. *Environ. Pollut.* **18,** 289–303.

O'Gara, P. J. (1922) Abstract of paper: sulphur dioxide and fume problems and their solutions. *Ind. Eng. Chem.* **14,** 44.

Oshima, R. J. (1975) Development of a system for evaluating and reporting economic crop losses caused by air pollution in California. III. Ozone dosage — crop loss conversion function — alfalfa, sweet corn. IIIA. Procedures for production, ozone effects on alfalfa, sweet corn and evaluation of these systems. California Air Resources Board, Sacramento, CA.

Oshima, R. J. (1978) The impact of sulfur dioxide on vegetation: a sulfur dioxide-ozone response model. California Air Resources Board Agreement. A6-162-30.

Oshima, R. J., Poe, M. P., Braegelmann, P. K., Baldwin, D. W., and Van Way, V. (1976) Ozone dosage-crop loss function for alfalfa: a standardized method for assessing crop losses from air pollutants. *JAPCA* **26,** 861–865.

Pagnotti, V. (1990) Seasonal ozone levels and control by seasonal meteorology. *J. Air Waste Manage. Assoc.* **40,** 206–210.

Parry, M. A. J. and Day, W. (1988) A comparison of indices that describe the relationship between exposure to ozone and reduction in the yield of agricultural crops [Comments on article by Lefohn et al. (1988a)]. *Atmos. Environ.* **22,** 2057–2058.

Pedersen, B. S and Cahill, T. A. (1989) Ozone at a remote, high-altitude site in Sequoia National Park, California. In *Effects of Air Pollution on Western Forests* (eds., Olson, R. K. and Lefohn, A. S.). Air and Waste Management Association, Pittsburgh, 207–220.

Peterson, D. L., Arbaugh, M. J., Wakefield, V. A., and Miller, P. R. (1987) Evidence for growth reduction in ozone-injured Jeffrey pine (*Pinus jeffreyi* Grev. and Balf.) in Sequoia and Kings Canyon National Parks. *JAPCA* **37,** 908–912.

Pinkerton, J. E. and Lefohn, A. S. (1986) Characterization of ambient ozone concentrations in commercial timberlands using available monitoring data. *Tappi J.* **April,** 58–62.

Pinkerton, J. E. and Lefohn, A. S. (1987) The characterization of ozone data for sites located in forested areas of the eastern United States. *JAPCA* **37,** 1005–1010.

Prinz, B. and Krause, G. H. M. (1988) State of scientific discussion about the causes of the novel forest decline in the Federal Republic of Germany and surrounding countries. Presented at the 15th Int.l Meet. Specialists in Air Pollution Effects on Forest Ecosystems, October 2–8, 1988. IUFRO. Air Pollution and Forest Decline. Interlaken, Switzerland.

Prinz, B., Krause, G. H. M. and Jung, K. D. (1985) Untersuchungen der LIS zur problematik der waldschaden. In *Waldschaden — Theorie und Praxis auf der Suche nach Antworten*. R. Oldenbourg Verlag, Munchen, Wien. 143–194.

Reagan, J. (1984) Air quality interpretation. In National Crop Loss Assessment Network (NCLAN) 1982 Annual Report, EPA-600/3-84-049, U.S. Environmental Protection Agency, Corvallis, OR, 198–219.

Reinert, R. A. and Nelson, P. V. (1979) Sensitivity and growth of twelve elatiorbegonia cultivars to ozone. *Hort. Sci.* **14,** 747–748.

Runeckles, V. C. (1974) Dosage of air pollutants and damage to vegetation. *Environ. Conserv.* **1,** 305–308.

Runeckles, V. C. (1988) A comparison of indices that describe the relationship between exposure to ozone and reduction in the yield of agricultural crops [Comments on article by Lefohn et al. (1988a)]. *Atmos. Environ.* **22,** 1241–1242.

Runeckles, V. C. and Wright, E. F. (1989) Exposure-yield response models for crops. In *Proc. 82nd Annual Meeting of the Air and Waste Management Association*, June 25–30, 1989, Anaheim, CA. Air and Waste Management Association, Pittsburgh.

Shadwick, D. S. (1989) Personal communication. The standard errors were determined for the seasonal average of the daily maximum 7-h and 12-h concentrations, W126, and 7-month accumulation of the number of hourly average concentrations equal to or greater than 0.08 ppm, using a combination of simulation and standard statistical estimation. ManTech Environmental Technology, Inc., Research Triangle Park, NC.

Showman, R. E. (1991) A comparison of ozone injury to vegetation during moist and drought years. *JAPCA* **41,** 63–64.

Smith, L. A. and Lefohn, A. S. (1991) Co-occurrence of ozone and wet deposited hydrogen ion in the United States. *Atmos. Environ.* **25A,** 2707–2716.

St. John, D. S., Bailey, S. P., Fellner, W. H., Minor, J. M., and Snee, R. D. (1981) Time series search for trend in total ozone measurements. *J. Geophys. Res.* **86,** 7299–7311.

Stan H. J., Schicker S. and Kassner H. (1981) Stress ethylene evolution of bean plants — a parameter indicating ozone pollution. *Atmos. Environ.* **15,** 391–395.

Stasiuk, W. N. and Coffey, P. E. (1974) Rural and urban ozone relationships in New York State. *JAPCA* **24,** 564–568.

Stratmann, H. (1963) Freilandversuche zur Schwefeldioxidwirkungen auf die vegetation. II. Teil: Messung und Bewertung der SO2-Emissionen (Field experiments for the determination of the effects of sulfur dioxide on vegetation. Part II: measurement and assessment of SO2 emissions). Forschungsberichte des landes Nordrhein-Westfalen no. 1184. Westdeutscher Verlag, Koln and Opladen.

Taylor, G. E., Jr., and Norby, R. J. (1985) The significance of elevated levels of ozone on natural ecosystems of North America. In *Evaluation of the Scientific Basis for Ozone/Oxidants Standards* (ed., Lee, S. D.). Air Pollution Control Association, Pittsburgh, 152–175.

Taylor, G. E., Jr, McLaughlin, S. B., and Shriner, D. S. (1982) Effective pollutant dose. In *Effects of Gaseous Air Pollution in Agriculture and Horticulture* (eds., Unsworth, M. H. and Ormrod, D. P.). Butterworth Scientific, London, 458–460.

Thompson, C. R., Kats, G., and Cameron, J. W. (1976) Effect of photochemical air pollutants on two varieties of alfalfa. *ES&T* **10**, 1237–1241.

Thompson, C. R., Olszyk, D. M., Kats, G., Bytnerowicz, A., Dawson, P. J., and Wolf, J. W. (1984) Effects of ozone or sulfur dioxide on annual plants of the Mojave Desert. *JAPCA* **34**, 1017–1022.

Tingey, D. T. (1984) The effects of ozone on plants in the United States. In *Proc. International Workshop on the Evaluation of the Effects of Photochemical Oxidants on Human Health, Agricultural Crops, Forestry, Materials and Visibility* (ed., Grennfelt P.). Swedish Environmental Research Institute, Göteborg, 60–71.

Tingey, D. T. and Taylor, G. E., Jr. (1982) Variation in plant response to ozone: a conceptual model of physiological events. In *Effects of Gaseous Air Pollution in Agriculture and Horticulture* (eds., Unsworth, M. H. and Ormrod, D. P.). Butterworth Scientific, London, 113–138.

Tingey, D. T., Hogsett, W. E., and Lee, E. H. (1989) Analysis of crop loss for alternative ozone exposure indices. In *Atmospheric Ozone Research and its Policy Implications* (eds., Schneider, T., Lee, S. D., Wolters, G. J. R., and Grant, L. D.). Elsevier, Amsterdam, 219–225.

Tingey, D. T., Reinert, R. A., Dunning, J. A., and Heck, W. W. (1971) Vegetation injury from the interaction of NO_2 and SO_2. *Phytopathology* **61**, 1506–1551.

Tonneijck, A. E. G. (1984) Effects of peroxyacetyl nitrate (PAN) and ozone on some plant species. In *Proc. International Workshop on the Evaluation of the Effects of Photochemical Oxidants on Human Health, Agricultural Crops, Forestry, Materials and Visibility* (ed., Grennfelt, P.). Swedish Environmental Research Institute, Göteborg, 118–127.

U.S. EPA. (1978) Altitude as a factor in air pollution. EPA-600/9-78-015, U.S. Environmental Protection Agency, Environmental Criteria and Assessment Office, Research Triangle Park, NC. Available from National Technical Information Service, Springfield, VA.

U.S. EPA. (1986) Air quality criteria for ozone and other photochemical oxidants. Vol III. EPA-600/8-84/020cF, U.S. Environmental Protection Agency, U.S. Environmental Criteria and Assessment Office, Research Triangle Park, NC.

U.S. EPA. (1988a) Summary of selected new information on effects of ozone on health and vegetation: Draft supplement to air quality criteria for ozone and other photochemical oxidants. EPA-600/8-88/105A, U.S. Environmental Protection Agency, Office of Health and Environmental Assessment, Washington, D.C.

U.S. EPA. (1988b) Review of the national ambient air quality standards for ozone assessment of scientific and technical information. U.S. Environmental Protection Agency, Office of Air Quality Planning and Standards, Research Triangle Park, NC.

U.S. EPA. (1989c) National air quality and emissions trends report, 1987. EPA-450/4-89-001, U.S. Environmental Protection Agency, Office of Air Quality Planning and Standards, Research Triangle Park, NC.

U.S. EPA. (1990) National air quality and emissions trends report, 1988. EPA-450/4-90-002, U.S. Environmental Protection Agency, Office of Air Quality Planning and Standards, Research Triangle Park, NC.

U.S. EPA. (1991) National air quality and emissions trends report, 1989. EPA-450/4-91-003, U.S. Environmental Protection Agency, Office of Air Quality Planning and Standards, Research Triangle Park, NC.

Wakim, P. G. (1989) Temperature-adjusted ozone trends for Houston, New York and Washington. In *Proc. 82nd Annual Meeting of the Air and Waste Management Association*. Air and Waste Management Association, Pittsburgh.

Walker, H. M. (1985) Ten-year ozone trends in California and Texas. *JAPCA* **35**, 903–912.

Winner, W. E., Lefohn, A. S., Cotter, I. S., Greitner, C. S., Nellessen, J., McEvoy, L. R., Jr., Olson, R. L., Atkinson, C. J., and Moore, L. D. (1989) Plant responses to elevational gradients of O_3 exposures in Virginia. *Proc. Natl. Acad. Sci. U.S.A.* **86,** 8828–8832.

Wolff, G. T., Lioy, P. J., and Taylor, R. S. (1987) The diurnal variations of ozone at different altitudes on a rural mountain in the Eastern United States. *JAPCA* **37,** 45–48.

Yarbrough, J. W. (1988) Recent ozone trends in Dallas and an evaluation of possible causative factors. In *Proc. 81st Annual Meeting of the Air Pollution Control Association.* Air Pollution Control Association, Pittsburgh.

CHAPTER 4

Experimental Methodology for Studying the Effects of Ozone on Crops and Trees

William J. Manning, Department of Plant Pathology, University of Massachusetts, Amherst, MA

Sagar V. Krupa, Department of Plant Pathology, University of Minnesota, St. Paul, MN

4.1 INTRODUCTION

Ozone (O_3) is an all-pervasive air pollutant. Since the pioneering studies of Haagen-Smit (1952), Richards et al. (1958), and Heggestad and Middleton (1959), O_3 has been recognized as the most important regional-scale phytotoxic air pollutant in the U.S. Increasing numbers of reports of O_3 injury on sensitive and/or indicator plants from countries such as Australia, Canada, Greece, India, Israel, Japan, Mexico, the Netherlands, Spain, the U.K., and Germany indicate that O_3 is also of increasing concern on a worldwide scale (Krupa and Manning, 1988).

Numerous investigators have studied the adverse effects of O_3 on crop growth and productivity (Heck et al., 1988; and Environ. Pollut., Vol. 53, 1988). Ozone has also been implicated as a major factor in forest decline (Krause et al., 1983; Ashmore et al., 1985; de Bauer et al., 1985; McLaughlin, 1985; Prinz, 1987; Linzon and Chevone, 1987).

Presently there is international concern regarding the depletion of stratospheric O_3 (Krupa and Manning, 1988; Krupa and Kickert, 1989). It is tropospheric O_3, however, that has direct effects on vegetation. For a detailed treatment of tropospheric O_3 production and destruction, the reader should consult Altshuller (1986), Demerjian (1986), Finlayson-Pitts and Pitts (1986), Krupa and Manning (1988), and Chameides and Lodge (Chapter 2, this volume).

4.1.1 Characteristics of Ambient Ozone

As with many air pollutants, there is a significant temporal and spatial variability in surface concentrations of ambient O_3. Ozone concentrations vary with altitude, latitude, season, and hour of the day (Pruchniewicz, 1973; U.S. EPA, 1986; Legge and Krupa, 1989). According to Altshuller (1986), surface O_3 concentrations measured at nonurban locations are influenced by (a) transport of O_3 formed in the stratosphere into the free troposphere and subsequent transport down into the planetary boundary layer (PBL); (b) photochemical O_3 formation within the free troposphere and the clean PBL; (c) photochemical O_3 formation within the polluted PBL, especially during the passage of warm high-pressure systems; and (d) O_3 formation within single or superimposed plumes. Thus, surface O_3 concentrations are governed by both natural and anthropogenic processes.

As previously stated, there is a seasonality in the patterns of surface O_3 concentrations. Depending on the geographic location and the type of atmospheric processes governing the O_3 concentrations in the tropospheric boundary layer, maximum hourly concentrations, or any modified numerical expressions thereof, may be observed during early spring, summer, or early fall (Singh et al., 1978; Logan, 1985; Lefohn et al., 1990).

Where local photochemistry is the dominant process, where transport distances of O_3 and/or its precursors are fairly short (few hours), or where previous day precursors are transported in sufficient amounts to distant geographic locations, daily maximum O_3 concentrations occur during the daylight hours soon after the incidence of the highest surface solar radiation (Pratt et al., 1983; Legge and Krupa, 1989). Where stratospheric intrusion (relatively infrequent) and certain types of long-range transport processes are involved, maximum O_3 concentrations may occur during hours of the day not expected to be governed by direct photochemistry (Finlayson-Pitts and Pitts, 1986; Legge and Krupa, 1989).

Based on the information presented above, we can conclude that seasonal and daily patterns of surface O_3 are highly site specific. This is an extremely important factor in understanding cause-effect relationships in vegetation-effects research at a given geographic location. One way to address this issue is to examine data for surface O_3 measurements gathered over several years at the effects study location, or in its vicinity, prior to the initiation of a field effects study.

Ambient O_3 exposure-vegetation response relationships are inherently stochastic (Krupa and Teng, 1982). At many rural locations, surface O_3 exposure profiles consist of 1 or more days of relatively low hourly O_3 concentrations interspersed at random with 1 or more days of O_3 episodes with relatively high hourly O_3 concentrations. When annual or seasonal hourly O_3 concentrations are examined for their frequency distributions, they generally do not follow a normal distribution (Figure 4.1) (Pratt et al., 1983). They are best described by the numerical functions of the Weibull distribution (Lefohn and Benedict, 1982; Legge and Krupa, 1989). These types of frequency distributions exhibit a peak to the left at relatively low hourly O_3 concentrations, with a long tail to the right towards high hourly O_3 concentrations (Figures 4.1 and 4.2). Krupa and Kickert (1987), Lefohn and Runeckles (1987), and Lefohn et al. (1989) have discussed the importance of the timing of the O_3 episodes or stress (tail part of the Weibull distribution) relative to the varying sensitivity of the changing growth stages of a given plant species. This subject is discussed in greater detail in other parts of this chapter.

4.1.2   Types of O_3 Effects on Crops and Trees

The effects of O_3 on plants can be observed at the cell, organ, whole plant, and population levels. Depending on the response variable measured and the level at which it is measured, the influence of O_3 can be positive, no measurable effect, or negative. In the end, it is the adverse effects of O_3 on vegetation that are of concern to most investigators. The emphasis of what is to follow in the rest of this chapter is on whole plants and populations. For a treatment of the effects of O_3 on plant biochemistry and physiology, the reader is referred to Unsworth and Ormrod (1982), Koziol and Whatley (1984), Schulte-Hostede et al. (1988), and Runeckles (Chapter 5, this volume).

Episodic, acute O_3 exposures (relatively high O_3 concentrations from a few to several hours on 1 or more days) can result in injury symptoms (visible evidence) on sensitive plant species under appropriate environmental conditions. Such symptoms on broadleaved plants consist of chlorosis, fleck, stipple, or uni- and bifacial necrosis. On conifers, acute response consists of mottle, banding, and tip chlorosis and necrosis spreading downward (Krupa and Manning, 1988). Plants have the ability to compensate for acute effects, depending on the respite time between acute exposures, and on plant phenology when the initial stress occurred (Lefohn and Runeckles, 1987). Historically, much of the research on acute effects has been conducted under controlled environment and greenhouse conditions (Heck et al., 1979; U.S. EPA, 1986).

Chronic effects are caused by exposure to frequent, relatively low hourly O_3 concentrations, with periodic, random, intermittent peaks of relatively high hourly O_3 concentrations on 1 or more days. Chronic effects can lead to changes in plant growth, productivity, and quality. Such changes may or may not involve symptom production. Where symptoms are present, they include chlorosis, delayed early season growth, premature senescence, and leaf abscission. In reality, chronic exposure may be considered as a series of acute exposures interspersed with nonepisode days of O_3. Since plants can compensate for stress (O_3 episode) during respite periods, the frequency of O_3 episodes and the time interval between such episodes are critical in evaluating and modeling plant response (Krupa and Kickert, 1989; Lefohn and Runeckles, 1987). Much of the

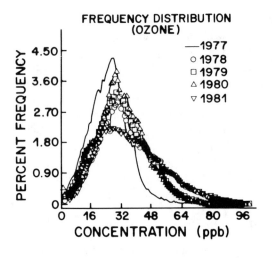

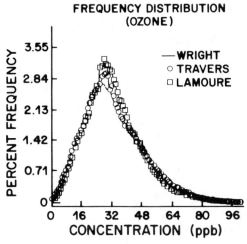

Figure 4.1A. Frequency distributions for ambient ozone. Above, the frequency distributions for ozone are broken down by year, whereas at the bottom, they are broken down by site. Among years, 1978, and particularly 1980, had more high O_3 concentrations (Pratt et al. 1983).

research on chronic exposures has been conducted under greenhouse and field conditions.

With annuals, O_3 exposure and plant response are compressed within one growth cycle. With perennials during a given year, if the influence of carry-over of the products of physiological and biochemical processes from the previous years are accounted for, then they may be considered similar to annuals in a given year. For example, most tree species are considered to be sensitive to acute O_3 exposures during the 8–13 week period after the onset of spring growth (Davis and Wood, 1972, 1973). In the end, after many years of O_3 exposure and

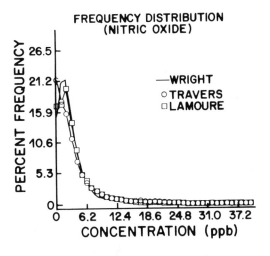

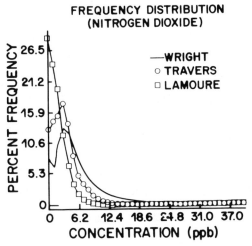

Figure 4.1B. Frequency distribution for nitric oxide and nitrogen dioxide. Only minor differences were observed among sites. The top shows the frequency distribution for nitric oxide, broken down by site. The concentrations were similar at all three sites. The bottom shows the frequency distribution for nitrogen dioxide, broken down by site. Major differences in the concentrations of NO_2 occurred at the different sites, with a gradient extending out from the Minneapolis-St. Paul, MN, urban area (Pratt et al., 1983).

response, final tree growth change should be considered as a product of its individual annual responses to stress. Since such a relationship is considered to be stochastic in nature, cause-effect relationships will vary in space and from season to season or year to year. This particular issue has led to a great deal of uncertainty in modeling regional scale perennial crop or tree responses to O_3 (Krupa and Teng, 1982; Heck et al., 1988).

Many of the important considerations described in the preceding narrative are

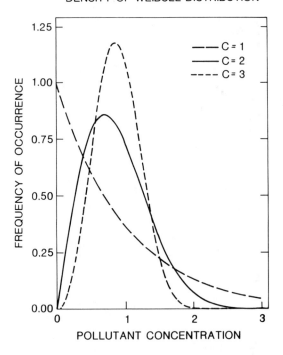

DENSITY OF WEIBULL DISTRIBUTION

Figure 4.2.   Theoretical distributions illustrating the Weibull family of frequency distributions.

critical in selecting the best experimental approach, design, and data-analysis strategy. However, it is important to note that these issues have been the subject of much controversy (Krupa and Nosal, 1989; Lefohn et al., 1989).

4.2   RATIONALE FOR THE SELECTION OF THE EXPERIMENTAL METHODOLOGY

The rationale or the objectives of a given study should dictate the experimental methodology used. Accepting the fact that O_3 is a regional scale phytotoxic air pollutant of worldwide concern, there are a number of specific reasons for conducting O_3 exposure-plant response studies:

1. Confirmation of O_3 as the primary causal agent of an observed problem on a given plant species or a number of species in a given geographic area. This involves artificially reproducing the O_3 injury symptoms on healthy plants under controlled environment and/or greenhouse conditions.
2. Description of O_3 dose-response relationships of plant species considered responsive to O_3 exposures. Depending on the objective, these types of studies have been conducted under all three conditions: growth chamber,

greenhouse, and field. In addition to providing assistance in developing regulatory policies, these types of studies are required for an understanding of the interactions between various environmental variables and O_3 exposure characteristics that lead to an O_3 effect.

3. Understanding the mechanism of O_3 injury and/or stress repair on selected or example plant species. Historically, stress physiology and stress repair have been the subject of attention of plant physiologists. Understanding the mechanism of O_3 injury can provide information as to why some genera, species, varieties, or cultivars are O_3 sensitive while others are tolerant. Certainly this has implications in plant breeding.

4. Screening for differential response of a number of genera, species, varieties, or cultivars. Plant pathologists and entomologists typically screen plants for their responses to biotic pathogens and pests. Similarly, a number of scientists have examined the responses of crop cultivars to O_3 exposure (U.S. EPA, 1986). This has also been the case with urban tree species, and to a limited extent, with native plant species (U.S. EPA, 1986). While the identification and use of tolerant cultivars of a given crop species are feasible, such an approach is impractical with natural ecosystems where numerous components are involved within a given ecosystem. Screening studies with crops should be closely integrated with ozone-injury-mechanism studies to gain the most knowledge and application. An important consideration is the behavior of a given receptor to O_3 exposures in the context of its response to other biotic and abiotic stress factors.

5. Crop-yield loss analysis, including quality. A key issue in the context of regional scale O_3 research has been the effects on crop yields and quality. Yield loss and stress mitigation translate to economics (cost-benefit analysis). No doubt this is a critical component in developing air quality regulatory policies. In comparison to yield studies, much less information has been gathered on the nutritional quality of the consumed plant part (U.S. EPA, 1986).

6. Growth analysis of tree species. As stated previously, chronic exposure to O_3 can result in altered tree growth, vigor, and reproductivity. When such effects are observed within the major components of an ecosystem, it can lead to alterations within other components in the system, as illustrated in the San Bernardino forest in Southern California (U.S. EPA, 1986).

7. Understanding of nutrient, material balance, and other properties of small scale, reconstructed ecosystem components or small samples of the natural ecosystem itself.

8. Development of a database for setting ambient air quality standards. While all the previously mentioned types of studies are important in setting ambient air quality standards, specific research programs, such as the U.S. National Crop Loss Assessment Network (NCLAN), have been conducted to expressly address the issue. At least in the U.S., the motivation for such studies has been the need for national-level assessment of O_3 impacts and a need to identify O_3 exposure statistics that are effective and yet simple to implement as an ambient air quality standard.

4.3 REQUIREMENTS OF THE EXPERIMENTAL METHODOLOGY

There are several requirements that should be considered in selecting the proper experimental method that would satisfactorily address the objectives of the study. In addition, the experimental and the numerical modeling steps in response assessment are closely related, since the data collected will often dictate the type of model that can be developed. Specifying the use of a given model at the onset of the study, however, will influence the experimentation. In O_3 research it is common to find these two steps treated separately and even conducted by different researchers, a situation that invariably leads to confusion and conflict between the biological validity of a model and its mathematical representation (Krupa and Teng, 1982).

It is beyond the scope of this chapter to provide descriptions of experimental designs and statistical data analysis methods. The reader is referred to the work of, among others, Conover (1971), Cochran (1973), Little and Hills (1978), Snedecor and Cochran (1978), Montgomery (1984), and France and Thornley (1984). However, in the following narrative some of the issues relevant to the selection and/or the development of experimental approaches are discussed.

When attempting to generate field data for quantifying the effect of O_3 dose on plant response, a major consideration is whether the method used will allow significant accounting of all other sources of variation except the O_3. There is an implicit acknowledgment, when using traditional experimental designs and analysis of variance, that this is possible. In practice, with field experiments using different O_3 exposure regimes, variation caused by "interfering" factors is a problem. If the ideal situation of noninterference is desired, then an experimental approach that can satisfy this need should be selected, or the model used in the data analysis must incorporate these sources of variation.

The relationships between pollutant dose and response should be viewed as a three-dimensional surface where response is affected differently by the O_3 dose at different growth or developmental stages (Krupa and Kickert, 1987). When attempting to generate data to model this surface, it is more meaningful to create a broad range of exposure regimes rather than to intensively replicate a few exposure regimes. Most published literature has provided data that only enables an understanding of part of the O_3 dose-response surface (Heck et al., 1988; Environ. Pollut. Vol. 53, 1988). This is largely attributable to the experimental designs that have been used — randomized block, factorial — which facilitate comparison of exposure regimes by emphasizing replication, but do not facilitate defining the response surface. Defining a response surface demands appropriate experimental designs, such as response surface methodology, which, in general, emphasizes the number of treatments over the number of replications (Myers, 1971).

Accuracy (exactness of a measurement) and precision (reproducibility of a measurement) are critical in the selection of the experimental approach. While these issues can be satisfactorily addressed in controlled experiments, the size of the receptors used in field studies is an important determinant of the accuracy of dose-response equations. There is a "representational error" incurred whenever an equation developed using a small unit is applied to the estimation of effects on a larger unit. For example, this occurs when results from small-plot fumigations are extrapolated to regional effects. The representational error is a reflection of different amounts of variation that can be accounted for in samples of different sizes. In general, decreasing the size of the receptor unit in an exposure experiment increases the amount of uncertainty associated with the subsequent dose-response function.

As precision increases, the ability to detect small differences between treatments will increase. This issue can be addressed reasonably well under controlled environment and certain types of greenhouse conditions. However, it is not logical to assume that we can improve the precision of current O_3 exposure-response models in field studies until the quality of experimental data is improved to reflect the dynamic interaction between O_3 dose and plant response.

Any experimental design and method used must have diversity and broadness in establishing cause-and-effect relationships. Furthermore, the approach should have general applicability and realism relative to ambient O_3 exposure regimes in a given geographic locality. In this context, one major source of variation is the reference response level to which treatment responses are compared. The so-called background levels of O_3 will vary according to the locality, and it may be necessary to assume that vegetation in a geographic area has sufficiently adapted to this "background level" to show no response. The related question is whether it is logical to consider the zero response as being equivalent to zero ozone dose. Regardless of which response level is taken as the reference point, the level itself is determined by other environmental and experimental variables.

When all the factors discussed in the preceding section are considered, every available experimental method has advantages and disadvantages. This subject is discussed in greater detail in the following sections. Many of the disadvantages can be minimized by a combination of improved experimental design; more intensive, diverse, and in-depth data gathering; and more innovative and powerful modeling or numerical analysis. No doubt, many studies have been limited by the costs for technology and labor. Nevertheless, until atmospheric scientists, biologists, and mathematicians develop fully cooperative and integrated studies, any critical program on ambient O_3 exposure and plant response will continue to stimulate controversy (Lefohn et al., 1989). Any integrated studies would require exchange of scientific knowledge between the participants from the various disciplines **before** the study is conducted, and not during or at the end. This is essential for developing a scientifically sound experimental strategy at the outset, which maintains continued involvement by all participants on an intensive basis until the study is fully completed.

4.4 SELECTION AND CARE OF PLANT MATERIAL

There are many approaches for the study of O_3 effects on vegetation. None of them, however, will provide useful results unless careful consideration is also given in advance to the biological, cultural, and physical factors that affect plant responses to O_3.

Guderian et al. (1985) point out that there is no absolute resistance (immunity) to air pollutants. With the appropriate concentration and duration of exposure, every plant species can be injured. However, this must be interpreted within the context of the ambient conditions at a geographic location. Relative resistance or tolerance to O_3 is based on inherent genetic characters, that may change during different stages of plant development. Plant tolerance to O_3 is also determined by both previous environmental conditions and those that occur during, and sometimes after, exposure to O_3 episodes.

Important biological, cultural, and physical factors that affect plant responses to O_3 are summarized in Tables 4.1 and 4.2. All these factors interact to affect plant responses to O_3, but they are separated here for ease of presentation. For more extensive consideration of these factors, the reader should consult Brennan et al. (1964), Darrall (1989), Guderian et al. (1985), Heck (1968), Heggestad and Heck (1971), Leone and Brennan (1979), Manning and Keane (1988), and Ormrod et al. (1973).

Physical factors that affect plant responses to O_3 can be controlled, to one extent or another, in most chamber systems. Most chambers have air movement sufficient to alter boundary-layer resistance. If air movement is too fast, soil moisture may be depleted rapidly, affecting plant responses to O_3. Light intensity, quality, and photoperiod are frequent problems in chambers. Atypical growth will occur if light conditions are not appropriate for the plants being exposed to O_3. Temperature and relative humidity can usually be regulated in

Table 4.1. Physical Factors Affecting Plant Responses to Ozone

Factors	Responses
Air movement	Must be sufficient to alter boundary layer resistance to allow O_3 uptake
Light Intensity, photoperiod, and quality	Ideal value varies for each plant; less or more than ideal value for each plant reduces O_3 sensitivity
Temperature	Injury increases in a range from 3 to 30°C
Water Relative humidity	Controls stomatal opening and gas exchange; uptake of O_3 and injury should increase as relative humidity increases
Soil moisture	Water stress increases O_3 tolerance due to stomatal closure

Table 4.2. Biological and Cultural Factors Affecting Plant Responses to Ozone

Factors	Responses
Biological	
Genetic diversity	Homogeneous plants give uniform responses; species, clones, cultivars and provenances react differently to O_3
Stage of plant development	Plant developmental stages and leaf age affect responses to O_3
Cultural	
General cultural practices	Optimal or usual practices will result in "typical" responses to O_3
Growth media	Soilless media allow uniformity and reproducibility, but are less relevant than natural soils
Nutrients	Optimal levels usually result in optimal injury from O_3
Pesticides	Variable effects, ranging from none to protection, from injury to reduced tolerance to joint effects with O_3

chambers. Problems with soil moisture can occur in closed and open-top chambers, due to high air-flow volume or rain shadow effects.

Air pollution scientists are sometimes not familiar enough with plant biology to select the most-appropriate plants for experiments or the most-appropriate stage of plant development or leaf age. Genetic diversity in test plants can lead to the masking of O_3 effects, or to the selection of tolerant or susceptible individuals. Genetic diversity in many tree species means that clones must be used to minimize variability, or large numbers of seedlings with numerous replications must be used. However, consideration must be given to the changing tolerance (or susceptibility) of plants and leaves as they age and senesce.

Cultural practices appropriate for test plants need to be followed if realistic O_3 responses are to be obtained. Soilless growth media, ranging from gravel in a styrofoam cup amended with a nutrient solution to peat-vermiculite mixtures with slow-release fertilizers, are frequently used to provide uniformity and reproducibility in experiments. However, they usually lack relevance. This problem can be resolved by using natural soils. Optimal nutrient levels usually result in optimal O_3 injury in plants. Ormrod et al. (1973) also found that temperature and N and P nutrition interacted to affect radish response to O_3.

The effects of pesticides on plant responses to O_3 is seldom considered by investigators. Air pollution scientists are often single-parameter oriented, and they frequently use excessive applications of fungicides, insecticides, and miticides on a preventive schedule (Manning and Keane, 1988). Guderian et al. (1985) report that a number of common pesticides can alter the response of plants

Table 4.3. Response of Plants to Ozone as Influenced by Selected Pesticides

Pesticides	Plants	Responses
Fungicides		
Contact		
dithiocarbamates (maneb, zineb, thiram, and ferbam)	herbaceous crops	protection
Systemic		
benomyl	herbaceous and woody plants	protection
carboxin	herbaceous crops	protection
Herbicides		
chloramben	tobacco	reduced tolerance
diphenamid	tobacco	protection
isopropalin	tobacco	protection
Insecticides		
diazinon	bean	protection
methomyl	bean	more than additive injury
Nematicides		
aldicarb fensulfothion oxamyl phenamiphos	bean and tobacco	reduced tolerance

to O_3 (Table 4.3). Antioxidants are also used as additives in many insecticides. More consideration should be given to interactive effects between pesticides and plants in relation to their responses to O_3.

4.5 EXPERIMENTAL METHODOLOGY

In this section we discuss the principal methods used to study the effects of O_3 on plants. Sufficient detail is given to acquaint the reader with each method, but additional details must be obtained from the original papers and reports. For each method, many investigators have also developed their own modified versions. These are almost as numerous as the number of investigators. For more extensive coverage of methodology, the reader is referred to Guderian et al. (1985), Heagle (1989), Heagle et al. (1988), Hogsett et al. (1987a,b), Krupa (1984), Krupa and Nosal (1989), Last (1986), and Ormrod et al. (1988).

Each method has its strengths and weaknesses, and these are pointed out here. The choice of method will depend on the objectives of the experiment and the

Table 4.4. Summary of Controlled Environment Exposure Systems Used to Determine the Effects of Ozone on Plants

Systems	References
Modified greenhouses	Darley and Middleton (1961), Menser et al. (1966), Heggestad et al. (1967), Berry (1970), Manning and Feder (1976)
Modified growth chambers	Wood et al. (1973)
Experimental chambers (used in greenhouses, growth rooms, or growth chambers)	
Cuvettes	Olszyk and Tibbitts (1981), Noble and Jensen (1983)
Rectangular chambers	Cantwell (1968), Heck et al. (1968), Heagle and Philbeck (1974), Keane and Manning (1988)
Round chambers Continuous stirred tank reactors (CSTRs)	Rogers et al. (1977), Heck et al. (1978)

equipment and finances available for the work. For some investigators, the methods are more important than the results, which leads to problems with data relevancy. As long as the limitations and advantages of each method are understood in advance, there are usually several methods appropriate for any objective. However, it is critical that the results be interpreted within the context of the methodology used, rather than extrapolation of such results to inappropriate situations.

4.5.1 Controlled Environment Exposure Systems

Investigators are likely to choose controlled environment exposure systems for experiments involving considerable control over O_3 concentrations and environmental variables. (Table 4.4). Modified greenhouses have the least control, but are useful for large plants and studies of long duration (i.e. several months). Smaller and usually fewer plants can be included in experimental chambers, and experiment durations are also usually shorter than in modified greenhouses. A large number of chamber designs have been used, and their descriptions have been published (Hogsett et al., 1987a,b).

4.5.1.1 Modified Greenhouses

Traditionally, greenhouses have been used to grow large numbers of plants

Table 4.5. Yields of Potato Cultivars Grown in Charcoal-Filtered or Nonfiltered Air in Greenhouses, Beltsville, MD, 1971

| | Yields (g) | | |
| | Charcoal-filtered | Nonfiltered | NF × 100 |
Cultivars	(CF)	(NF)	CF
Katahdin	218	279	128
Penn 71	251	269	107
Pungo	282	287	102
Norgold Russett	251	252	100
Superior	318	304	96
Kennebec	369	339	92
Wauseon	273	218	80
Norchip	466	346	74
La Chipper	380	276	73
Alamo	410	272	66
Haig	295	187	63
Norland	401	199	50

[a] LSD, 0.05 = 71 g for cultivars within an air regime, 81 g between air regimes.
[b] Adapted from Heggestad (1973).

under conditions where temperature and day length can be partially manipulated. The discovery by Darley and Middleton (1961) that activated-charcoal filters could protect plants from O_3 injury in Los Angeles provided the basis for the first experimental systems for the study of the effects of O_3 on plants in greenhouses.

Menser et al. (1966) used charcoal filters to demonstrate that O_3 in ambient air caused weather fleck of tobacco at the USDA facilities in Beltsville, MD. Heggestad et al. (1967) developed a system to cool greenhouse air with pad coolers, which allowed the use of carbon-filtered air (CF) or nonfiltered (NF) air in the greenhouses during the warm summer growing months, without excessive heat build-up. This system has been used successfully at Beltsville for many years. An example of results obtained from CF and NF greenhouses is given in Table 4.5. Berry (1970) developed an alternatively heated or cooled lean-to greenhouse with humidity control to expose tree seedlings to O_3 and SO_2.

Feder utilized small plastic greenhouses as O_3 fumigation chambers (Manning and Feder, 1976). Incoming air was either charcoal filtered or nonfiltered. Mixtures of both CF and NF could also be achieved. Ozone was also artificially generated and added to the air within a chamber, and the air recirculated with small additions of CF or NF as desired. Manning (unpublished) improved the design by generating O_3 from O_2 and introducing it into the incoming air stream, after the charcoal filter. The incoming air was moved through a polyethylene tube or sleeve along the top of the chamber. The tube had perforations that pointed downward at a 45° angle, facilitating the downward flow of air over the plants on the benches below. This is a pass-through system with continual introduction of O_3 on a programmable basis.

Glass greenhouses are expensive to build and maintain. Plastic greenhouses are cheaper, easier to use, have a better design for fumigation, and allow for greater replications of treatments. Unless cooled, greenhouses cannot be used during the warm-to-hot growing season, and supplementary lighting is required during the cooler months.

4.5.1.2 Modified Growth Chambers

Wood et al. (1973) modified a commercially available growth chamber to be corrosion resistant for exposure of plants to O_3 and other pollutants. Temperature, light, relative humidity, and O_3 concentrations were carefully controlled and programmed using the growth chamber controls. In these chambers, air containing O_3 moved downward over the plants and recirculated into the chamber. This type of chamber has been used extensively by Davis and Wood (1972, 1973) and others at Penn State University.

4.5.1.3 Other Indoor Chambers

These types of systems are used to study the effects of carefully controlled and defined concentrations of O_3 on plants, with minimal interference from other environmental factors that affect plant growth. Single or multiple groups of chambers can be used in greenhouses or growth rooms, resulting in varying degrees of control over light, temperature, relative humidity, etc. Care must be taken to use inert materials to construct the chambers, to avoid release of phytotoxic compounds during experiments, which will confound the results (Hardwick et al., 1984). Many different chamber designs have been developed and are described elsewhere (Heagle and Philbeck, 1979; Hogsett et al., 1987a,b). Results from these chambers are relevant to the conditions under which they were obtained, and extrapolation to ambient or real-world conditions is usually not possible.

4.5.1.4 Cuvettes

Cuvettes are small chambers that are used to enclose part or all of a leaf attached to a plant, to allow monitoring gas fluxes on a leaf-area basis. The nature and use of cuvettes has been reviewed in more detail by Bennett (1979) and Hogsett et al. (1987a).

The simplest types of cuvettes are used to document the effects of very-short-term exposures to O_3 on gas exchange rates. These measurements of gas exchange rates are performed before or after, but not during, exposure of the leaf to O_3. Gas fluxes are calculated from measures of the flow rate, differential gas concentrations between inlet and outlet ports, and leaf area. Olszyk and Tibbitts (1981) used this system to determine the effects of O_3 on stomatal responses of Scots pine (*Pinus sylvestris*) needles.

A cuvette can be converted to a dynamic exposure system when O_3 is introduced directly into the cuvette while gas exchange rates are being determined. This is usually accomplished by injection of O_3 in the line above the inlet port of a pass-through cuvette (Noble and Jensen, 1983). In more-sophisticated versions (Figure 4.3), O_3 is the only variable, with temperature, light, air flow rates, and other environmental factors held constant throughout the length of the exposure, which can vary from a few minutes to several hours (Legge et al., 1979). Taylor et al. (1982) used a method like this to determine O_3 flux in soybean leaves of different ages. Cuvettes are often used periodically during long-term studies to determine the cumulative effects of O_3 exposure on gas exchange rates.

To minimize the effects of temperature and humidity increases, simple-type cuvettes are frequently used with plants inside growth chambers or controlled environment rooms, where control of light and temperature are better than under ambient conditions. Problems with low air movement in the cuvette can lead to increases in humidity and leaf-surface temperature. Increases in boundary-layer resistance are more difficult to manage, and dictate very short leaf exposure periods in these simple cuvettes. In larger cuvettes, fans are used to increase air movement, and liquid coolants are used to control the temperature.

Ennis et al. (1990a,b) developed a system of branch chambers, or large fan-driven cuvettes, to expose individual branches on potted red spruce trees to separate or various combinations of CO_2, H_2O, O_3, and H_2O_2. The branch chambers allowed determination of gaseous flux for a variety of gas exposures for single branches.

Branch chambers were placed over individual branches on a potted tree inside an environmental chamber, where light, temperature, and humidity were controlled (Figure 4.4). Mass flow controllers were used to deliver gas mixtures to the branch chambers. Details of the construction of a branch chamber are shown in Figure 4.5.

4.5.1.5 Rectangular Chambers

Cantwell (1968) placed a glass fumigation chamber inside a commercially available growth chamber, utilizing the light, temperature, and humidity controls of the growth chamber. Air in the growth chamber was charcoal filtered. Ozone was introduced from the outside into the growth chamber, and then, introduced into the fumigation chamber, circulated, and then vented to the outside.

A free-standing simple chamber for exposure studies with O_3 and other pollutants was described by Heck et al. (1968) for use singly or in connected groups or banks, with a single air-management system. Chambers of two different sizes were developed to allow their use in greenhouses, growth rooms, or growth chambers. Chambers were constructed of wood frame, finished with white glass enamel, and covered with a mylar film. Filtered air from a plenum moved through piping into the bottom of each chamber through perforations in the floor. Temperature and relative humidity of this air were preconditioned. Ozone, or other pollutants, were added through ports in the inlet duct of each chamber. An exhaust blower maintained negative pressure in the system and provided a single-pass flow through the chambers. In a sunny greenhouse, chamber air temperatures were 2.2 to 3.3°C above the temperature within the

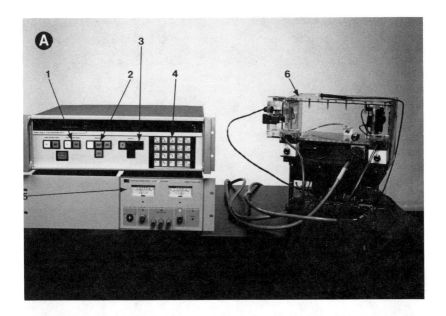

Figure 4.3.   A cuvette system for pollutant exposure-response studies. (Photographs courtesy of A. H. Legge, University of Calgary.) (A) A cuvette with its environmental control and power supply units: (1) wind and temperature controls, (2) temperature select section, (3) humidity select section, (4) key pad, (5) power supply. (B) Close-up view of the cuvette: (1) cone-shaped container for inserting the seedling or a branch, (2) humidity sensor, (3) air outlet, (4) air inlet, (5) quantum sensor, (6) fans, (7) wind speed sensor.

Figure 4.4. Environmental chamber and branch chambers used to measure gas flux measurements on individual red spruce branches (Ennis et al., 1990a).

greenhouse, at the rate of one exchange of air per minute squared within the chambers.

For long-term exposure of tree seedlings, and other plants of similar size, to steady or variable concentrations of O_3, alone or in combination with other

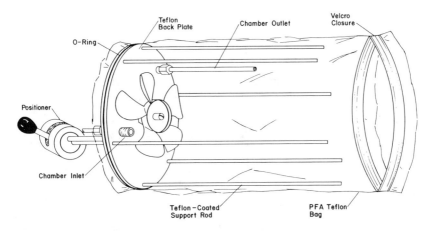

Figure 4.5. Details of construction of a branch exposure chamber, as used by Ennis et al. (1990a).

pollutants, Plotkin and Manning (unpublished) developed a system of eight chambers in pairs for use in a greenhouse or controlled-environment room. Two sets of eight clear-plexiglas chambers (each $72 \times 45 \times 45$ cm), in pairs of two each, were constructed in a greenhouse and connected by means of valves and a plenum box to a closed pass-through air exposure system (Figure 4.6). All incoming air was charcoal filtered, using an air filtration and movement system developed by Mandl et al. (1973). Ozone was generated from tanks of dry O_2, utilizing an OREC (Ozone Research & Equipment Corp., U.S.) O_3 generator, and added via teflon tubing to the plenum boxes. Air from the plenum boxes entered each chamber via the valve and was dispersed through a plexiglas tube at the top, with perforations at $45°$ angles. The air passed over the plants and exited through an opening in a corner at the bottom. Each pair of chambers can be a separate treatment, with some receiving only charcoal-filtered air. Valve adjustments were used to equalize air flow to all chamber pairs. By use of a damper in the incoming air line and valve adjustments, different rates of air flow can be maintained in each eight-chamber unit. Each chamber was provided with its own source of light, a 400-W HID lamp. Keane and Manning (1988) used these chambers to assess the long-term effects of O_3 on growth and mycorrhizal formation in white birch seedlings (*Betula papyrifera*).

4.5.1.6 Round Chambers: Continuous Stirred Tank Reactors (CSTRs)

Rectangular chambers of various types are useful under controlled-environment conditions. They may, however, have "dead spots" in the corners where pollutant concentrations and other environmental factors may vary from those in the rest of the chamber.

Therefore, Rogers et al. (1977) adapted a system, used previously for the

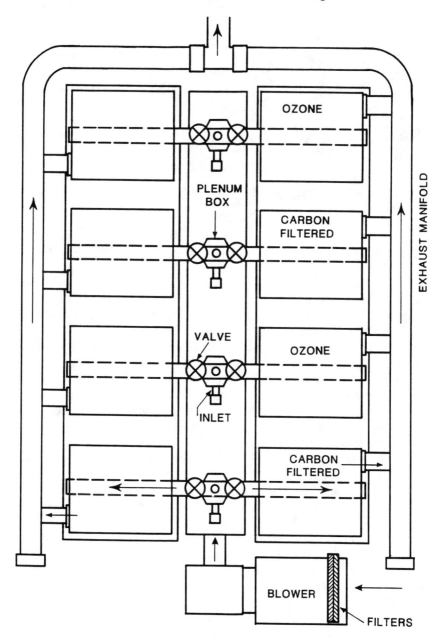

Figure 4.6.   Schematic diagram of closed pass-through multiple-chamber fumigation system (viewed from above), as used by Keane and Manning (1988).

study of chemical reactions, for investigations of gas exchange by plants in response to O_3. The chamber was cylindrical and the air inside the chamber was continuously stirred by an impeller, hence the name continuous stirred tank reactor (CSTR). Thorough mixing in the chamber ensures uniform air distribution in the cylinder and low leaf-boundary-layer resistance. A CSTR is essentially a very large cuvette. Ozone flux, net photosynthesis, and transpiration rates can be calculated from measurements of inlet and outlet concentrations of O_3, CO_2, and water vapor. This basic CSTR design was adapted by Heck et al. (1978) into multiple units for use in growth rooms and greenhouses (Figure 4.7).

The CSTRs are cylinders of varying sizes. The original version was 106.7 cm in diameter × 122 cm high, enclosed with a clear teflon film (Figure 4.8). The chamber had baffles along the sides and a motor-driven aluminum impeller at the top. Air entered the CSTR at the level of the impeller, was blown downward through the chamber, and exited in a lower corner of the chamber. Some investigators have CSTR systems where temperature, humidity, and light are also controlled.

At the present time, CSTRs are widely used, especially for studies on the physiology and biochemistry of plant responses to O_3. They are expensive, and a minimum of four units is required to achieve reproducibility of treatments. They must be built to exact specifications or they do not work properly. CSTRs in greenhouses cannot be used in the summer unless the greenhouses are air conditioned or pad cooled.

4.5.2 Field Exposure Systems

Controlled-environment exposure systems allow the greatest control over O_3 concentrations, environmental variables, and plant growth. Results from these systems usually are relevant only to the conditions under which they were obtained, and their relevancy to ambient conditions is questionable. As a result, many investigators have developed field exposure systems for more natural ambient environmental conditions. Ozone concentrations can still be regulated, but to a lesser degree. Control of environmental variables varies from method to method. These field exposure methods are summarized in Table 4.6, and some examples are given here.

4.5.2.1 Open-Air: Plumes

Plume systems are used where large-scale fumigations of plants in the field are desired, with artificially generated pollutant concentrations added above the normal ambient concentrations. They were originally developed for work with SO_2. The artificially generated pollutant is injected into ambient air above the plant canopy, and prevailing winds diffuse and disperse the pollutant as a plume over the plant canopy. Changes in wind speed and direction cause a high degree of variability in pollutant concentrations. However, there are no other microcli-

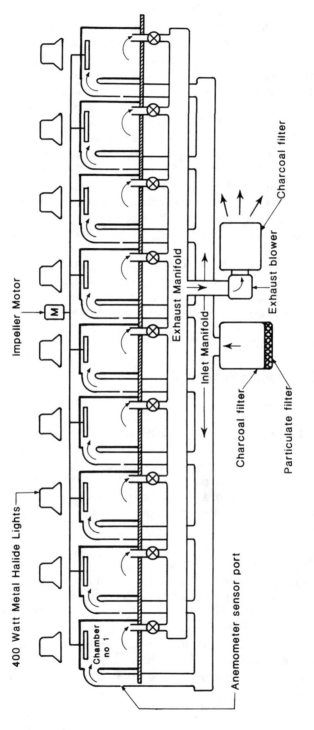

Figure 4.7. A bank of 9 CSTRs for use in a greenhouse (Heck et al., 1978).

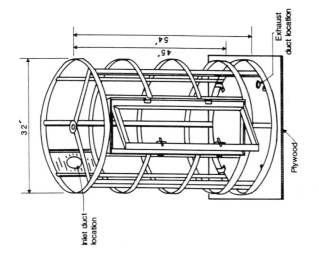

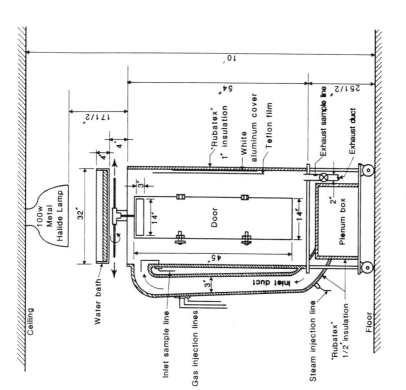

Figure 4.8.   Diagram of a single CSTR chamber, as used in a phytotron (Heck et al., 1978).

Table 4.6. Summary of Field Exposure Systems Used to Determine the Effects of Ozone on Plants

Systems	References
Open-air	
Plumes [Circular, grid, linear, square, zonal (ZAPS)]	Lee and Lewis (1978), Greenwood et al. (1982), Thompson et al. (1984), McLeod et al. (1985), Runeckles et al. (1990)
Air exclusion Pollutant removal	Jones et al. (1977), Shinn et al. (1977), Olszyk et al. (1986a,b)
Linear gradients	Shinn et al. (1977), Laurence et al. (1982), Reich et al. (1982), Olszyk et al. (1986)
Field chambers	
Branch exposure chambers	Evans and Dougherty (1986)
Closed chambers	Roberts (1981), Ashenden et al. (1982), Musselman et al. (1986)
Closed greenhouses	Thompson and Taylor (1969)
Cuvettes	Legge et al. (1979), McLaughlin et al. (1982)
Down-draft chambers	Runeckles et al. (1978)
Open-top chambers	Mandl et al. (1973), Heagle et al.(1973, 1979), Nystrom et al. (1982)
Self-ventilating chambers	Miller and Yoshiyama (1973)

mate effects. Different types of these plume systems have been described more extensively by Hogsett et al. (1987a,b) McLeod et al. (1985, 1988), and Runeckles et al. (1990).

ZAPS, the zonal air pollution system, was originated by Lee and Lewis (1978). A network of aluminum pipes was suspended above a native grassland ecosystem. Sulfur dioxide was released from openings at regular intervals along the pipes. A series of plots, each receiving a different initial SO_2 concentration, were established, and plant response results were compared to those receiving no additional SO_2. The rate of SO_2 release was constant, causing problems when winds changed and during periods when no appreciable winds occurred. Greenwood et al. (1982) developed a simplified rectangular version of this system, using a microcomputer to continually acquire data on SO_2 concentrations and

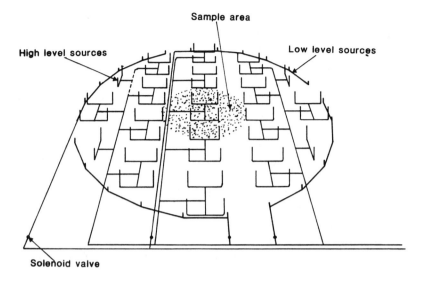

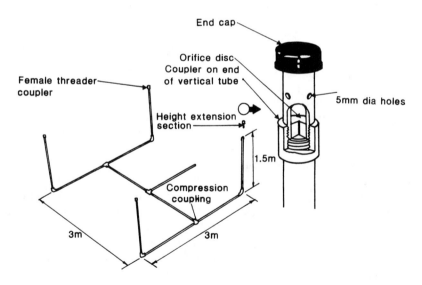

Figure 4.9. Diagram of a circular plume system and details of emitter head for release of SO_2 (McLeod et al., 1985, as described in Hogsett et al., 1987b).

wind speed and to control SO_2 release accordingly. McLeod et al. (1985, 1988) have developed a circular system for use with SO_2 and cereals (Figure 4.9).

Runeckles et al. (1990) have developed the newest modification of the ZAPS system. By this approach, twelve different concentrations and patterns of O_3 were obtained, allowing simulation of a wide range of ambient O_3 concentrations. This was facilitated by different rates of discharge of O_3 through manifolds suspended above individual plots. The plots were in blocks of four, and wind speed and direction determined the actual exposure patterns received in any plot. The overall O_3 concentration supplied to the manifolds was proportional to that in the ambient air. O_3 was supplied from 0700 to 2059 h. The concentration of O_3 addition was reduced automatically at low and high wind speeds.

Of all the methods, plume systems, at the present time, represent the ultimate forms of dispensing O_3 to plants at **above ambient concentrations**, with the least amount of environmental disturbance. However, they lack a great deal of precision in maintaining control over O_3 concentrations, and this becomes an increasing problem with changes in wind speed and direction. Extensive monitoring is also required to determine the O_3 concentrations in the study plots and to allow computerized control of O_3 release.

4.5.2.2 Air Exclusion Systems: Pollutant Removal

Open-top chambers (OTCs) represent the application of engineering technology to air pollution research in the field. Building on the OTC experience, several others have extended the concept to design air exclusion systems for field use without enclosures or chambers. Like all open-air systems, wind turbulence greatly affects system effectiveness.

Jones et al. (1977) placed perforated PVC (polyvinyl chloride) ducts between rows of soybeans and used a high-pressure blower to inflate the ducts with charcoal-filtered (CF) air. The CF air rising through the canopy excluded or minimized SO_2 contact with the soybean leaves. This was done only during SO_2 episodes and was 85% effective. A similar system was developed by Shinn et al. (1977) (Figure 4.10), except that a short fiberglass wall was erected around the plot to reduce the incursion of ambient air.

Olszyk et al. (1986a) developed a more elaborate and better-controlled version of the system of Jones et al. (1977). The length of the ducts, the hole diameter, and the spacing of holes on the ducts can be adjusted as required for either air exclusion or for artificial introduction of pollutants in gradient studies. There was a 6% average loss in the quantity of light, but relative humidity was comparable to ambient air. With low ambient wind speed and high ambient O_3, up to 75% O_3 exclusion could be achieved with this system. This compares favorably with results achieved with OTCs and closed-top chambers (CTCs).

Field tests were conducted to compare alfalfa growth in the air exclusion system, in an OTC, and in a closed chamber (Musselman et al., 1986). Plant growth in the exclusion system was comparable to that in ambient air. Growth was somewhat reduced in the OTC and was reduced even more in the CTC,

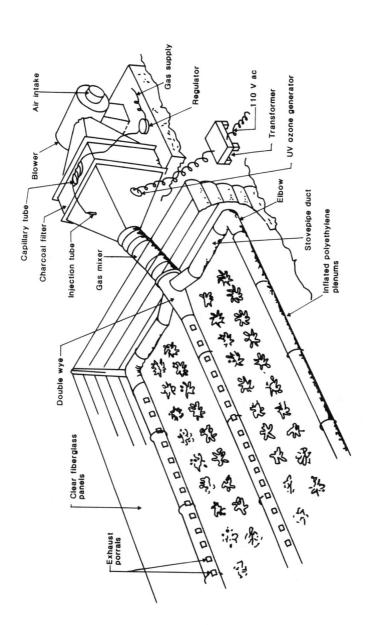

Figure 4.10.   Air exclusion system for exposure of plants to charcoal-filtered air, O_3, or other pollutants in the field (Shinn et al., 1977, as described in Hogsett et al., 1987b).

compared to ambient air. During fall and winter, growth was stimulated in the OTC, but not in the exclusion system.

4.5.2.3 Linear Gradients

The same apparatus described in the previous section can be used to expose plants in the field to an artificially introduced pollutant or such a treatment plus ambient or carbon-filtered air in various proportions. Typically, the introduced pollutant is supplied at a high concentration and then is allowed to diffuse down the pipes or tubing or is vented with a blower. Concentration gradients are established, allowing approximations of dose-response studies in the field without a chamber. Problems occur when moderate-to-strong winds affect air-flow rates. Careful monitoring of ambient air and air in the pipes or ducts for O_3 concentrations is essential. Shinn et al. (1977) have used this system with a variety of crops. Reich et al. (1982) and Laurence et al. (1982) have used a linear gradient system with soybeans and O_3, SO_2, and HF. Olszyk et al. (1986) have the most sophisticated new design and have achieved good results with alfalfa.

Both pollutant removal and linear gradient systems could be useful for long-term field studies with low-growing row crops in areas where turbulent ambient winds are not a problem. Even small changes in wind patterns can affect O_3 concentrations in these systems. Extensive O_3 monitoring is required. However, conditions closely approximate those of ambient microclimate, more so than OTCs and CTCs.

4.5.2.4 Field Chambers: Branch Exposure Chambers

Hernandez (1981) developed a simple wind-activated branch exposure chamber (BEC) to enclose pine branches in Ajusco, Mexico (Figure 4.11). The chamber was a clear-plastic-covered cylinder with 0, 50, or 100% activated charcoal in the end panels to allow varying amounts of ambient O_3 to pass through the chambers via wind movement. These BECs were used to assess ozone-induced changes in pigments of *Pinus hartwegii* needles from trees at Ajusco (Hernandez et al., 1986).

Evans and Dougherty (1986) have developed newer mechanized BECs to determine the effects of ambient air pollutants, including O_3, acidic fog, and above ambient concentrations of O_3, on needle physiology and growth responses of branches, on large or mature conifer trees. Teflon-covered frames were placed over branches and incoming air was either not filtered, charcoal-filtered, or charcoal-filtered with artificially added O_3. Branches could be enclosed and studied for a whole season.

Mattson et al. (1990) have reviewed current results from a number of BEC studies with ponderosa pine and red spruce. They concluded that success with BECs depends on (1) autonomy of mature branches in satisfying their own carbon requirements from photosynthesis, (2) comparability in environmental conditions between BECs and the ambient environment, and (3) good control of introduced pollutants (O_3) in the BECs. $CO_2{}^{14}$ pulse-trace experiments with

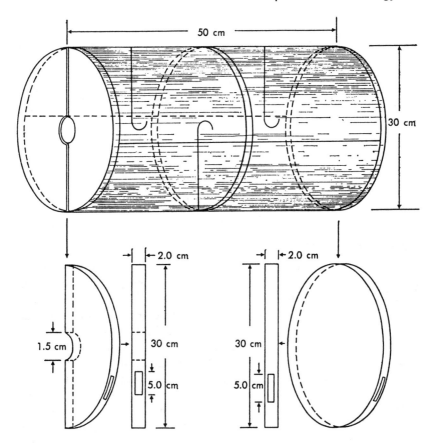

Figure 4.11. Construction details of a branch chamber used to enclose branches of *Pinus hartweggi* in Ajusco, Mexico (Hernandez, 1981).

ponderosa pine branches showed a high degree of autonomy. Less than 1 to 2.4°C temperature differences in BECs were reported, as was a 10% decrease in light intensity, compared to the ambient. O_3 distribution in the BECs was very good. Charcoal filtration efficiencies of 77 to 88% have been reported for such systems.

4.5.2.5 Closed-Top Chambers

Several investigators have developed closed-top chambers (CTCs) for experiments that require more precise control of O_3 than can be accomplished in open-top chambers (OTCs). Relevance to ambient environmental conditions is not required in these experiments. Rain does not enter the chambers, nor is dew formation possible. Air flow-through rates in CTCs are usually high to help minimize temperature increases. However, this may lead to increased water usage by plants. Artificial soil mixes, with automatic watering and fertilization systems, are often used for plants grown in pots or on the ground, within these chambers.

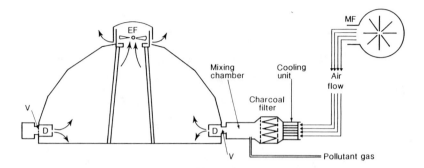

Figure 4.12. Diagram of a "solar dome" closed-top chamber developed at the University of Lancaster, England (as described in Hogsett et al., 1987b).

4.5.2.6 Hemispherical Domes

T. A. Mansfield et al., at the University of Lancaster in England, adapted a commercially available glass and aluminum frame hemispherical dome ("solar dome") for use as a CTC for air pollution research (Ashenden et al., 1982). Lucas, from Mansfield's group, has described the use of the solar dome with O_3 (Figure 4.12) (for more information, the reader should refer to Hogsett et al., 1987b). Roberts (1981) has also developed a similar CTC.

Each hemispherical dome is 4.6 m in diameter and 2.0 m high, with a volume of 20 m^3 and a plant growth area of 12.6 m^2. Charcoal-filtered or unfiltered air is introduced at the chamber base and distributed around the base of the chamber via a duct. Air is pulled out of the chamber by an exhaust fan at the top. Air moves at 40 to 60 m^3 min^{-1}, with 2 to 3 air exchanges per minute. Cooling units are needed to reduce summer air temperatures in the CTC to near ambient. Other environmental factors are not controlled.

4.5.2.7 Octagonal Chambers

Musselman et al. (1986) developed a new design for an octagonal CTC for use in southern California (Figure 4.13). Each CTC is 2.1 m high × 2.5 m across and constructed with flat aluminum members covered with teflon film. Teflon remains clear longer than plastic and minimizes differences between chamber and ambient environmental conditions. The CTC is fixed permanently in place, with underground air ducts connected to two blowers: one with a charcoal filter (CF), and one without (NF). CF and NF air can be introduced into the CTC in any combination or with artificially regulated O_3 concentrations, at adjustable air flow rates. Pollutant concentrations are stable in time and space, as are chamber temperatures. Quantity of light was 11% less than ambient, and midday temperatures could be higher within the chamber. All natural soil within the chambers was replaced with an artificial soilless solid growing medium commonly used in greenhouses. Drip irrigation, an automatic timer, and a fertilizer proportioner were used.

CTCs are useful for replicating closely controlled experiments in the field.

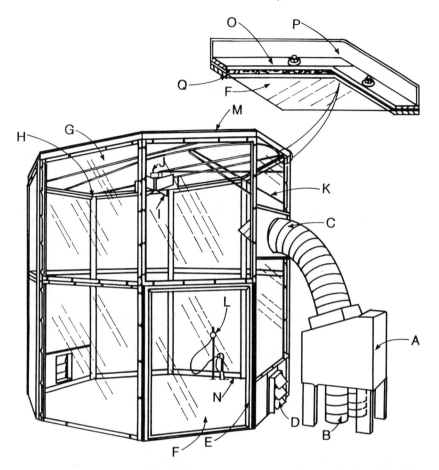

Figure 4.13. Diagram of a closed-top chamber: (A) air inlet mixing box; (B) Air inlets from filtered and nonfiltered blowers with butterfly valves to adjust flows; (C) Aluminum inlet duct to chamber from mixing box; (D) louvers for air-exiting chamber; (E) chamber door; (F) teflon film on chamber wall; (G) chamber teflon top panel; (H) aluminum conduit post for attaching wall panels; (I) impeller; (J) impeller motor and mount; (K) impeller support frame; (L) pollutant air sample tube; (M) chamber top frame support; (N) thermocouple sensor; (O) aluminum bar to secure teflon film to aluminum angle; (P) aluminum wall frame; (Q) weather strip (reprinted from Musselman et al., 1986, as described in Hogsett et al., 1987b).

They are essentially small greenhouses with uniform conditions. They are used to determine the effects of O_3 on plants without regard to the relevance of such results to ambient field conditions. Musselman et al. (1986) designed their chambers to determine how the ambient pollutants in Riverside, CA affect crop yield when all other factors are held constant.

4.5.2.8 Closed Greenhouses

The first versions of field exposure chambers were represented by small, closed greenhouses, with and without charcoal filters, placed over crops or trees in the field. Thompson and Taylor (1969) used greenhouses to enclose 16-year-old citrus trees in Riverside, CA. Ambient air reduced yields of oranges and lemons up to 50% due to premature leaf and fruit drop. Environmental conditions in these greenhouses were quite different from the ambient, therefore, results should be considered as not very relevant to natural conditions.

4.5.2.9 Cuvettes

Small clip-on cuvettes can be used in the field to determine the short-term effects of O_3 on gas exchange. They can also be used to expose single leaves to CO_2 containing C^{11} and C^{14} isotopes. Measurements of radioactivity from decay of these isotopes can be used to determine the rates of photosynthesis and translocation of photosynthate to other locations in the plant. McLaughlin et al. (1982) used this method to determine how chronic exposure to ambient air pollutants affected photosynthesis and carbon allocation in natural stands of white pine (Pinus strobus).

Legge et al. (1979) developed a much larger cuvette or leaf chamber (13 cm × 31.3 cm × 19.7 cm, with an internal volume of 4.9 l) (Figure 4.3). The cuvette can be heated or cooled, and air circulation is also provided. It is portable and can be used indoors or in the field. In the field the temperature-controlling system adjusts the cuvette temperature to ambient and prevents heat build-up. This allows exposures of needles or leaves for as much as 8 h at a time.

4.5.2.10 Down-Draft Chambers

Runeckles et al. (1978) developed a down-draft chamber (Figure 4.14) for use in the field to more closely mimic the downward flow of ambient air over vegetation (as opposed to the upward flow of introduced air in open-top chambers). Air is introduced (with or without filters) into the chamber from its top. This air moves downward with considerable turbulence, exiting the chamber around the edges, just above the soil level. Within the chamber there was a small increase in temperature and some reduction in net radiation. Air distribution within the chamber was more uniform than in an open-top chamber. Ambient rain is excluded from the chamber.

4.5.2.11 Open-Top Chambers

In response to the observed excessive increase in daytime temperature and the lack of exposure to ambient rainfall in closed field chambers, Heagle et al. (1973) and Mandl et al. (1973) developed large cylindrical OTCs. Charcoal-filtered or dust-filtered ambient air is blown into the bottom of the OTC at a velocity that permits it to rise within the chamber and exit through the open top. This reduces ingress of ambient air from above the OTC and prevents problems with daytime increases in temperature. In addition, ambient rainfall enters the chamber through the open top.

The Heagle OTC design, in use for more than 15 years, has been the most popular with investigators, including the U.S. Environmental Protection Agency

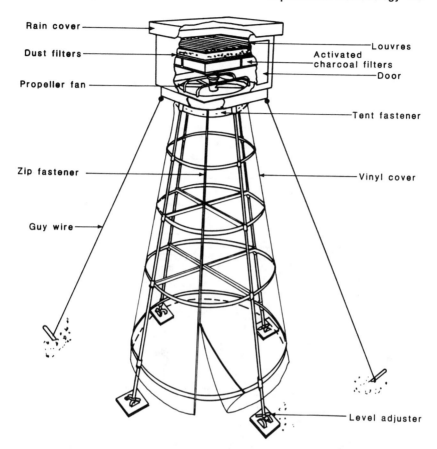

Figure 4.14. Diagram of a down-draft field chamber for exposing plants to O_3, charcoal-filtered air, or ambient air (Runeckles et al., 1978, as described in Hogsett et al., 1987b).

National Crop Loss Assessment Network (NCLAN) (Heagle et al., 1973; Heagle et al., 1988; Heck et al., 1982, 1984, 1988). Similar OTCs have been used to study O_3 effects on plants as diverse as cereals, grapevines, and large trees (Figures 4.15 and 4.16). Design variations in OTCs have been described in detail by Hogsett et al. (1987a,b) and Last (1986). With the addition of a rain cap (Figure 4.17), OTCs can also be used to study precipitation effects and soil moisture regime interactions (Hogsett et al., 1985).

The Heagle OTC is shown in Figure 4.18. It is a large cylinder (3.05 m in diameter × 2.44 m high), made of channeled aluminum framework, and covered with upper and lower panels of clear, flexible polyvinyl chloride film. The upper panel is a single layer. The lower panel is double layered, with six rows of 2.5 cm diameter holes at 17 to 19 cm intervals on the inner layer of the panel. An air inlet duct is attached to the lower double panel. Air entering from a fanbox is blown into this lower panel via the inlet duct. This inflates the lower double panel; the air passes through the holes of the inner panel into the chamber, and

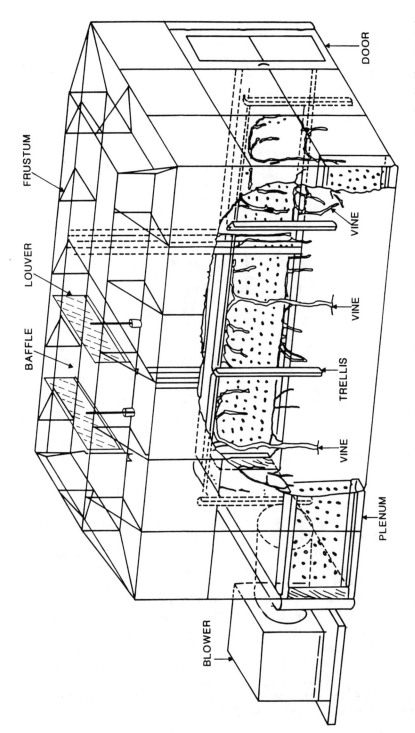

Figure 4.15.   Diagram of a large rectangular open-top field chamber for use with grapevines (Laurence and Kohut, 1984, as described in Hogsett et al., 1987b).

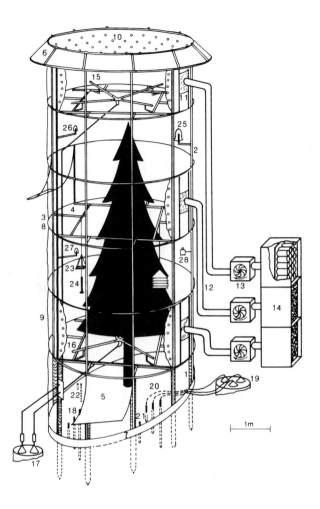

1. Wooden posts
2. Frame of coated iron retangular tube (30x30x2, 5 cm)
3. Coated band iron hoop (30 3, 5 mm)
4. Cross-struts of coated iron retangular tubes (30x30x2, 5cm)
5. Foil-door
6. Frustrum
7. Chamber rim PVC - Strips (200, 3 mm)
8. Foil tension ring, PVC - hoop (30, 5 mm)
9. Chamber skin of PE-foil SPR 3, 0.2 mm
10. Ventilation ring channel of perforated PE-foil SPR 3, 0.2 mm
11. Connection piece for ventilation tubes
12. Air duct tubes from blower, PVC 0 160 mm
13. Radial compressor 220 volts, 550 watt, 28 m^3 per minute
14. Filter element with particle filter and activated charcoal cartridges

15. Sampling rim for rain solution, PVC - tubes 0 32 mm
16. Sampling rim for canopy leachate, plexiglass - tubes, 0 40 mm
17. PE- small samplers for canopy leachate and rain solution, 5 l
18. Ceramic sond for soil tension in the depth of 15, 45 and 90 cm
19. Glass-sampler for rain solution. 2 l
20. Ceramic candles for soil solution in the depth of 15, 45 and 90 cm
21. Rhizoscope - tube, glass, 50 cm
22. Sensor for soil temperature in the depth of 15 and 40 cm
23. SAM - sampler for dry SO$_2$ - deposition
24. PICHE - evaporimeter
25. Active sampler for aerosols
26. Glass head for air sampling
27. Shelter for RH - and T - sensor
28. PAR - quantum sensor

Figure 4.16.   Open-top chamber used to enclose one Norway spruce (*Picea abies*) tree at the Hohenheim University research plot at Edelmannshof, Federal Republic of Germany.

then moves upward within and out of the chamber through the open top. Approximate velocity of air movement through the chamber is 1500 cfm, with two or three exchanges per minute. Incoming air is passed through a particle

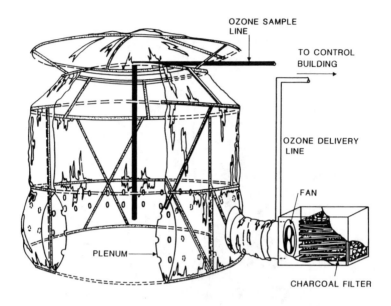

Figure 4.17.   An open-top chamber with frustrum and ambient rain-exclusion cap (Hogsett et al., 1985, from Hogsett et al., 1987b).

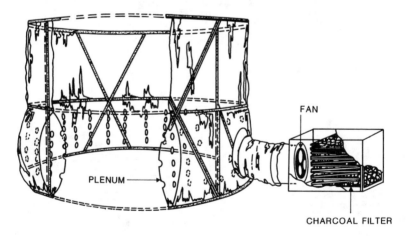

Figure 4.18.   Diagram of the original open-top field chamber (Heagle, 1973, as described in Hogsett et al., 1987b).

(dust) filter and is designated as nonfiltered air (NF). Activated-charcoal elements can be added to the fan box to produce charcoal-filtered air (CF).

Under conditions of low wind velocity, the charcoal filters may remove 50 to 60% of the ambient O_3. Downdraft incursions of ambient air into OTCs increase

Table 4.7. Some Advantages and Disadvantages of the Use of Open-Top Chambers (OTCs)

Advantages	Disadvantages
Widely used with 15+ years historical record, especially for CF/NF comparisons	Problems with comparisons of results from CF, NF, and AA[a]
Crops can be grown to maturity in the field	Limited space in chambers. Long-term use may mask effects in perennial plants and trees
Dose-response studies at concentrations above ambient can be made by O_3 additions	Microclimate effects may affect results, e.g., soil moisture problems, changes in insect and disease incidence
Each OTC is cost effective, portable,and durable	The many OTCs required for field work are expensive and labor-intensive. Each chamber requires a 20-amp circuit.
	Use leads to increased plant growth in cool seasons and winter, compared to AA plants

[a] CF = charcoal filtered; NF = nonfiltered; AA = chamberless ambient air plot.

Table 4.8. Effects of Ozone on Yields of Vona Winter Wheat in Open-Top Chambers, Ithaca, NY[a]

Treatments	Seed weight kg ha[-1]	% loss	100-seed weight weight (g)	% loss
Filtered air	5331.0	—	3.26	—
Ambient air	4049.8	24	2.32	29
Nonfiltered air	3552.1	33	2.47	24
Nonfiltered +0.03 ppm O_3	2322.3	56	1.77	46
Nonfiltered +0.06 ppm O_3	1698.8	68	1.41	57
Nonfiltered +0.09 ppm O_3	1430.0	73	1.30	69

[a] Adapted from Kohut et al. (1987).

with wind velocity, causing problems. Several designs involving additions of a baffle or frustrum to the top portion of the OTCs to reduce downdrafts have been evaluated (Buckenham et al., 1981; Davis and Rogers; 1980, Kats et al., 1976; Kohut et al., 1978). The use of a truncated cone-shaped frustrum (Kohut et al., 1978) increases the removal efficiency of ambient O_3 to 75% (Davis and Rogers, 1980). The frustrum reduced the OTC opening from 3.05 to 2.1 m in diameter. The ingress of ambient rain and solar radiation are reduced, but the uniformity of distribution of introduced O_3 within the OTC is increased (Davis and Rogers, 1980).

SO$_2$-O$_3$ EXCLUSION AND INCLUSION

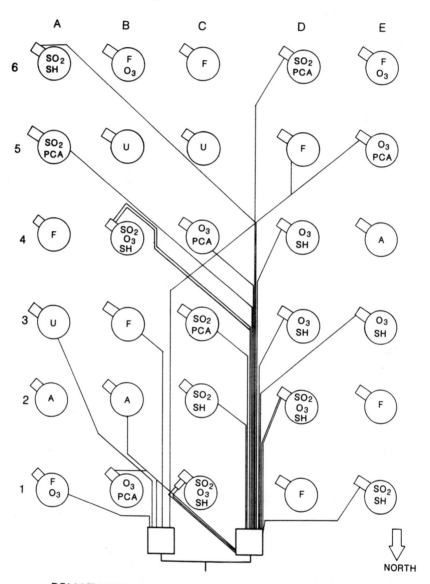

POLLUTANT DISPENSING AND MONITORING SYSTEMS

Figure 4.19. Schematic representation of open-top field fumigation chamber arrangement and the pollutant dispensing/monitoring lines. Chamber treatment codes are: A = ambient field plot with no chamber; U = unfiltered ambient air; FO$_3$ = ambient air filtered with type "AC" charcoal to remove O$_3$ only; F = ambient air filtered with type "CH" charcoal to remove both O$_3$ and SO$_2$; SH = same as F, but O$_3$, SO$_2$ or both artificially added to simulate ambient concentrations at the field site; PCA = same as F but O$_3$ or SO$_2$ added to randomly exceed state of Minnesota Ambient Air Quality Standards (O$_3$, 70 ppbh^{-1}, and SO$_2$, 250 ppbh^{-3}) (Nystrom et al., 1982).

Table 4.9. Open-Top Chamber Effects: Physical Changes

Parameters measured	Changes
Air filtration effects	
Activated charcoal (CF)	50–75% reduction in ambient O_3, increase in NO over ambient, and considerable reductions in NO_2, PAN, and SO_2
Particles (dust) (NF)	5–10% reduction in ambient O_3, some reduction in SO_2
Gas exchange	
Canopy resistance	Similar to ambient air
CO_2 uptake	Inconsistent effects
Leaf boundary layer resistance	Less in OTCs
Stomatal conductance	Variable, ranging from 0 to 22% decrease
Microclimate	
Air turbulence (due to ingress of ambient air)	Greater in downwind half of the chamber
Dewpoint	0.5–2.0°C higher
Light	Decrease of 12–20%, especially with dirty plastic, sun shadows possible
Relative humidity	5-10% increase or decrease, usually less than 5%
Temperatures	
Air	2.0–3.7°C increase, usually less than 2.0
Leaf	Slight increase
Windspeed	Decrease

Table 4.10. Open-Top Chamber Effects: Biological Changes

Factors	Changes
Microclimate effects (not defined)	Plants in NF may yield less than in AA
	Plants in NF are often taller than in AA; chambers can delay leaf senescence and shoot maturity in grapevines
Pesticide usage (usually increased)	Unusually low occurrences of diseases, insects and mites
Pollination problems	Fewer bees in chambers reduces pod set and seed yields in broadbeans
Position effects in chambers	Higher yields may occur in northern rather than southern half of chambers

OTCs have been used for many years, allowing the accumulation of considerable information about their performance under a variety of conditions. Some advantages and disadvantages of the use of OTCs are summarized in Table 4.7. More extensive consideration of OTCs has been provided elsewhere (Heagle, 1989; Heagle et al., 1988; Hogsett et al., 1987a,b; Krupa, 1984; Krupa and Nosal, 1989; Ormrod et al., 1988; Unsworth et al., 1984a,b).

OTCs were originally used to compare plant responses to CF and NF under field conditions (Heagle, 1989). CF was found to increase yields of plants, such as bean, cotton, field corn, peanut, potato, soybean, and wheat, when comparisons were made with yields from NF. This allowed the determination of possible ambient O_3 effects on plants in any given area. Results varied from year to year, as weather conditions and ambient O_3 concentrations and exposure patterns varied.

Proportional or fixed concentrations of O_3, separately or combined with other pollutants, can be added to OTCs. This allows dose-response studies in the field, with a range of concentrations of O_3 above the ambient (Table 4.8) (Heagle, 1989; Heagle et al., 1979a,b). Hogsett et al. (1987) and Nystrom et al. (1982) developed computerized systems to maintain OTCs and dispense O_3 in patterns that simulate fluctuating ambient air concentrations (Figure 4.19). Data from OTC dose-response studies have been used to predict how changes in future, elevated ambient O_3 concentrations will affect plant yields.

Howell et al. (1974) expanded the CF vs. NF experimental approach to include comparisons with a comparably sized, nonenclosed ambient air (AA) or chamberless plot. They compared the yield of four soybean cultivars in CF, NF, and AA for a 3-year period. Comparisons between CF and NF showed seed-yield increases for CF up to 20%. Average seed yields from AA, however, were higher than NF, suggesting that yield differences between CF and NF were due to chamber effects.

Chamber effects have been intensively investigated and described in detail by Ashmore et al. (1988), Colls et al. (1988), Heagle (1989), Heagle et al. (1988), Manning and Keane, (1988), Musselman et al. (1978), Olszyk et al. (1980, 1989), and Weinstock et al. (1982). A condensed summary of chamber effects, based on all of the previously cited work, is presented in Tables 4.9 and 4.10.

In terms of plant response, decreases in light (or sun shadow) and changes in rainfall distribution (or rain shadow) seem to be the most important variables. If chamber position is not a factor, then chamber effects, like its influence on light and rainfall, should be comparable between CF and NF chambers, and these should not be a factor when yield effects due to O_3 are determined. Problems may occur, however, when yield comparisons are made among CF, NF, and AA. Plants in AA are exposed to different environmental conditions, and their responses may be different from those in CF and NF. Yields may be higher, lower, or the same when NF and AA are compared, and these relationships may change from year to year (Heagle et al., 1989). Musselman et al. (1978) found that grapevines grew less in CF and NF than in AA and exhibited delayed shoot maturity and leaf senescence. Broad bean *(Vicia faba)* yields in NF were reduced

Table 4.11. Summary of the Effects of Open-Top Chambers (Unfiltered Air) on Broadbeans (*Vicia Faba*) Compared to Results from Ambient Air (Outside Plots)[a]

Parameters	Treatments[b]		
	Unfiltered air (NF)		Outside plot (AA)
Total bean fresh wt. (g)	27.6		37.7**
Number of mature beans	11.4		14.7*
Number of mature pods		ns	
Dry wt. of mature pods (g)	9.7		13.8***
Mean dry wt. per pod (g)	2.24		2.92***
Mean fresh wt. per bean (g)		ns	
Number of immature pods		ns	
Dry wt. of immature pods (g)		ns	
Number of side shoots	1.30		1.60*
Dry wt. of leaf and stem (g)	9.7		13.0**
Root/shoot ratio		ns	
Leaf area ratio		ns	

[a] From Ashmore et al. (1988b).

[b] Tabulated values are the mean values per plant in outside plots and in unfiltered air chambers. ns indicates differences were not significant at p = 0.05. ***, ** and * indicate significance levels of p = 0.001, 0.01, and 0.05, respectively.

by poor pollination due to low numbers of bees (Table 4.11) and soil moisture stress (Ashmore et al. 1988b; Colls et al., 1988). However, Heggestad et al. (1980) reported no measurable chamber effects for yields of snapbeans over a 5-year period.

Another possible reason why results from CF, NF, and AA may not be comparable is the composition of the air that the plants are exposed to as a result of charcoal filtration (CF), particle (dust) filtration (NF), and no filtration (AA). An important consideration here is that, frequently, investigators do not specify the type of charcoal used for filtration. Different types of charcoals provide different degrees of filtration of different pollutants. Nystrom et al. (1982) have used this feature to advantage. Most recently, Olszyk et al. (1989) examined air composition in CF, NF, and AA in the South Coast Air Basin in California and compared the efficiency of charcoal and particle filters (Figure 4.20). Concentrations of NO in CF were higher than in NF and AA, probably due to the conversion of NO_2 to NO on the charcoal filter. Concentrations of O_3, NO_2, PAN, NO_3, SO_4^{2-}, and NH_4^+ were greatly reduced by charcoal filtration. To the contrary, plants in NF were exposed to near ambient concentrations of O_3, NO_2, and PAN.

OTCs are useful in assessing ambient O_3 effects on vegetation and simulations of ambient effects, using comparisons between CF and NF. Dose-response studies, with elevated O_3 concentrations added to CF and NF, can be accomplished with OTCs. Problems may develop, however, when comparisons are made between CF, NF, and AA. Chamber effects need to be investigated in a

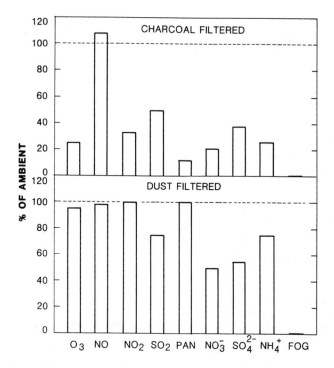

Figure 4.20. Effects of charcoal-filtration and particle (dust) filtration of ambient air on photochemical oxidant pollutants in open-top field chambers in the South Coast Air Basin of California (Olszyk et al., 1989).

systematic way to determine their influence on plant responses in CF and NF, and differences between results in NF and AA.

4.5.2.12 Self-Ventilating Chambers

All of the chambers previously described in this chapter depend on the introduction of forced air for their operation. In remote areas, particularly in forests, the electricity required to power the blowers or fans that force air into chambers may not be available, and installation of such chambers is either impossible or prohibitively expensive. Miller and Yoshiyama (1973) developed an inexpensive, small, self-ventilated, portable chamber for use in such areas. They successfully used it to determine the effects of ambient O_3 on tree seedlings in an area of the San Bernardino mountains where high O_3 concentrations were known to occur. Comparisons were made between chambers with and without charcoal filters.

Each chamber consisted of a basal plant enclosure area (3×3 ft. square) with a pyramid-shaped roof with an opening for an 8-ft-tall 1×1 ft metal stack with internal damper, capped by a rotating ventilator. The chambers were painted white on the exterior. Incoming air was either filtered or not filtered, passing into the chamber by the natural convection of warmer air from the bottom of the plant enclosure, upward through the cooler stack, and through the wind-driven rotating ventilator, enhancing air flow through the chamber in the process. The temperature within the chambers was about 4°F higher than the outside, insuring airflow. Air exchange rates ranged from 0.5 min^{-1} at night to 3.5 min^{-1} in the afternoon. No ambient air plots were used for comparison of growth effects due to the chambers.

4.5.3 Field Plot Systems

When investigators wish to determine the effects of ambient O_3 on plants, they may use a field plot system. No chambers or exclusion systems are involved, and there are no microclimate effects. Plants are exposed to normal fluctuations in ambient O_3 and changes in environmental factors that influence plant responses. Continual monitoring of ambient O_3 concentrations and environmental variables is required. Results at a given location will vary with time and reflect real-world conditions. The investigator has no control over O_3 concentrations or environmental variables. Lack of absolute control plots for sake of comparison is considered a criticism of these methods. However, numerical techniques (e.g., response surface methodology) (Myers, 1971) are available to determine what the control would be in a particular study. Field-plot systems are summarized in Table 4.12, and some examples are given here.

4.5.3.1 Ambient Exposures

The ideal way to conduct O_3/plant response studies in ambient air in field plots would be to have locations where there are defined gradients in O_3 concentrations or differing exposure patterns over distance. The value of this method is that O_3 exposures occur within a range of normal environmental variables that fluctuate independently from O_3 concentrations. These natural gradients are difficult to find due to the all-pervasive nature of O_3 in most locations. In the South Coast Air Basin of California, however, there are gradients in O_3 concentrations, starting from low concentrations near the coast and increasing to high concentrations in the mountains (Oshima et al., 1976, 1977a,b). At locations in the U.S., there are other O_3 gradients, but they do not have intermediate points in the gradient that are distinct enough to provide statistically different exposure regimes (Krupa, 1984). Ashmore et al. (1988) have also reported an O_3 gradient in southern England.

Oshima et al. (1976) made good use of the O_3 gradient in Southern California by developing a standardized plant culture method for alfalfa (*Medicago sativa*) at several locations on the gradient. They used a standardized soilless mixture for potted alfalfa plants and standardized cultural conditions. By using regression analysis, Oshima et al. (1976) were able to account for most of the variability in alfalfa yields at each point along the gradient and were able to attribute it primarily to the effects of ambient O_3. This work was continued with tomato plants in 13 field plots along the gradient (Oshima et al., 1977b). It was determined that yield reductions (Table 4.13) at locations of high O_3 were due

Table 4.12. Summary of Field Plot Systems Used to Determine the Effects of Ozone on Plants

Systems	References
Ambient exposures (natural differences in concentrations and duration)	Oshima et al. (1976,1977a,b), Krupa and Nosal (1989)
Cultivar comparisons	Heggestad (1973), Manning et al. (1974), Karnosky (1976)
Indicator plants	Heck and Heagle (1970), Manning and Feder (1980), Posthumus (1982), Steübing and Jäger (1982), Guderian et al. (1985)
Long-term growth reductions	Miller (1983), Skelly et al. (1983), Peterson et al. (1987)
Protective chemicals	Koiwai et al. (1974), Manning et al. (1974), Carnahan et al. (1978), Kender and Forsline (1983), Beckerson and Ormrod (1986), Smith et al. (1987)

to reductions of fruit size, as detected by size distributional analysis. Regression analyses comparing O_3, air temperatures, and relative humidity indicated that effects were due primarily to O_3.

Krupa and Nosal (1989) grew alfalfa in a common soil type at nine sites in Minnesota over a 2-year period. Using time-series and spectral-coherence analyses, they found that at low O_3 flux density, median O_3 concentration provided the best coherence with height growth of alfalfa, while at high O_3 flux density, cumulative integral of exposure was the best.

One of the drawbacks of using differences in ambient O_3 concentrations to determine plant response is that plants are grown in pots with artificial growing medium, and cultural conditions (i.e. watering and fertilization) are the same at each location. The "location effect", however, cannot be determined. Oshima et al. (1977a) addressed this in a second tomato experiment. Five cultivars, varying in foliar injury response to O_3, were planted in field plots at Riverside, where ambient O_3 concentrations were high, and at Irvine, where O_3 concentrations were low. These are the extremes of the O_3 gradient in the South Coast Air Basin. Tomato yields were smaller at Riverside (Figure 4.21) due to depressed early-season production and reduced fruit size. Again, regression analysis indicated that differences in ambient O_3 concentrations, not differences in air temperature and relative humidity, were responsible for the difference in the yield of tomato fruits. Krupa and Nosal (unpublished) did not standardize the cultural conditions in their study. On the contrary, they included several sets of ambient temperature data, precipitation data, etc. in their modeling of O_3 exposure and plant response.

Table 4.13. Relationship of Percent Marketable Tomato Fruit to O_3 Dose and Other Environmental Variables at Eleven Locations along a Natural O_3 Gradient in the South Coast Air Basin in California[a]

Location	Marketable fruit	Seasonal values for monitored variables			
		Avg max temp (°C)	Avg min temp (°C)	RH (%)[b]	Ozone dose[c]
1	99.5	27.4	15.8	67.9	70
2	94.8	35.1	11.4	38.7	164
3	99.4	27.5	14.4	55.4	169
4	98.1	28.7	14.7	59.3	455
5	97.1	30.2	15.8	54.1	630
6	91.5	34.0	12.8	38.7	1670
7	82.9	31.2	13.0	49.8	2347
8	88.2	31.2	12.9	49.4	2382
9	70.4	33.1	14.8	38.0	2429
10	78.6	31.7	13.7	41.7	3085
11	78.1	32.9	14.3	41.7	3085

[a] Adapted from Oshima et al. (1977b).
[b] Average hourly values of relative humidity between 0700 and 2000 (Pacific Standard Time).
[c] Calculated from pphm-hr greater than 10 pphm between June 20 to final harvest, 1975.

While no adverse O_3 effects were observed by Krupa and Nosal in their study, O_3 exposure appeared to be the most important independent variable regulating alfalfa height growth.

The ambient exposures or natural gradient method provides information about the effects of ambient O_3 on plants with no effects through changes in the natural environment. Many replications are possible, and resistant and susceptible cultivars and protectant chemicals could also be integrated into the approach. Intensive monitoring of O_3 and climate variables, however, is required at, or as close as possible to, each site. Unless artificial soil mixtures are used, consideration must be given to the effects of the soil condition in explaining the results. Defined natural O_3 gradients are few, but as more extensive ambient air monitoring is undertaken, additional gradients most likely will be found.

4.5.3.2 Cultivar Comparisons

The ideal way to determine the effects of ambient O_3 on plants in the field would be to compare the responses of two isogenic lines that differ from each other only in their tolerance or responsiveness to O_3. Presently no such lines are readily available, but it is well known that plant species, cultivars, lines, and clones vary in their sensitivity to O_3 (Brennan et al., 1964; Genys and Heggestad, 1978; Heggestad, 1973; Houston, 1974; Karnosky, 1976; Keller, 1988; Reinert et al., 1984). Some examples of useful plants that have cultivars or clones with known tolerance or sensitivity to O_3 are given in Table 4.14.

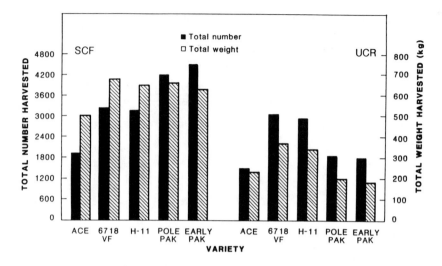

Figure 4.21. Comparison of yields for five tomato cultivars at Irvine (SCF) (low ambient O_3) and Riverside (UCR) (high ambient O_3) (Oshima et al., 1977a).

Unless the tolerance of a plant to O_3 has been previously established, it is necessary to screen populations of the plant to a range of O_3 concentrations (Drummond and Pearson, 1979) under carefully controlled conditions. Plant responses must then be verified under field conditions where O_3 concentrations and environmental conditions are variable. Most screening tests are done to achieve acute O_3 injury symptoms. Frequently there is little agreement between results obtained under controlled conditions and those from field screenings. Taylor (1974), however, found good correlation (0.78) between O_3 sensitivity in tobacco seedlings and mature plants. Plants resistant as seedlings were also resistant in the field. Acute injury screening was used to identify O_3-tolerant and -sensitive bean cultivars that were also tolerant and susceptible in the field. Relative yield responses could also be predicted for cultivars that were very tolerant or very sensitive to O_3 in an acute screening program (Heck et al., 1988). These bean cultivars could then be used in field research programs to determine the effects of O_3 on yields.

Even when plant response to O_3 has been defined, care must be taken to select cultivars that are closely related to each other in as many ways as possible, before response and yield comparisons are made. Heck et al. (1988) determined that bean cultivar BBL 290 is sensitive to O_3, and BBL 274 is tolerant. We have found this to be the case in our work (Kostka-Rick and Manning, unpublished), but plant sizes, pod sizes, and days-to-pod maturity dates for the two cultivars are different. BBL 290 is smaller and matures more rapidly than BBL 274 and, hence, is exposed to less O_3. In this case, these cultivars would be more useful if they were considered separately, and a protectant chemical, such as EDU (ethylene diurea), was used to provide the comparisons within each cultivar

Table 4.14. Some Plants That are Known to Have Cultivars or Clones with Known Tolerance or Sensitivity to O_3

Plants	Latin names	References
Bean	Phaseolus vulgaris	Manning and Feder (1976); Heggestad et al. (1976, 1980); Reinert et al.(1984); Heck et al. (1988)
Morning glory	Ipomoea purpurea	Manning (1977)
Potato	Solanum tuberosum	Rich and Hawkins (1970); Heggestad (1973)
Tobacco	Nicotiana tabacum	Heggestad and Menser (1962); Taylor (1974)
Trembling aspen	Populus tremuloides	Karnosky (1976); Keller (1988)
White pine	Pinus strobus	Genys and Heggestad (1978); Houston (1974)

instead of between them. Treated BBL 274 plants would also serve as a double control.

4.5.3.3 Indicator Plants

Plant responses, notably visible, characteristic symptoms, have long been used as indications of a number of air and soil pollutants. It is not surprising then that unusual foliar injury symptoms on vegetation in the Los Angeles area, beginning around 1942, served to indicate that a new type of air pollutant was present (Middleton et al., 1950). A number of investigators began to use a variety of plants to detect photochemical oxidants — principally O_3 and peroxyacyl compounds (Heck, 1966). As O_3 became an air pollution problem in most industrial nations, interest in using indicator plants has increased greatly. Results of this work is considered in detail elsewhere (Guderian et al., 1985; Manning and Feder, 1980; Posthumus, 1982; Steübing and Jäger, 1982).

Plants used as bioindicators of O_3 must be easy to grow under standardized cultural systems, easily obtained, sensitive to O_3 (with distinct symptoms), have indeterminate growth, and have few confounding disease or pest problems (Cole, 1958; Manning and Feder; 1980, Ormrod and Adedipe, 1975). Standardized methods for evaluating and quantifying O_3-induced foliar injury must also be developed (Gumpertz et al., 1982). Some common bioindicators of O_3 are given in Table 4.15.

Noble and Wright (1958) developed a standardized assay for ambient O_3 in Los Angeles in the 1950s. Annual bluegrass (*Poa annua*) responds to O_3 with

Table 4.15. Some Common Indicator Plants for Use Where Ozone is the Principal Phytotoxicant

Plants	Latin Names	References

Grass
    Annual
    bluegrass
	Poa annua	Noble and Wright (1958)

Morning glory	*Ipomoea purpurea*	Manning (1977); Nouchi and Aoki (1979)
Pinto bean cv. 111	*Phaseolus vulgaris*	Oshima (1974)
Spinach	*Spinacea oleracea*	Posthumus (1982)

Tobacco
    Bel-W3 (sensitive)
    Bel-C (intermediate)
    Bel-B (tolerant)
	Nicotiana tabacum	Heck and Heagle (1970);
		Heggestad and Menser (1962)

Trembling aspen clones	*Populus tremuloides*	Karnosky (1976); Keller (1988)

characteristic leaf injury. A 15-station network of controlled chambers was established in the Los Angeles area. Plants were kept in carbon-filtered air chambers where relative humidity, daylength, and nutrition were controlled. Plants were exposed to ambient air for 1-d periods and then evaluated for extent of foliar injury. Because annual bluegrass is also sensitive to peroxyacyl compounds, the effect of a photochemical oxidant mixture was measured, rather than O_3 alone.

An automated air pollution biological indicator (AMBI) system for O_3 was developed in Southern California by Oshima (1974). Instead of *Poa annua*, he used 10 to 12-day-old Pinto bean (cv 111) seedlings. Seedlings at that age are no longer sensitive to peroxyacyl compounds, but are quite sensitive to O_3. Like Noble and Wright (1958), Oshima developed a multistation network of AMBIs along the O_3 gradient in the Los Angeles Basin. The AMBIs were self-contained, and watering was automated. Plants were evaluated weekly by referring to standardized reference photographs of categories of O_3 injury on Pinto bean leaves (Table 4.16). Significant correlation between average weekly O_3 dosage and average weekly plant injury indices was obtained.

Heggestad and Middleton (1959) demonstrated that O_3 caused weather fleck on tobacco on the east coast of the U.S. In the course of investigations with tobacco, three very useful lines of tobacco were discovered. Bel-W3 is quite sensitive to O_3 and has become the standard for indicating O_3 in ambient air all over the world. Bel-C is intermediate in sensitivity, and Bel-B is quite tolerant. Bel-W3 and Bel-B are often used together with Bel-W3 for comparison purposes (Ashmore et al., 1980; Manning and Feder, 1980).

Table 4.16. Ozone Injury Ratings Used to Evaluate Field Response of Pinto Beans[a]

Rating	Description
0	No visible injury
1	Traces of ozone stipple visible
2	Ozone stipple scattered over most of the primary leaf
3	Dense ozone stipple covering all or almost all of the primary leaf. Veins may be in sharp contrast with the leaf surface
4	Stipple present with severe bifacial injury occurring in patches, or areas of dead tissue

[a] From Oshima (1974).

Heck and Heagle (1970) used Bel-W3 tobacco to assess the frequency of occurrence of phytotoxic concentrations of O_3 around Cincinnati, OH. They did not find a consistent relationship between ambient O_3 concentrations and plant injury. Jacobson and Feder (1974) used Bel-W3 tobacco to biomonitor O_3 in all the northeastern states of the U.S. Kelleher and Feder (1978) used Bel-W3 to demonstrate long-range transport of O_3 from the New York area to Nantucket Island off the coast of Massachusetts. Good correlation between the extent of O_3 injury on Bel-W3 and ambient O_3 concentrations was found in Israel (Naveh et al., 1978). As Guderian et al. (1985) point out, the response of Bel-W3 tobacco is usually the first indication that a country or region has developed an O_3 problem. Bel-W3 is also useful for monitoring the success of pollution abatement programs.

Manning (1977) evaluated the response of ten cultivars of morning glory (*Ipomoea purpurea*) to ambient O_3 in Massachusetts. Only cultivars with cordate, nonpubescent leaves, such as "Heavenly Blue," showed injury symptoms. Newer cultivars, like "Scarlet O'Hara," with smaller, thicker, deeply lobed pubescent leaves, did not exhibit injury symptoms. In Japan, however, Nouchi and Aoki (1979) found O_3 injury on "Scarlet O'Hara." Using both exposure chambers and observations in the field, they "calibrated" plant responses so that intensity of responses could be used to infer the average ambient O_3 concentration during the time of plant exposure to ambient air.

Little work has been done with trees as bioindicators. Clones of trembling aspen (*Populus tremuloides*) are known to differ in sensitivity to O_3 (Karnosky, 1976), and Keller (1988) proposed that these differential responses to ambient O_3 can be used to indicate O_3 problems.

Indicator plants for O_3 have been widely studied, and much is known about them. As such, they also make good subjects for experiments where predictable plant responses to O_3 are required.

Like any method, the use of indicator plants is not without problems. Poor plant culture, failure to adhere to schedules, and the use of more than one person to evaluate plant responses are common factors that produce poor results. Indicator plants must also be grown as close as possible to electronic O_3 monitors, otherwise correlation with ambient O_3 episodes is not possible.

4.5.3.4 Long-Term Growth Reductions

Calculation of long-term reductions in the radial growth of trees can be accomplished from interpretation of data obtained from tree cores. In even-aged stands of trees of the same species, such as red spruce (*Picea rubens*), it is possible to determine changes in mean basal-area increments from an analysis of ring-width measurements of core samples (Federer and Hornbeck, 1987). Most forests, however, are mixed stands of uneven age. These factors, plus a range of fluctuating environmental factors, make it difficult to unequivocally attribute long-term growth reductions to the effects of O_3 (Innes and Cook, 1989).

Miller (1983) determined the effects of 30 years of exposure to ambient O_3 on Ponderosa pine (*Pinus ponderosa*) and other trees in the San Bernardino Mountains of Southern California. Susceptible trees were weakened by ambient O_3 through foliar injury and premature needle drop. Over time, this resulted in reduced radial growth of branches and trunks. Weakened trees were more readily invaded by bark beetles (*Dendroctonus breviconus*) and by the opportunistic root-rot fungus Fomes annosus (*Heterobasidion annosum*). This pattern of long-term growth decline is now beginning to develop for Jeffrey pines (*Pinus jeffreyi*) in Sequoia and Kings Canyon National Parks, north of Los Angeles and east of San Francisco (Peterson et al., 1987). Decline of growth of white pines (*Pinus strobus*) in the Blue Ridge Mountains of Virginia has also been documented. Weakened trees there became more susceptible to root rot caused by *Verticicladiella procera* (Skelly et al., 1983).

Determination of long-term growth reductions in trees depends on the analysis and interpretation of changes in tree rings. It is often difficult to separate the factors that may be responsible for changes in tree ring widths. Drought, other environmental factors, and stand composition can affect growth and ring widths and mask possible O_3 effects. There is also considerable debate about how to take core samples, how many should be taken, and how to interpret them (Cook, 1987; Innes and Cook, 1987; McLaughlin et al., 1987).

4.5.3.5 Protective Chemicals

For more than 30 years, many investigators have evaluated diverse groups of chemical compounds, including antioxidants, antisenescence agents, antitranspirants, dusts, growth regulators, growth retardants, and pesticides (Table 4.17), to determine whether these compounds will protect plants from O_3 injury. Reports of the activity of these compounds have been reviewed by Bialobok (1984), Guderian et al. (1985), Kender and Forsline (1983), Ormrod et al. (1988), and others. The advantages and disadvantages of the use of protective chemicals are given in Table 4.18.

While there have been many reports of chemicals that were successful in preventing O_3 injury to plants, most of them are of a preliminary nature, and further experiments need to be conducted. The development of the open-top chamber method in the 1970s also contributed to a decrease in interest in protective chemicals. A great deal of work, however, has been done with the systemic fungicide benomyl, and the antioxidant EDU. Several chemical industries have recently shown a renewed interest in, among other substances, ascorbic acid (Table 4.17).

Benomyl is a useful compound for protecting plants from O_3. This compound can be applied to soil, or used as a foliar spray. Benomyl is systemic when applied to soil and systemic in leaves when applied as a spray. The active component is the benzimidazole moiety, which is known to delay senescence in plants. Benomyl is also a fungicide. Critics of chemical protection claim that it is not possible to separate the fungicidal from antisenescence effects. Manning et al. (1974) used benomyl to suppress O_3 injury symptoms on an O_3-sensitive bean cultivar in the field. An O_3-tolerant cultivar was included for comparison. Benomyl-sprayed, O_3-sensitive bean plants exhibited higher yields than nonsprayed plants. Benomyl sprays had no effects on yields of the tolerant bean cultivars or on fresh and dry weights of the plants. Incidence of diseases in roots, stems, shoots, leaves, and pods for either bean cultivar were not affected by benomyl. The use of benomyl on grape leaves, however, could lead to problems. Benomyl could decrease O_3 stipple, but also decrease powdery mildew, a common grape disease. This would make it difficult to clearly determine the effects of O_3 on grape growth and yield. Where crop biology and disease incidence are well known, benomyl could be used successfully to protect plants from O_3 and to determine yield losses.

EDU (ethylene diurea) is currently the best known systemic antioxidant. It has been used extensively, and many reports are available for many crops and trees (Manning, 1992). EDU was developed specifically as an antioxidant to protect plants (Carnahan et al., 1978). Contrary to Ormrod et al. (1988), EDU is not a fungicide. At appropriate concentrations it retards senescence in plants. Heagle (1989) points out that its mode of action is not completely understood. One of the reasons for this is that we do not have a clear understanding of the complete mechanism for O_3 injury at the molecular level. EDU has excellent potential as a tool for use in research on the mechanisms or the mode of action of O_3.

Carnahan et al. (1978) clearly demonstrated that EDU could protect bean leaves from O_3 injury. A large quantity of EDU was produced by the duPont Chemical Company (U.S.) and distributed to many investigators. Many trials showed that EDU could be used to suppress O_3 injury and, most likely, yield depressions (Table 4.19) (Clarke et al.; 1983, Legassicke and Ormrod; 1981, Manning, 1990; Smith et al., 1987). In many cases, however, the experiments were not well planned, and assumptions were made without verification. The principal assumptions were that EDU itself had no adverse effects on plants and

Table 4.17. Examples of Chemicals Used to Protect Plants from Ozone Injury

Chemicals	Formulas	Plants	References
Antioxidants			
Ascorbic acid	K-ascorbate, N-ascorbate	bean, celery, citrus, lettuce, petunia	Dass and Weaver (1968), Freebairn and Taylor (1960)
Butox	piperonyl butoxide	bean, tobacco	Koiwai et al. (1974), Rubin et al. (1980)
DPA	diphenylamine	apple	Elfving et al. (1976)
EDU	ethylene diurea N-[2-(2-oxo-1-imidazolidinyl) ethyl]-N'- phenylurea	many plants	Carnahan et al. (1978), Manning (1991)
NBC	nickel-N-dibutyl dithiocarbomate	bean	Dass and Weaver (1968)
Santoflex 13	N-(1,3-dimethyl butyl)-N'-phenyl-p-phenylene-diamine	apple, bean, melon, tobacco	Gilbert et al. (1977)
Antisenescence agents			
Polyamines	putrescine, spermidine, spermine	tomato	Ormrod and Beckerson (1986)
Antitranspirants			
Folicote	paraffinic hydrocarbon waxes	solaneous crops	Knapp and Fieldhouse (1970)
Wilt Pruf (also with DPA)		apple	Elfving et al. (1976)
Dusts	charcoal, diatomaceous earth, ferric oxide, kaolin	tobacco	Bialobok (1984), Jones (1963)

Growth regulators			
cytokinins			
BA	6-benzylamine purine	radish	Adedipe and Ormrod (1972)
Kinetin	N-6 benzyladenine	bean	Tomlinson and Rich (1973)
Growth retardants			
CBBP	2,4-dichloro-benzyl tributyl phosphonium chloride	petunia	Cathey and Heggestad (1972)
SADH	succinic acid 2, 2-dimethyl hydrazide	petunia	Cathey and Heggestad (1972)
Pesticides			
Fungicides			
benomyl	methyl-1-butyl-carbamyl-2 benzimidazole	many plants	Manning et al. (1974), Moyer et al. (1974)
carboxin	5,6-dihydro-2-methyl-1, 4-oxathiin-3-carboxanilide	bean	Rich et al. (1974)
dithiocarbamates maneb (manganese) zineb (zinc)	ethylene bis dithiocarbamates	many plants	Kendrick et al. (1962)
Herbicides			
diphenamid	N,N-dimethyl-2, 2-diphenyl acetamide	tobacco	Reilly and Moore (1982)
isopropalin	2,6-dinitro-N, N-dipropyl cumidine	tobacco	Reilly and Moore (1982)
Insecticides			
Spectracide 25	diazinon	bean	Tess et al. (1979)

Table 4.18. Some Advantages and Disadvantages of the Use of Protective Chemicals

Advantages	Disadvantages
No chambers required, ambient conditions used	O_3 dose/response studies cannot be done unless coupled with some other exposure methods
No microclimate effects	Ambient O_3 concentrations and environmental conditions must be measured
Plant numbers and plot sizes can be varied according to the requirements of the experiment	Repeat applications of chemicals needed
High degree of replication possible	Extensive plant toxicology studies needed before start of experiments
Low costs for materials and labor	Results vary from year to year (also considered an advantage)
Protocol is simple and uncomplicated, requiring little equipment	

that the rate used in pot tests for suppression of acute injury from one exposure to O_3 (500 ppm actual EDU) could also be used in the field, with multiple applications of EDU to suppress both acute and chronic injury. Failure to verify these assumptions has cast doubt on some results and is used by critics to discredit the use of EDU (Heagle, 1989; Ormrod et al., 1988).

Basic plant toxicology studies with EDU, with and without exposure to O_3 episodes, need to be conducted to find the proper rates and number of applications for each plant species to provide protection against O_3 without adversely affecting plant growth. OTCs could be useful for doing this. With bean and radish, we have found that 150-ppm EDU is sufficient for protection against O_3 without adversely affecting plant growth. Two applications of EDU are sufficient for radish, with three applications required for bean (Kostka-Rick and Manning, 1989, unpublished).

EDU was developed as a commercial product by the duPont Chemical Company (U.S.). When it was decided that there was no market for the product, its production was discontinued. Consequently, interest in EDU declined when it was no longer available. OTCs became the method of choice for field studies on the effects of O_3 on plants. The high costs and labor needed to operate the numbers of OTCs required for field research, plus increasing doubts about the

Table 4.19. Examples of the Effects of EDU Applications on Yield of Ozone-Sensitive "Tiny Tim" Tomatoes in the Field at Harrow, Ontario[a]

EDU treatment	Number of fruit per plant	Avg fruit wt (g)	% ripe fruit	Yield per plant (g)
Control	317	5.85	67	1856
1.0 gl⁻¹, foliar spray	413	5.89	67	2431
0.5 g per plant, soil drench	369	6.26	61	2312

[a] Adapted from Legassicke and Ormrod (1981).

validity of the data from some of the studies conducted in OTCs, have led to renewed interest in EDU. The duPont Co. has funded the manufacture of a new supply of EDU, which will be made available to investigators. EDU will then become a useful research tool to investigate the effects of ambient concentrations of O_3 on plants. Proximity to an electronic O_3 monitor is necessary to determine the ambient O_3 concentrations. As environmental conditions and O_3 concentrations in the field vary from year to year, results from EDU studies will vary from year to year also. Heagle (1989) sees this variability as a criticism of the use of EDU. Others see it as an accurate and a more desirable method because it reflects real-world conditions.

EDU alone cannot be used to do O_3 dose/response studies in the field (Heagle, 1989). It was not developed for that purpose. However, EDU coupled with, for example, plume systems can be used to do more definitive O_3 dose/response studies under ambient conditions.

EDU has a great potential as a chemical protectant to prevent O_3 injury to plants. Some of the other chemical compounds may also have this property, but not much is known about most of them. Continued research and toxicology studies with EDU will lead to the development of a valuable verified method for the study of the effects of ambient, or above ambient, concentrations of O_3 on plants.

4.6 CONCLUSIONS

Descriptions of 18 different O_3 exposure systems or methodologies for plant-response studies have been presented in this chapter. All the approaches have advantages and disadvantages and vary in costs and ease of implementation. Investigators should carefully define the objectives of their work and then select the appropriate system or method to achieve their objectives. Useful results can be obtained when the appropriate system or method is used, and its advantages and disadvantages are well understood. In some cases, parts of two or more approaches can be combined to achieve greater scientific strength. It is critical, however, that results be interpreted within the context of the approach used, rather than inappropriate extrapolations of such results to other situations. Significant uncertainties exist in the transfer of plot-level research results to regional-scale assessment of O_3-induced plant effects. This uncertainty must be defined in any cost-benefit analysis. The design and initiation of a study are inseparable from any data analysis strategy. Therefore, the two steps must be integrated at the **outset** and not **during** or at the **end** of an investigation. Many studies to date, including some large research programs, have missed this very important requirement. The use of results from such investigations in developing air quality regulatory policies can lead to scientific controversy. Future studies must consider this important issue in any experimentation.

REFERENCES

Adedipe, N. O. and Ormrod, D. P. (1972) Hormonal regulation of ozone phytotoxicity in *Raphanus sativus Z. Pflanzenphysiol.* **68**, 254–258.

Altshuller, A. P. (1986) The role of nitrogen oxides in non-urban ozone formation in the planetary boundary layer over North America, Western Europe and adjacent areas of ocean. *Atmos. Environ.* **20**, 245–268.

Ashenden, T. W., Tabner, P. W., Williams, P., Whitmore, M. E., and Mansfield, T. A. (1982) A large-scale system for fumigating plants with SO_2 and NO_2. *Environ. Pollut.* **B3**, 21–26.

Ashmore, M. R., Bell, J. N. B., and Mimmack, A. (1988) Crop growth along a gradient of ambient air pollution. *Environ. Pollut.* **53**, 99–121.

Ashmore, M. R., Bell, J. N. B., and Reily, C. L. (1980) The distribution of phytotoxic ozone in the British Isles. *Environ. Pollut.* **B1**, 195–216.

Ashmore, M. R., Bell, N., and Rutter J. (1985) The role of ozone in forest damage in West Germany. *Ambio* **14**, 81–87.

Bennett, J. H. (1979) Foliar exchange of gases. In *Methodology for the Assessment of Air Pollution Effects on Vegetation* (eds., Heck, W. W., Krupa, S. V., Linzon, S. N., and Frederick, E. R.). Air Pollution Control Association, Pittsburgh.

Berry, C. R. (1970) A plant fumigation chamber suitable for forestry studies. *Phytopathology* **60**, 1613–1615.

Bialobok, S. (1984) Controlling atmospheric pollution. In *Air Pollution and Plant Life* (ed., Treshow, M.). John Wiley & Sons, New York, 451–478.

Brennan, E., Leone, I. A., and Daines, R. H. (1964) The importance of variety in ozone plant damage. *Plant Dis. Rept.* **48**, 923–924.

Buckenham, A. H., Parry, M. A., Whittingham, C. P., and Young, A. T. (1981) An improved open-topped chamber for pollution studies on crop growth. *Environ. Pollut.* **B2**, 475–482.

Cantwell, A. M. (1968) Effect of temperature on response of plants to ozone as conducted in a specially designed plant fumigation chamber. *Plant Dis. Rept.* **52**, 957–960.

Carnahan, J. E., Jenner, E. L., and Wat, E. K. W. (1978) Prevention of ozone injury to plants by a new protectant chemical. *Phytopathology* **68**, 1225–1229.

Carney, A. W., Stephenson, G. R., Ormrod, D. P., and Ashton, G. C. (1973) Ozone-herbicide interactions on crop plants. *Weed Sci.* **21**, 508–511.

Cathey, H. M. and Heggestad, H. E. (1972) Reduction of ozone damage to *Petunia hybrida* Vilm. by use of growth regulating chemicals and tolerant chemicals. *J. Am. Soc. Hort. Sci.* **97**, 685–700.

Clarke, B. B., Henninger, M. R., and Brennan, E. (1983) An assessment of potato losses caused by oxidant air pollution in New Jersey. *Phytopathology* **73**, 104–108.

Cochran, W. G. (1973) Experiments for non-linear functions. *J. Am. Stat. Assoc.* **68**, 771–781.

Cole, G. A. (1958) Air pollution with relation to agronomic crops: III. Vegetation survey methods in air pollution studies. *Agron. J.* **50**, 553–555.

Colls, J. J., Sanders, G. E., and Clark, A. G. (1988) Open-top chamber experiments on field-grown *Vicia faba* L. at Sutton Bonington (1987–1988). In *Proc. 3rd Workshop on Open-Top Chambers, The European Community Project on Open-Top Chambers, Results on Agricultural Crops, 1987-1988* (ed., Bonte, J.). 43–72.

Conover, W. J. (1971) *Practical Nonparametric Statistics.* John Wiley & Sons, New York, 462.

Cook, E. R. (1987) The decomposition of tree-ring series for environmental studies. *Tree-Ring Bull.* **47**, 37–59.

Darley, E. F. and Middleton, J. T. (1961) Carbon filter protects plants from damage by air pollution. *Florist's Rev.* **127**, 15–16, 43, 45.

Darrall, N. M. (1989) The effect of air pollutants on physiological processes in plants. *Plant Cell Environ.* **12**, 1–30.

Dass, H. C. and Weaver, G. M. (1968) Modification of ozone damage to *Phaseolus vulgaris* by antioxidants, thiols and sulfhydryl reagents. *Can. J. Plant Sci.* **48**, 569–574.

Davis, D. D. and Wood, F. A. (1972) The relative susceptibility of eighteen coniferous species to ozone. *Phytopathology* **62**, 14–19.

Davis, D. D. and Wood, F. A. (1973) The influence of plant age on the sensitivity of Virginia pine to ozone. *Phytopathology* **63**, 381–388.

Davis, J. M. and Rogers, H. H. (1980) Wind tunnel testing of open-top field chambers for plant effects assessment. *JAPCA* **30**, 905–908.

de Bauer, M. L., Tejeda, T. H. and Manning, W. J. (1985) Ozone causes needle injury and tree decline in *Pinus hartwegii* at high altitudes in the mountains around Mexico City. *JAPCA* **35**, 838.

Demerjian, K. L. (1986) Atmospheric chemistry of ozone and nitrogen oxides. In *Air Pollutants and Their Effects on the Terrestrial Ecosystem* (eds., Legge A. H. and Krupa S. V.). John Wiley & Sons, New York, 105-127.

Drummond, D. B. and Pearson, R. G. (1979) Screening of plant populations. In *Methodology for the Assessment of Air Pollution Effects on Vegetation* (eds., Heck, W. W., Krupa, S. V., Linzon, S. N., and Frederick E. R.). Air Pollution Control Association, Pittsburgh.

Elfving, D. C., Gilbert, M. D., Edgerton, L. F., Wilde, M. H., and Lisk, D. J. (1976) Antioxidant and antitranspirant protection of foliage against ozone injury. *Bull. Environ. Contam. Toxicol.* **15**, 336–341.

Ennis, C. A., Lazrus, A. L., Kok, G. L., and Zimmerman, P. R. (1990a) A branch chamber system and techniques for simultaneous pollutant exposure experiments and gaseous flux determinations. *Tellus* **42B**, 170–182.

Ennis, C. A., Lazrus, A. L., Zimmerman, P. R., and Monson, R. K. (1990b) Flux determinations and physiological response in the exposure of red spruce to gaseous hydrogen peroxide, ozone, and sulfur dioxide. *Tellus* **42B**, 183–189.

Evans, L. S. and Dougherty, P. (1986) Exposure systems and physiological measurements: Forest Response Program Quality Assurance Methods Manual. EPA/600/x-86/193a, p. 99.

Federer, C. A. and Hornbeck, J. W. (1987) Expected decreases in diameter growth of even-aged red spruce. *Can. J. For. Res.* **17**, 266–269.

Finlayson-Pitts, B. J. and Pitts, J. N., Jr. (1986) *Atmospheric chemistry: Fundamentals and Experimental Techniques*. John Wiley & Sons, New York, 1098.

France, J. and Thornley, J. H. M. (1984) *Mathematical Models in Agriculture*. Butterworth Scientific, London. 335.

Freebairn, H. T. and Taylor, O. C. (1960) Prevention of plant damage from airborne oxidizing agents. *Proc. Am. Soc. Hort. Sci.* **76**, 693–699.

Genys, J. B. and Heggestad, H. E. (1978) Susceptibility of different species, clones and strains of pines to acute injury caused by ozone and sulfur dioxide. *Plant Dis. Rept.* **62**, 687–691.

Gilbert, M. D., Elfving, D. C., and Lisk, D. J. (1977) Protection of plants against ozone injury using the antioxidant N-(1,3-dimethylbutyl)-N[1]-phenyl-p-phenelene-diamine. *Bull. Environ. Contam. Toxicol.* **18**, 783–785.

Greenwood, P., Greenhalgh, A., Baker, C., and Unsworth, M. (1982) A computer controlled system for exposing field crops to gaseous air pollutants. *Atmos. Environ.* **16**, 2261–2266.

Guderian R., Tingey, D. T., and Rabe, R. (1985) Effects of photochemical oxidants on plants. In *Photochemical Oxidants* (ed., Guderian R.). Springer-Verlag, New York.

Gumpertz, M. L., Tingey, D. T., and Hogsett, W. E. (1982) Precision and accuracy of visual foliar injury assessments. *J. Environ. Qual.* **11**, 549–553.

Haagen-Smit, A. J. (1952) Chemistry and physiology of Los Angeles smog. *Ind. Eng. Chem.* **44**, 1342–1346.

Hardwick, R. C., Cole, R. A., and Fyfield, T. P. (1984) Injury to and death of cabbage (*Brassica oleracea*) seedlings caused by vapours of dibutyl pthalate emitted from certain plastics. *Ann. Appl. Biol.* **105**, 97–105.

Heagle, A. S. (1989) Ozone and crop yield. *Ann. Rev. Phytopathol.* **27**, 397–423.

Heagle, A. S. and Philbeck, R. B. (1979a) Exposure techniques In *Methodology for the Assessment of Air Pollution Effects on Vegetation* (eds., Heck, W. W., Krupa, S. V., Linzon, S. N., and Frederick, E. R.). Air Pollution Control Association, Pittsburgh.

Heagle, A. S., Body, D. E., and Heck, W. W. (1973) An open-top field chamber to assess the impact of air pollution on plants. *J. Environ. Qual.* **2**, 365–368.

Heagle, A. S., Philbeck, R. B., Rogers, H. H., and Letchworth, M. B. (1979b) Dispensing and monitoring ozone in open-top field chambers for plant effects studies. *Phytopathology* **69**, 15–20.

Heagle, A. S., Kress, L. W., Temple, P. J., Kohut, R. J., Miller, J. E., and Heggestad, H. E. (1988) Factors influencing ozone dose-yield response relationships in open-top field chamber studies. In *Assessment of Crop Loss from Air Pollutants* (eds., Heck, W. W., Taylor, O. C., and Tingey, D. T.). Elsevier, London, 141–179.

Heck, W. W. (1966) The use of plants as indicators of air pollution. *Air Water Pollut. Int. J.* **10**, 99–111.

Heck, W. W. (1968) Factors influencing expression of oxidant damage to plants. *Ann. Rev. Phytopathol.* **6**, 165–188.

Heck, W. W. and Heagle, A. S. (1970) Measurement of photochemical air pollution with a sensitive monitoring plant. *JAPCA* **20**, 97–99.

Heck, W. W., Dunning, J. A., and Johnson, H. (1968) Design of a simple plant exposure chamber. U.S. Department of Health, Education, and Welfare, Publ. APTD-68-6, Cincinnati.

Heck, W. W., Philbeck, R. B., and Dunning, J. A. (1978) A continuous stirred tank reactor (CSTR) system for exposing plants to gaseous air contaminants. Principles, specifications, construction and operation. Agric. Res. Serv., USDA, ARS-S-181.

Heck, W. W., Taylor, O. C., and Tingey, D. T. (eds.) (1988) *Assessment of Crop Loss from Air Pollutants*. Elsevier, London, 552.

Heck, W. W., Krupa, S. V., Linzon, S. N., and Frederick, E. R. (eds.) (1979) *Methodology for the Assessment of Air Pollution Effects on Vegetation.* Air Pollution Control Association, Pittsburgh, 1.1 to 18.3 .

Heck, W. W., Taylor, O. C., Adams, R., Bingham, G., Miller, J., Preston, E., and Weinstein, L. H. (1982) Assessment of crop loss from ozone. *JAPCA* **32**, 353–361.

Heck, W. W., Cure, W. W., Rawlings, J. O., Zaragoza, L. J., Heagle, A. S., Heggestad, H. E., Kohut, R. J., Kress, L. W., and Temple, P. J. (1984) Assessing impacts of ozone on agricultural crops. I. Overview. *JAPCA* **34**, 729–735.

Heck, W. W., Dunning, J. A., Reinert, R. A., Prior, S. A., Rangappa, M., and Benepal, P. S. (1988) Differential responses of four bean cultivars to chronic doses of ozone. *J. Am. Soc. Hort. Sci.* **113**, 46–51.

Heggestad, H. E. (1973) Photochemical air pollution injury to potatoes in the Atlantic coastal states. *Am. Potato J.* **50**, 315–328.

Heggestad, H. E. and Heck, W. W. (1971) Nature, extent and variation of plant response to air pollutants. *Adv. Agron.* **23**, 111–145.

Heggestad, H. E. and Menser, H. A. (1962) Leaf-spot sensitive tobacco strain Bel-W3: a biological indicator of the air pollutant, ozone. *Phytopathology* **52**, 735 (Abstr.).

Heggestad, H. E. and Middleton, J. T. (1959) Ozone in high concentrations as cause of tobacco leaf injury. *Science* **129**, 208–210.

Heggestad, H. E., Craig, W. L., and Meiners, J. P. (1976) Response in Beltsville greenhouses of dry bean cultivars and breeding lines to oxidant air pollution. *Proc. Am. Phytopathol. Soc.* **3**, 326 (Abstr.).

Heggestad, H. E., Menser, H. A., and Bailey, W. A. (1967) A system for filtering air through activated carbon to be used with fan pad cooling in greenhouses. Preprint 67–38h, Air Pollution Control Association Meeting, Cleveland.

Heggestad, H. E., Heagle, A. S., Bennett, J. H., and Koch, E. J. (1980) The effects of photochemical oxidants on the yield of snapbeans. *Atmos. Environ.* **14**, 317–326.

Hernandez, T. T. (1981) Recononocimiento y evaluacion del dano por gases oxidantes en pinos y avena del Ajusco, D. F. Tesis, professional, Departamento de Fitotecnia, UAch, Chapingo, Mexico, 90.

Hernandez, T. T., de Baurer, L. I., and Delgado, M. L. O. (1986) Identificacion y determinacion de los principales pigmentos fotosinteticos de hojas de *Pinus hartwegii* afectadas por gases oxidantes. *Agrosciencia* **66**, 71–82.

Hogsett, W. E., Tingey, D. T., and Holman, S. R. (1985) A programmable exposure control system for determination of the effects of pollutant exposure regimes on plant growth. *Atmos. Environ.* **19**, 1135–1145.

Hogsett, W. E., Olszyk, D., Ormrod, D. P., Taylor, G. E., Jr., and Tingey, D. T. (1987a) Air pollution exposure systems and experimental protocols: Vol. 1: a review and evaluation of performance. U.S. EPA Pub. 600/3-87/037a.

Hogsett, W. E., Olszyk, D., Ormrod, D. P., Taylor, G. E., Jr., and Tingey, D. T. (1987b) Air pollution exposure systems and experimental protocols: Vol. 2: description of facilities. U.S. EPA Pub. 600/3-87/0376.

Houston, D. B. (1974) Response of selected *Pinus strobus L.* clones to fumigations with sulphur dioxide and ozone. *Can. J. For. Res.* **41**, 65–68.

Howell, R. K., Koch, E. J., and Rose, L. P., Jr. (1979) Field assessment of air pollution-induced soybean yield losses. *Agron. J.* **71**, 285–288.

Innes, J. L. and Cook, E. R. (1989) Tree-ring analysis as an aid to evaluating the effects of pollution on tree growth. *Can. J. For. Res.* **19**, 1174–1189.

Jacobson, J. S. and Feder, W. A. (1974) A regional network for environmental monitoring of atmospheric oxidant concentrations and foliar injury to tobacco indicator plants in the eastern United States. *Mass. Agric. Exp. Sta. Bull.* 604.

Jones, H. C., Lacasse, N. L., Liggett, W. S., and Weatherford, F. (1977) Experimental air exclusion system for field studies of SO_2 effects on crop productivity. U.S. EPA-600/7-77-122.

Karnosky, D. F. (1976) Threshold levels for foliar injury to *Populus tremuloides* by sulphur dioxide and ozone. *Can. J. For. Res.* **6**, 166–169.

Kats, G., Thompson, C. R., and Kuby, W. C. (1976) Improved ventilation of open-top chambers. *JAPCA* **26**, 1089–1090.

Keane, K. D. and Manning, W. J. (1988) Effects of ozone and simulated acid rain on birch seedling growth and mycorrhizal associations. *Environ. Pollut.* **50**, 55–65.

Kelleher, T. J. and Feder, W. A. (1978) Phytotoxic concentrations of ozone on Nantucket Island: long-range transport from the Middle Atlantic states over the open ocean confirmed by bioassay with ozone-sensitive tobacco plants. *Environ. Pollut.* **17**, 187–194.

Keller, Th. (1988) Growth and premature leaf fall in American aspen as bioindications for ozone. *Environ. Pollut.* **52**, 183–192.

Kender, W. J. and Forsline, P. J. (1983) Remedial measures to reduce air pollution losses in horticulture. *Hort. Sci.* **18**, 680–684.

Knapp, C. E. and Fieldhouse, D. J. (1970) Alar and Folicote sprays for reducing ozone injury on four solonaceous genera. *Hort. Sci.* **5**, 338.

Kohut, R. J., Krupa, S. V., and Russo, F. (1978) An open-top chamber study to evaluate the effects of air pollution on soybean yield. *Proc. 4th Joint Conference on Sensing Environmental Pollutants,* American Chem. Society, Washington, D.C., 71–73.

Kohut, R. J., Amundson, R. G., Laurence, J. A., Colavito, L., vanLeuken, P., and King, P. (1987) Effects of ozone and sulfur dioxide on yield of winter wheat. *Phytopathology* **77**, 71–74.

Koiwai, A., Kitano, H., Fukuda, M., and Kisaki, T. (1974) Methylene dioxphenyl and its related compounds as protectants against ozone injury to plants. *Agr. Biol. Chem.* **38**, 301–307.

Koziol, M. J. and Whatley, F. R. (eds.) (1984) *Gaseous Air Pollutants and Plant Metabolism.* Butterworth Scientific, London, 466.

Krause, G. H. M., Prinz, B., and Jung, K. D. (1983) Forest effects in West Germany. In *Air Pollution and the Productivity of the Forest* (eds., Davis, D. D., Miller, A. A., and Dochinger, L.). Izaak Walton League of America, Washington, D.C., 297–332.

Krupa, S. V. (1984) Field exposure methodology for assessing the effects of photochemical oxidants on crops. In *Proc. 77th Annual Meeting Air Pollution Control Association.* Preprint no. 88-1042. Air Pollution Control Association, Pittsburg.

Krupa, S. V. and Kickert, R. N. (1989) The greenhouse effect: impacts of ultraviolet (UV)-B radiation, carbon dioxide (CO_2) and ozone (O_3) on vegetation. *Environ. Pollut.* **61**, 263–393.

Krupa, S. V. and Manning, W. J. (1988) Atmospheric ozone: formation and effects on vegetation. *Environ. Pollut.* **50**, 101–137.

Krupa, S. V. and Nosal, M. (1989) Effects of ozone on agricultural crops. In *Atmospheric Ozone Research and Its Policy Implications* (eds., Schneider, T., Lee, S. D., Wolters, G. J. R., and Grant, L. D.). Elsevier, Amsterdam, 229–238.

Krupa, S. V. and Teng, P. S. (1982) Uncertainties in estimating ecological effects of air pollutants. In *Proc. 75th Ann. Meetings Air Pollution Control Association,* 82-6.1, Air Pollution Control Association, Pittsburgh, 1–10.

Last, F. T. (1986) Microclimate and plant growth in open-top chambers. Comm. European Communities Air Pollution Research Report 5, p. 340.

Laurence, J. A., MacLean, D. C., Mandl, R. H., Schneider, R. E., and Hansen, K. S. (1982) Field tests of a linear gradient system for exposure of row crops to SO_2 and HF. *Wat. Air Soil Pollut.* **17**, 399–407.

Lee, J. J. and Lewis, R. A. (1978) Zonal air pollution system: design and performance. U.S. EPA pub. 600/3-78-021.

Lefohn, A. S. and Benedict, H. M. (1982) Development of mathematical index that describes ozone concentration, frequency and duration. *Atmos. Environ.* **16**, 2529–2532.

Lefohn, A. S. and Runeckles, V. C. (1987) Establishing a standard to protect vegetation-ozone exposure/dose considerations. *Atmos. Environ.* **21**, 561–568.

Lefohn, A. S., Krupa, S. V., and Winstanley, D. (1990) Surface ozone exposures measured at remote locations around the world. *Environ. Pollut.* **63**, 189–224.

Lefohn, A. S., Runeckles, V. C., Krupa, S. V., and Shadwick, D. S. (1989) Important considerations for establishing a secondary ozone standard to protect vegetation. *JAPCA* **39**, 1039–1045.

Legassicke, B. C. and Ormrod, D. P. (1981) Suppression of ozone-injury on tomatoes by ethylene diurea in controlled environments and in the field. *Hort. Sci.* **16**, 183–184.

Legge, A. H. and Krupa, S. V. (eds.) (1989) *Acidic Deposition: Sulphur and Nitrogen Oxides.* Lewis Publishers, Chelsea, MI. 659.

Legge, A. H., Savage, D. J., and Walker, R. B. (1979) A portable gas-exchange leaf chamber. In *Methodology for the Assessment of Air Pollution Effects on Vegetation* (eds., Heck, W. W., Krupa, S. V., Linzon, S. N., and Frederick, E. R.). Air Pollution Control Association, Pittsburgh.

Leone, I. A. and Brennan, E. J. (1979) Plant growth and care. In *Methodology for the Assessment of Air Pollution Effects on Vegetation* (eds., Heck, W. W., Krupa, S. V., Linzon, S. N., and Frederick, E. R.). Air Pollution Control Association, Pittsburgh.

Linzon, S. N. and Chevone, B. I. (1987) Tree decline in North America. *Environ. Pollut.* **50**, 87–99.

Little, T. M. and Hills, F. J. (1978) *Agricultural Experimentation.* John Wiley & Sons, New York, 350.

Logan, J. A. (1985) Tropospheric ozone: seasonal behavior, trends and anthropogenic influence. *J. Geophys. Res.* **90**(10), 463–482.

Mandl, R. L., Weinstein, L. H., McCune, D. C., and Keveny, M. (1973) A cylindrical, open-top chamber for the exposure of plants to air pollutants in the field. *J. Environ. Qual.* **2**, 371–376.

Manning, W. J. (1977) Morning glory as an indicator plant for oxidant air pollution: cultivar sensitivity. *Proc. Am. Phytopathol. Soc.* **4**, 192.

Manning, W. J. (1992) EDU: a research tool for assessment of the effects of ozone on vegetation. *Environ. Pollut.* (in press).

Manning, W. J. and Feder, W. A. (1976) Effects of ozone on economic plants. In *Effects of Air Pollutants on Plants* (ed., Mansfield, T. A.), Cambridge University Press, Cambridge, 209.

Manning, W. J. and Feder, W. A. (1980) *Biomonitoring Air Pollutants with Plants.* Applied Science Publishers, Elsevier London.

Manning, W. J. and Keane, K. D. (1988) Effects of air pollutants on interactions between plants, insects and pathogens. In *Assessment of Crop Loss from Air Pollutants* (eds., Heck, W. W., Taylor, O. C., and Tingey, D. T.). Elsevier, London, 365–386.

Manning, W. J., Feder, W. A., and Vardaro, P. M. (1974) Suppression of oxidant injury by benomyl: effects on yields of bean cultivars in the field. *J. Environ. Qual.* **3**, 1–3.

Mattson, K. G., Arnaut, L. Y., Reams, G. A., and Cline, S. P. (1990) Response of forest trees to sulfur, nitrogen, and associated pollutants. U.S. EPA/USDA-Forest Service Response Program, Major Program Output No. 4, p. 108.

McLaughlin, S. B. (1985) Effects of air pollution on forests. *JAPCA* **35**, 512–534.

McLaughlin, S. B., McConathy, R. K., Duvick, D., and Mann, L. K. (1982) Effects of chronic air pollution stress on photosynthesis, carbon allocation, and growth of white pine trees. *For. Sci.* **28**, 60–70.

McLaughlin, S. B., Dawing, D. J., Blasing, T. J., Cook, E. R., and Adams, H. S. (1987) An analysis of climate and competition as contributors to decline of red spruce in high elevation Appalachian forests of the eastern United States. *Oecologia* **72**, 487–501.

McLeod, A. R. and Baker, C. K. (1988) The use of open-field systems to assess yield response to gaseous pollutants. In *Assessment of Crop Loss from Air Pollutants* (eds., Heck, W. W., Taylor, O. C., and Tingey, D. T.). Elsevier, London, 181–210.

McLeod, A. R., Fackrell, J. E., and Alexander, K. (1985) Open-air fumigation of field crops: criteria and design for a new experimental system. *Atmos. Environ.* **19**, 1639–1649.

Menser, H. A., Jr., Heggestad, H. E., and Grosso, J. J. (1966) Carbon filter prevents ozone fleck and premature senescence of tobacco leaves. *Phytopathology* **56**, 466-467.

Middleton, J. T., Kendrick, J. B., Jr., and Schwalm, H. W. (1950) Injury to herbaceous plants by smog or air pollution. *Plant Dis. Rept.* **34**, 245–252.

Miller, P. R. (1983) Ozone effects in the San Bernardino National Forest. In *Air Pollution and the Productivity of the Forest* (eds., Davis, D. D., Miller, A. A., and Dochinger, L.), Izaack Walton League of American, Washington, D.C., 161–197.

Miller, P. R. and Yoshiyama, R. M. (1973) Self-ventilated chambers for identification of oxidant damage to vegetation at remote sites. *Environ. Sci. Technol.* **7**, 66–68.

Montgomery, D. C. (1984) *Design and Analysis of Experiments.* John Wiley & Sons, New York, 538.

Moyer, J. W., Cole, H., Jr., and Lacasse, N. L. (1974) Suppression of naturally occurring oxidant injury on azalea plants by drench or foliar spray treatment with benzimidazole or oxathiin compounds. *Plant Dis. Rept.* **58**, 41–44.

Musselman, R. C., Kender, W. J., and Crowe, D. E. (1978) Determining air pollutant effects on the growth and productivity of "Concord" grapevines using open-top chambers. *J. Am. Soc. Hort. Sci.* **103**, 645–648.

Musselman, R. C., McCool, P. M., Oshima, R. J., and Teso, R. R. (1986) Field chambers for assessing crop loss from air pollutants. *J. Environ. Qual.* **15**, 152–157.

Myers, R. H. (1971) *Response Surface Methodology.* Allyn and Bacon, Boston, MA. 246.

Naveh, Z., Chaim, S., and Steinberger, E. H. (1978) Atmospheric oxidant concentrations in Israel as manifested by foliar injury in Bel W3 tobacco plants. *Environ. Pollut.* **16**, 249–262.

Noble, R. D. and Jensen, K. F. (1983) An apparatus for monitoring CO_2 exchange rates in plants during SO_2 and O_3 fumigations. *J. Exp. Bot.* **34**, 470–475.

Noble, W. M. and Wright, L. A. (1958) Air pollution with relation to agronomic crops. II. A bio-assay approach to the study of air pollution. *Agron. J.* **50**, 551–553.

Nouchi, I. and Aoki, K. (1979) Morning glory as a photochemical oxidant indicator. *Environ. Pollut.* **18**, 289–303.

Nystrom, S. D., Hendrickson, R. C., Pratt, G. C., and Krupa, S. V. (1982) A computerized open-top field chamber system for exposing plants to air pollutants. *Agric. Environ.* **7**, 213–221.

Olszyk, D. M. and Tibbitts, T. W. (1981) Stomatal response and leaf injury of *Pinus sylvestris* L. with SO_2 and O_3 exposures. *Plant Physiol.* **67**, 539–544.

Olszyk, D. M., Byternowicz, A., and Takemoto, B. K. (1989) Photochemical oxidant pollution and vegetation: effects of mixtures of gases, fog and particles. *Environ. Pollut.* **61**, 11–29.

Olszyk, D. M., Tibbitts, T. W., and Hertzberg, W. M. (1980) Environment in open-top field chambers utilized for air pollution studies. *J. Environ. Qual.* **9**, 610–615.

Olszyk, D. M., Kats, G., Dawson, P. J., Byternowicz, A., Wolf, J., and Thompson, C. R. (1986a) Characteristics of air exclusion systems vs. chambers for field air pollution studies. *J. Environ. Qual.* **15**, 326–334.

Olszyk, D. M., Byternowicz, A., Kats, G., Dawson, P., Wolfe, J., and Thompson, C. R. (1986b) Crop effects from air pollutants in air exclusion systems vs. field chambers. *J. Environ. Qual.* **15**, 417–422.

Ormrod, D. P. and Adedipe, N. O. (1975) Experimental exposures and crop monitors to confirm air pollution. *Hort. Sci.* **10**, 493-494.

Ormrod, D. P. and Beckerson, D. W. (1986) Polyamines. *Hort. Sci.* **21**, 1070–1071.

Ormrod, D. P., Adedipe, N. O., and Hofstra, G. (1973) Ozone effects on growth of radish plants as influenced by nitrogen and phosphorus nutrition and by temperature. *Plant Soil* **39**, 437–439.

Ormrod, D. P., Marie, B. A., and Allen, O. B. (1988) Research approaches to pollutant crop loss functions. In *Assessment of Crop Loss from Air Pollutants* (eds., Heck, W. W., Taylor, O. C., and Tingey, D. T.). Elsevier, London, 27–44.

Oshima, R. J. (1974) A viable system of biological indicators for monitoring air pollutants. *JAPCA* **24**, 576–578.

Oshima, R. J., Poe, M. P., Braegelmann, P. K., Baldwin, D. W., and VanWay, V. (1976) Ozone dosage-crop-loss function for alfalfa: a standardized method for assessing crop losses from air pollutants. *JAPCA* **26**, 861–865.

Oshima, R. J., Braegelmann, P. K., Baldwin, D. W., Van Way, V., and Taylor, O. C. (1977a) Responses of five cultivars of fresh market tomato to ozone: a contrast of cultivar screening with foliar injury and yield. *J. Am. Soc. Hort. Sci.* **102**, 286–289.

Oshima, R. J., Braegelmann, P. K., Baldwin, D. W., Van Way, V., and Taylor, O. C. (1977b) Reduction of tomato fruit size and yield by ozone. *J. Am. Soc. Hort. Sci.* **102**, 289–293.

Peterson, D. L., Arbaugh, M. J., Wakefield, V. A., and Miller, P. R. (1987) Evidence of growth reduction in ozone-injured Jeffrey pine (*Pinus jeffreyi* Grev. and Balf.) in Sequoia and Kings Canyon National Parks. *JAPCA* **37**, 906–912.

Posthumus, A. C. (1982) Biological indicators of air pollution. In *Effects of Gaseous Air Pollution in Agriculture and Horticulture* (eds., Unsworth, M. H. and Ormrod, D. P.). Butterworth Scientific, London, 115–120.

Pratt, G. C., Hendrickson, R. C., Chevone, B. I., Christopherson, D. A., O'Brien, M., and Krupa, S. V. (1983) Ozone and oxides of nitrogen in the rural upper midwestern U.S.A. *Atmos. Environ.* **10**, 2013–2023.

Prinz, B. (1987) Ozone effects on vegetation. In *Proc. Advanced Research Workshop on Tropospheric Ozone* (ed., Isaksen, I. S. A.), D. Reidel Publishing, Dordrecht, Holland, 425.

Pruchniewicz, P. G. (1973) The average tropospheric ozone content and its variation with season and latitude as a result of the global ozone circulation. *Pure Appl. Geophys.* **106–108**, 1058–1073.

Reich, P. B., Amundson, R. G., and Lassoie, J. P. (1982) Reduction in soybean yield after exposure to ozone and sulfur dioxide using a linear gradient exposure technique. *Wat. Air Soil Pollut.* **17**, 29–36.

Reilly, J. J. and Moore, L. D. (1982) Influence of selected herbicides on ozone injury in tobacco (*Nicotiana tabacum*). *Weed Sci.* **30**, 260–263.

Reinert, R. A., Dunning, J. A., Heck, W. W., Benepal, P. S., and Rangappa, M. (1984) Screening of bean (*Phaseolus vulgaris*) for sensitivity to ozone. *Hort. Sci.* **19**, 86–88.

Rich, S. and Hawkins, A. (1970) The susceptibility of potato varieties to ozone in the field. *Phytopathology* **60**, 1309.

Rich, S., Ames, R., and Zuckel, J. W. (1974) 1,4-oxanthiin derivatives protect plants against ozone. *Plant Dis. Rept.* **58**, 162–164.

Richards, B. L., Middleton, J. T., and Hewitt, W. B. (1958) Air pollution with relation to agronomic crops. V. Oxidant stipple of grape. *Agron. J.* **50**, 559–561.

Roberts, T. M. (1981) Effects of stack emissions on agriculture and forestry. *CEGB Res.* **12**, 11–24.

Rogers, H. H., Jeffries, H. E., Stachel, E. P., Heck, W. W., Ripperton L. A. and Witherspoon A. M. (1977) Measuring air pollutant uptake by plants: a direct kinetic technique. *JAPCA* **27**, 1192–1197.

Rubin, B., Leavitt, J. R. C., Penner, D., and Saettler, A. W. (1980) Interaction of antioxidants with ozone and herbicide stress. *Bull. Environ. Contam. Toxicol.* **25**, 623–629.

Runeckles, V. C., Wright, E. F., and White, D. (1990) A chamberless field exposure system for determining the effect of gaseous air pollutants on crop growth and yield. *Environ. Pollut.* (in press).

Runeckles, V. C., Staley, L. M., Bulley, N. R., and Black, T. A. (1978) A down draft field chamber for studying the effects of air pollutants on plants. *Can. J. Bot.* **56**, 768–778.

Schulte-Hostede, S., Darrell, N. M., Blank, L. W., and Wellburn, A. R. (eds.) (1988) *Air Pollution and Plant Metabolism.* Elsevier, London, 379.

Shinn, J. H., Clegg, B. R., and Stuart, M. L. (1977) A linear-gradient chamber for exposing field plants to controlled levels of air pollutants. UCRL Reprint no. 80411. Lawrence Livermore National Laboratory, University of California, Livermore.

Singh, H. B., Ludwig, F. L., and Johnson, W. B. (1978) Tropospheric ozone: concentrations and variabilities in clean, remote atmospheres. *Atmos. Environ.* **12**, 2185–2196.

Skelly, J. M., Yang, Y. S., Chevone, B. I., Long, S. J., Nellessen, J. E., and Winner, W. E. (1983) Ozone concentrations and their influence on forest species in the Blue Ridge Mountains of Virginia. In *Air Pollution and the Productivity of the Forest* (eds., Davis D. D., et al.). Izaak Walton League of America, Washington, D.C., 143–159.

Snedecor, G. W. and Cochran, W. G. (1978) *Statistical Methods.* Iowa State University Press, Ames, 593.

Steübing L. and Jäger H. J. (1982) *Monitoring of Air Pollutants by Plants, Methods and Problems.* Dr. W. Junk, Pub., The Hague, 161.

Taylor, G. E., Jr., Tingey, D. T., and Ratsch, H. C. (1982) Ozone flux in *Glycine max* (L.) Merr.: sites of regulation and relationship to leaf injury. *Oecologia* **53**, 179–186.

Taylor, G. S. (1974) Ozone injury on tobacco seedlings can predict susceptibility in the field. *Phytopathology* **64**, 1047–1048.

Teso, R. R., Oshima, R. J., and Carmean, M. J. (1979) Ozone-pesticide interactions. *California Agriculture* **April**, 13–15.

Thompson, C. R. and Taylor, O. C. (1969) Effects of air pollutants on growth, leaf-drop, fruit-drop, and yield of citrus trees. *Environ. Sci. Technol.* **3**, 934–940.

Thompson, C. R., Olszyk, D. M., Kats, G., Bytnerowicz, A., Dawson, P. J., and Wolf, J. W. (1984) Effects of O_3 or SO_2 on annual plants of the Mojave Desert. *JAPCA* **34**, 1017–1022.

Tomlinson, H. and Rich, S. (1973) Relating ozone resistance and antisenescence in beans treated with benzimadozole. *Phytopathology* **63**, 208.

Unsworth, M. H. and Ormrod, D. P. (eds.) (1982) *Effects of Gaseous Air Pollution in Agriculture and Horticulture.* Butterworth Scientific, London, 532.

Unsworth, M. H., Heagle, A. S., and Heck, W. W. (1984a) Gas exchange in open-top field chambers. I. Measurement and analysis of atmospheric resistances to gas exchange. *Atmos. Environ.* **18**, 373–380.

Unsworth, M. H., Heagle, A. S., and Heck, W. W. (1984b) Gas exchange in open-top field chambers. II. Resistances to ozone uptake by soybeans. *Atmos. Environ.* **18**, 381–385.

U.S. EPA (1986) Air quality criteria for ozone and other photochemical oxidants. Vol. III. EPA-600/8-84/020CF. U.S. Environmental Protection Agency, Research Triangle Park, NC.

Weinstock, L., Kender, W. J., and Musselman, R. C. (1982) Microclimate within open-top air pollution chambers and its relation to grapevine physiology. *J. Am. Soc. Hort. Sci.* **107**, 923–929.

Westman, W. E. (1979) Oxidant effects on Californian coastal sage scrub. *Science* **197**, 960–964.

Wood, F. A., Drummond, D. B., Wilhour, R. G., and Davis, D. D. (1973) An exposure chamber for studying the effects of air pollutants on plants. Progress Report No. 335, Pennsylvania State University, State College.

CHAPTER 5

Uptake of Ozone by Vegetation

Victor C. Runeckles, Department of Plant Science, University of British Columbia, Vancouver, B.C., Canada

5.1 INTRODUCTION

All terrestrial green plants have the capacity to absorb and emit gases. Most of this gas exchange takes place in the foliage and predominantly concerns the uptake and release of carbon dioxide and oxygen, which are involved in the processes of photosynthesis and respiration, and the release of water vapor, the final stage of the process of transpiration.

Such gas exchange is the consequence of combinations of the physical processes of mass transfer, diffusion, and sorption, and the chemical reactions that either utilize or release the gases in question. Similarly, ozone and other gaseous pollutants that may be present in ambient air are sorbed onto foliar and other surfaces and move into the plant tissues. Surface deposition may lead to chemical changes in the cuticle and cuticular waxes on the leaf surface, but movement into the interior tissues is the essential prerequisite for most of the biochemical and physiological effects of exposure that have been observed. Without access to the internal sites of reaction, a gaseous pollutant such as ozone may be relatively harmless to vegetation.

Thus, uptake plays a key role in determining the effects on metabolism and physiology, and in the description and quantification of dose response.

5.2 THE NATURE OF THE ATMOSPHERE-PLANT-SOIL SYSTEM

Since the plant is part of the larger system involving the atmosphere, vegetation, and soil, it is appropriate to review first the features of this overall system as they relate to ozone pollution. The formation of ozone in the lower layers of the troposphere, together with its dispersal and transport, are discussed in Chapters 2 and 3. With regard to the potential impact of ozone on vegetation, the important features of the ambient air are the ozone concentrations present, the volume of the parcel of polluted air to which the vegetation may be exposed, and the turbulence and mixing that are occurring within this volume, since these define the reservoir from which ozone can pass to the plant canopy.

As depicted in the simple diagram of the overall system in Figure 5.1, vegetation provides only one of the major "sinks" for any ozone present in the ambient air. The pollutant may also be adsorbed by soil and other materials and be deposited in snow and bodies of water.

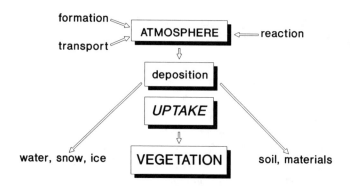

Figure 5.1. Overall scheme of gaseous pollutant uptake and deposition.

Figure 5.1 also recognizes that the system is dynamic: ozone is entering or being formed (and destroyed) within this reservoir and is leaving it via various routes. These routes of depletion are summarized in the term "deposition", a term that has, unfortunately, sometimes been used synonymously with "uptake". "Deposition" is best reserved for describing transfer from the ambient air, regardless of which sinks are involved, leaving "uptake" for when the focus is on the plant response. For uptake is concerned not only with sorption on the leaf and entry into its interior, but with transport to, and utilization at, the various reaction sites within it.

The relative importance of the three major sinks depicted in Figure 5.1 is obviously dependent on location, meteorology, and the extent and nature of the vegetative cover. Appreciable deposition of ozone may occur on soil surfaces (Macdowall, 1974; Turner et al., 1974). On the other hand, deposition on water bodies (or snow) is usually one order of magnitude less than on vegetation, although roughness of the water surface may double such deposition (Galbally and Roy, 1980). Transfer of ozone to foliage does not require rainfall, as in the "wet deposition" of several other air pollutants, although the presence of water on leaf surfaces can reduce uptake (Wesely et al., 1982).

With regard to uptake by plants, numerous studies have led to the description of more detailed models, which will be elaborated upon later, all of which recognize the coupling between the atmosphere and vegetation (Monteith, 1981). For the moment, however, it is appropriate to summarize the models in terms of three major components: the free air above the vegetation; the layer of air immediately above the vegetation (the boundary layer); and the vegetation itself, with its canopy of foliage.

As Chamberlain (1986) has pointed out, the processes at work in the first are almost purely meteorological in character. The second (which involves eddy and molecular diffusion to the surface of the vegetation) and the third (which involves molecular diffusion and other processes such as adherence to surfaces and solution within the tissues of the foliage) involve, to varying degrees, both the physics and chemistry of the gases of interest and of the vegetation, including its physiological state.

While these three components provide a simplistic view of uptake, their use can nevertheless provide insights into their individual importance to plant uptake and the conditions under which their roles may be modified. For example, Hicks and Matt (1988) have suggested that conditions above and within the canopy are more important than those within the boundary layer in affecting the uptake of gaseous sulfur dioxide; the same may well be true for ozone. They point out, though, that their treatment is based upon the "big leaf" concept, which regards the canopy as a homogeneous entity rather than as a heterogeneous assemblage of foliar and other surfaces and structures (O'Dell et al., 1977; Wesely and Hicks, 1977; Unsworth, 1981). However, recent progress has been made in relating overall canopy properties to those of individual leaves (Baldocchi et al., 1987).

Varying degrees of heterogeneity typify all canopies, and this heterogeneity takes many forms, ranging from the canopy level itself, e.g., the spatial arrangements and orientations of individual plants and leaves, to the heterogeneity of the structural and physicochemical features of tissues, cells, and their organelles. Nevertheless, the "big leaf" concept has provided a useful means of investigating uptake and its regulation, at the population and community levels.

5.3 THE CANOPY, THE PLANT, AND THE PATHWAYS OF UPTAKE

Since foliage is a major sink for ozone, the structure of the canopy and the density of the foliage within it will have a pronounced effect upon the concentration of ozone to which individual leaves will be exposed. Few studies have addressed this issue, but it has been clearly shown that the concentration of ozone (and other pollutants) diminishes as one passes down through a canopy to soil level (Bennett and Hill, 1973). As a result, leaves within a dense canopy will tend to be exposed to lower concentrations than those on the surface or at its edges. However, because the air movement within a dense canopy is reduced, the lower concentrations within it will be less prone to short-term fluctuations caused by turbulence than will the air above the canopy.

Other features of the environment of leaves within a canopy are also modified in relation to those on the outside, e.g., reduced light intensity, increased temperature, and increased humidity. Hence, the impact of a given ambient ozone concentration outside the canopy on such leaves will be a function of all of these factors.

In many crop situations the spacing of rows opens up the canopy to provide optimal leaf-area indices, permitting light penetration. However, such spacing also tends to reduce the resistance to the transfer of ozone between different layers of the canopy, thereby reducing the gradient within the crop and maintaining the potential for uptake down to the lowest leaves (Leuning et al., 1979a; Unsworth, 1981).

Canopy structure and density become particularly important when attempting to extrapolate to field situations from results obtained under laboratory or controlled conditions utilizing individual plants, though in some situations, comparable responses have been reported (Endress and Grunwald, 1988). Furthermore, canopy structure undoubtedly influences the magnitude of the response of different species in mixtures, whether of agricultural crops such as pasture and forage mixtures or of natural vegetation. In such mixed vegetation, some species may well "protect" others by virtue of their relative locations within the layers of the canopy. Their capacity to act as sinks (thereby influencing the ozone concentration) and to modify other environmental factors may influence the responses of the other species. This issue is discussed more fully in Chapter 6.

Within the canopy, it is the leaves that constitute the principal sites of uptake. It is therefore important to understand the features of leaf structure and anatomy that relate to the movement of a pollutant, such as ozone, from the outside air to the internal sites of reaction.

Sectional views of a leaf of a typical broad-leaved plant and of a pine needle are shown in Figure 5.2. On the outside of the leaf is the cuticle with or without distinct wax deposits whose thickness and chemistry vary considerably from species to species. The cuticle is attached to the outermost layer of cells, the epidermis. Most epidermal cells are devoid of pigments such as chlorophyll, the exceptions being the guard cells of the stomata, the elliptical pores that occur in most foliar epidermal surfaces. The guard cells of each stoma are capable of changing the aperture of the stomatal pore in response to a wide range of physical and chemical factors, as discussed in Chapter 6. In many desert plants and other xerophytes, the stomata are located in depressions or pits in the leaf surface.

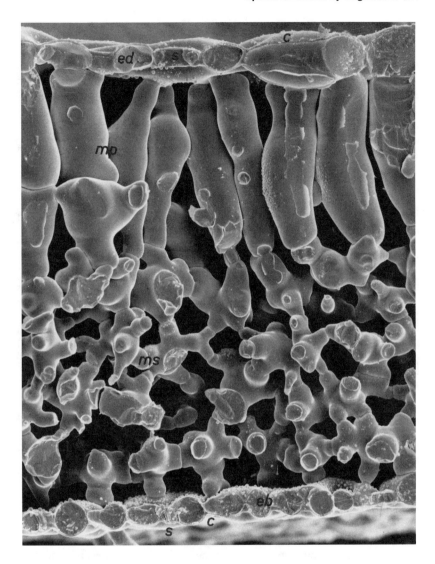

Figure 5.2A. Scanning electron micrograph of longitudinal section of a bean leaf (*Phaseolus vulgaris* L.). c = cuticle; e = epidermis (eb = abaxial; ed = adaxial); m = mesophyll (mp = palisade mesophyll; ms = spongy mesophyll); s = stomata. Micrograph courtesy of Dr. C.E. Jeffree, University of Edinburgh (reproduced from Jeffree et al., 1987, with permission).

The frequency of stomata is usually greater on the lower (abaxial) leaf surfaces of broad-leaved plants; in some species they may be virtually absent from the upper (adaxial) surface. On the other hand, in grasses and many plants with vertically orientated leaves, they tend to occur in equal numbers on both surfaces, while in conifers such as the pines, they may be uniformly distributed

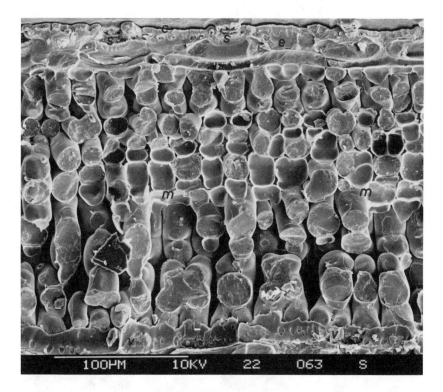

Figure 5.2B. Scanning electron micrograph of longitudinal section of a pine needle (*Pinus sylvestris* L.). c = cuticle; e = epidermis (eb = abaxial; ed = adaxial); m = mesophyll; s = stomata; v = vascular bundle. Micrograph courtesy of Dr. C.E. Jeffree, University of Edinburgh.

over the surface. The frequency of occurrence on the lower surface of leaves of most broad-leaved plants is in the range 40 to 300 mm^{-2}. The pores of open stomata account for 0.2 to 2% of the leaf surface area (Nobel, 1983).

Leaf hairs or trichomes and various glandular structures also occur on the surfaces of the leaves of many species, as shown in Figure 5.3 (a view of the abaxial surface of a birch leaf). The frequency of such structures influences the movement of air over the epidermis and to the stomata.

Within the leaf are the mesophyll cells, rich in chloroplasts that carry out the processes of photosynthesis. In the leaves of broad-leaved plants, the mesophyll typically consists of an upper palisade layer of somewhat densely packed, cylindrical cells, and a lower layer of spongy mesophyll consisting of loosely associated strands of cells, as shown in Figure 5.2A and Figure 5.3. In many monocotyledonous plants (especially the grasses), the mesophyll has a more uniform structure throughout the thickness of the leaf, without the marked distinction between palisade and spongy tissues. The same is true of the xerophytic pines, whose mesophyll cells have, typically, involuted walls. Towards the interior of the leaf are the vascular bundles containing the xylem and

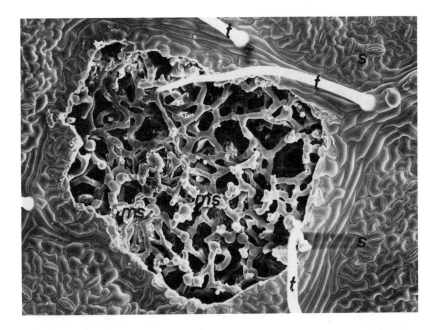

Figure 5.3.   Scanning electron micrographic view of the adaxial surface of a leaf of birch (*Betula pubescens* Ehrh.) with part of the epidermis removed, showing the large air spaces between the spongy mesophyll cells (ms), and stomata (s) and trichomes (t) on the surface. Micrograph courtesy of E.A. Neighbour and K. Oates, University of Lancaster.

phloem tissues responsible for the movement of water and nutrients into and out of the leaf.

In the C-4 plants, which include many members of the grass family, such as corn and sugarcane, an additional type of mesophyll can be distinguished: the bundle-sheath tissue surrounding the vascular bundles, typified by the high density of its cellular contents and the large numbers of chloroplasts in its cells.

In spite of the direct attachment of many of the mesophyll cells to each other, much of the leaf volume consists of air spaces between the cells. This is particularly true of the broad-leaved plants, as is shown in Figure 5.2A and Figure 5.3. In some species there are large, distinct air spaces beneath the stomata and within the palisade layer (the substomatal cavities). However, the occurrence of such cavities is perhaps not as ubiquitous as is generally assumed — witness the poorly defined cavity in the "typical" bean leaf depicted in Figure 5.2A, and the absence of any distinct cavity in the pine needle (Figure 5.2B). In many species, the stomata simply provide access to the general continuum of air spaces within the leaf.

The palisade mesophyll layer is generally 10 to 40% air by volume, and the spongy mesophyll, 50 to 80% (Nobel, 1983). Although the mesophyll cells of the pine are more closely packed, there are clear air spaces between many adjacent cells, as shown in the longitudinal section in Figure 5.2B. The interior air space also abuts some of the interior surfaces of the cells in the epidermal layers (in addition to the guard cells).

Much of the total surface of the mesophyll cells is exposed to this air space rather than being in contact with other cells. In typical broad-leaved mesophytes, although there is a lesser fraction of the internal air volume in the palisade tissue, palisade cells usually have a greater area of cell wall exposed to the internal air space than is the case with spongy mesophyll cells. In such plants, the ratio A^{mes}/A ranges from 10 to 40, where A^{mes} is the total area of exposed mesophyll cell walls and A is the area of one surface of the leaf (Nobel, 1983). Taylor et al. (1982) reported ratios ranging from 17.0 to 21.8 for leaves of different age classes of two soybean cultivars. In xerophytes such as pine, the ratios are somewhat higher: 20 to more than 50 (Nobel, 1983).

The exposed surface of the wall of a mesophyll cell is moist (or at least hydrated), since it interfaces on the exterior with the air in the adjacent air space that is close to saturation (Nobel, 1983), and on the interior with the cell wall that is in direct contact with the cell membrane or plasmalemma of the underlying cell itself.

This combination of leaf anatomy and physical conditions therefore leads to several possibilities with regard to the fate of ozone molecules in the ambient air surrounding a leaf:

1. Sorption and reaction on the cuticular surface
2. Passage through the cuticle into the epidermis
3. Reaction in the epidermal cells or passage to the interior of the leaf
4. Diffusion through the stomata into the substomatal cavities
5. Reaction with gases within the substomatal cavities and other air spaces
6. Sorption, partition into the liquid phase, and reaction on the epidermal and mesophyll cell surfaces within the substomatal cavities
7. Diffusion of ozone or its reaction products through the air spaces between the mesophyll cells
8. Following sorption and partition into the liquid phase, movement of ozone or products of its reaction through the cell walls to the cell membrane
9. Movement of ozone or its reaction products through the cell membrane to the interior of the cell and its organelles (nucleus, chloroplasts, mitochondria, vacuole)

It is not known whether ozone itself or the products of its reaction within the air spaces or with cell wall constituents and the cell membrane are the chemical species that are directly responsible for many of the various phytotoxic effects attributed to ozone. It may well be that some effects are caused directly by ozone, while others are caused by its reaction or decomposition products. This matter is discussed further, below and in Chapter 6, but the absence of information as to the precise nature of the chemical species involved does not preclude the development of overall models of ozone uptake. The relatively low solubility of ozone suggests that it is likely to penetrate deep into the interior air spaces, in contrast to the more soluble CO_2 (Taylor and Tingey, 1982; Taylor et al., 1988). This, together with the high palisade surface area interfacing the air spaces, may help explain why the acute effects of ozone injury that result in cell necrosis typically occur in the palisade layer (Evans and Ting, 1974). However, the high reactivity of ozone (discussed in Chapter 6) may, at the same time, serve to limit the penetration of ozone itself. Such reactivity is supported by the recent report that the computed concentrations of ozone within the intercellular space of leaves of the sunflower (*Helianthus annuus* L.) and *Perilla ocymoides* L., exposed briefly to concentrations as high as 1.5 ppmv, rapidly drop to zero (Laisk et al., 1989).

5.4 MODELS OF POLLUTANT UPTAKE

Following the early work of Gaastra (1959), numerous models of photosynthetic gas exchange based on the electrical circuit analogue of a network of resistances to gas flow have been described and reviewed (Jarvis, 1971; Cooke and Rand, 1980; Tenhunen et al., 1980; Monteith, 1981; Leuning, 1983). The concepts involved have been applied to the development of models of gaseous air pollutant uptake by numerous workers (Bennett et al., 1973; O'Dell et al., 1977; Wesely and Hicks, 1977; Taylor, 1978; Black and Unsworth, 1979; Unsworth, 1981, 1982; Taylor et al., 1982; Wesely et al., 1982; Bache, 1986; Hicks et al., 1987; Baldocchi et al., 1987, 1988; Hicks and Matt, 1988). The past decade has seen several reviews of the subject, covering both field and experimental situations (Heath, 1980; Unsworth, 1981, 1982; Hosker and Lindberg, 1982; Tingey and Taylor, 1982; Hosker, 1986; Cape and Unsworth, 1988; Luxmoore, 1988; Taylor et al., 1988).

A common feature of models applied to the uptake of ozone (or any gas) is that they are essentially extrapolations from the theory developed to describe the movement of water vapor out of transpiring leaves and, to a lesser degree, the exchange of CO_2 between the atmosphere and the leaf. The two sets of basic assumptions made are (1) that the eddy and molecular diffusivities of the gas of interest are inversely proportional to molecular weight: the greater the molecular weight, the lower the diffusivity; and (2) that the water vapor gradient starts from near saturation within the leaf's air spaces, particularly the substomatal cavities (Nobel, 1983). Hence, the values reported for the various resistances to the uptake of ozone are, for the most part, obtained by calculation from porometric measurements of water vapor transfer (e.g., see Taylor et al., 1982), not by direct observation.

Some of the uptake models described are concerned with overall dry deposition and, hence, incorporate soil and other sinks in addition to vegetation

(Unsworth, 1981, 1982; Baldocchi et al., 1987; Hicks et al., 1987; Baldocchi, 1988; Hicks and Matt, 1988). Most are based upon the electrical resistance analogue in which total pollutant flux, F_t (g m^{-2} s^{-1} units), onto and into the foliage, is treated as the result of the concentration difference between the ambient air, C_a (g m^{-3} units) and the internal sinks (C_i; usually C_i is considered to equal $C_o = 0$), such that

$$F_t = \frac{\left(C_a - C_o\right)}{r_t}$$

(5.1)

where r_t (s m^{-1} units) is the overall resistance to gas movement. This overall resistance may be viewed as comprising various series-parallel networks such as those depicted in Figure 5.4.

Although such models are conventionally described in terms of networks of resistances and therefore appear to describe the steady-state situation with regard to the fluxes resulting from a particular set of conditions occurring at a particular instant, the networks are dynamic and some or all of the resistances involved may vary appreciably over time.

Furthermore, although most models have utilized the concentration of the pollutant (expressed in g m^{-3} units) as the equivalent of the electrical potential across the network, the transfer process may also be viewed as being the outcome of differences in the partial pressure of the pollutant, or differences in mole fraction (Nobel, 1983). Since mole fraction (or mixing ratio) is dimensionless, resistance, on this basis, is expressed in m^2 s mol^{-1} units. Its reciprocal, conductance, has the same units as flux (mol m^{-2} s^{-1}) and is somewhat easier to visualize. The use of mole fractions offers the advantage that the resistance values obtained are independent of concentration and pressure and less dependent on temperature than those based on absolute concentrations (Nobel, 1983). Furthermore, mole fraction is equivalent to the widely used parts per million by volume (ppmv) or µl l^{-1} form of expressing concentration. At standard pressure (101.3 kPa) and 20°C, a resistance of 1 s m^{-1} corresponds to a resistance of 0.024 m^2 s mol^{-1}. In the following general discussion of the features of the various models that have been proposed, the choice of dimensions for the different components is unimportant, provided that they are self-consistent. However, the potential influence of temperature and especially pressure should be borne in mind and is discussed further in Section 5.6. The influence of temperature and pressure on the measurement of ambient concentration and the units in which it is expressed is discussed in Chapter 3 and by Lefohn et al. (1990).

The resistance, r_b, external to the leaf surface, limits transfer through boundary layers. The other resistances relate to diffusion through the stomata (r_s) and cuticle/epidermis (r_c) and to sorption and reaction on the leaf surface (r_p). Since the resistance analogue is strictly appropriate only for gas-phase processes, the resistances of the parts of the pathway that involve internal gas-phase diffusion, partition into the liquid phase, diffusion in the liquid phase, and the quasiresistances

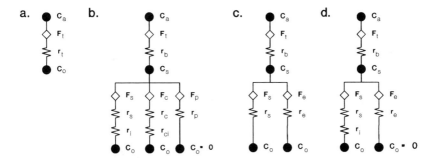

Figure 5.4.   Electrical resistance analogue models of ozone uptake.
(a) Overall model; (b) based on Unsworth, 1981; (c) based on Bennett et al., 1973; (d) based on Taylor et al., 1982. C_a is the ambient concentration, C_s is the concentration at the leaf surface, C_i is the concentration within the interior air spaces, and C_o is the final concentration, assumed to be zero. The fluxes and resistances are: F_t and r_t, total; F_c and r_c, through the cuticle; F_s and r_s, through the stomata; F_p and r_p, to the surface; and F_e and r_e, overall nonstomatal transfer. r_i and r_{ci} are internal resistances in the pathways through the stomata and cuticle, respectively.

of any chemical reactions that a pollutant enters into, are amalgamated for convenience into "residual" or "internal" resistances in the mesophyll (r_i) and in the cuticle/epidermis (r_{ci}) (Figure 5.4). However, these composite resistances may be subdivided into their physical and chemical components, as has been done for the influx and efflux of CO_2 in leaves (Tenhunen et al., 1980; Nobel, 1983).

In the models depicted in Figure 5.4c,d, no distinction is made between losses to the surface and passage through the epidermis, and the nonstomatal resistance paths are combined:

$$r_e = \left[\left(r_c + r_{ci} \right)^{-1} + r_p^{-1} \right]^{-1}$$

The model depicted in Figure 5.4c was developed in part to permit the estimation of the gas-phase ozone concentration within the leaf, C_i, rather than assuming a final sink concentration, $C_o = 0$.

When $C_i = C_o = 0$ is assumed, Equation 5.1 reduces to Equation 5.2:

$$F_t = \frac{C_a}{r_t} \tag{5.2}$$

where

$$F_t = F_p + F_c + F_s \quad \text{(cf Figure 5.4b), or}$$

$$F_t = F_e + F_s \quad \text{(cf Figure 5.4c,d),}$$

apportioned according to the values of r_p, r_c, r_{ci}, (or r_e), r_s and r_i, since

$$r_t = r_b + \left[\left(r_s + r_i \right)^{-1} + \left(r_c + r_{ci} \right)^{-1} + \left(r_p \right)^{-1} \right]^{-1} \quad \text{(cf Figure 5.4b),}$$

$$r_t = r_b + \left[r_s^{-1} + r_e^{-1} \right]^{-1} \quad \text{(cf Figure 5.4c), or}$$

$$r_t = r_b + \left[\left(r_s + r_i \right)^{-1} + \left(r_e \right)^{-1} \right]^{-1} \quad \text{(cf Figure 5.4d)}$$

Several models have incorporated the concept of deposition velocity, v_g (m s^{-1} units), defined as v_g = F/C, for an average driving concentration, C, expressed in absolute rather than mole fraction units. Deposition velocity is thus equivalent to conductance and is the reciprocal of the overall resistance:

$$v_g = \left[r_t \right]^{-1}$$

As Unsworth (1981) has pointed out, one should not assume that v_g is constant, since it clearly depends upon wind speed, physiological responses, and chemical reactivity and may vary by at least an order of magnitude, depending on the canopy environment (Fowler, 1981). Nevertheless, under standardized conditions, determination of v_g can reveal interspecific and intervarietal differences in uptake characteristics (Thorne and Hanson, 1972; Elkiey and Ormrod, 1981a,b).

Application of these models has led to certain general conclusions. With regard to uptake by foliage itself, the most important resistances are usually those affecting flux *to* the foliar surfaces (i.e., the boundary layer resistance, r_b, and the external aerodynamic resistances affecting transport to the boundary layer) and the flux *into* the foliage (i.e., the stomatal resistance, r_s). However, the composite internal resistances, r_i and r_{ic}, may also be significant (Tingey and Taylor, 1982; Taylor et al., 1988). Since the importance of r_s is probably overestimated in resistance analogue models of CO_2 uptake (Farquhar and Sharkey, 1982), the same may well be true of gaseous pollutant uptake (Taylor et al., 1988). On the other hand, Leuning (1983) has also pointed out that the net influx of a gas such as ozone is somewhat impeded by the outward flux of water vapor, as a result of intermolecular collisions, the net result of which would be to raise r_s by a few percent.

The composite internal resistances, r_i and r_{ci}, are impossible to determine directly, but one approach to their estimation (Black and Unsworth, 1979) is based on the rearrangement of the overall resistance equation to the form:

$$\left[r_t - r_b \right]^{-1} = \left[r_s + r_i \right]^{-1} + \left[r_c + r_{ci} \right]^{-1} + r_p^{-1} \tag{5.3}$$

which permits the iterative estimation of r_i and r_p by selecting a value for r_i such that $(r_t - r_b)^{-1}$ and $(r_s + r_i)^{-1}$ are linearly related.

Using this approach, Unsworth (1981) showed that the uptake of ozone by corn (*Zea mays*) leaves gave a straight-line relationship for the variables of Equation 5.3 when r_i was zero, over a range of r_s values from 300 to 1500 s m^{-1}, with $r_b = 10$ s m^{-1}. If $r_i = 0$, it is also probable that $r_{ci} = 0$. This suggests that the interior of the corn leaf at the time of the observations was a perfect sink for ozone either as a result of rapid chemical reaction within the leaf's air spaces or as a result of rapid sorption on the mesophyll cell walls, or both, as suggested by the conclusions of Laisk et al. (1989). The intercept for corn corresponded to a resistance, $r_c + r_p$, of 2180 s m^{-1}.

In contrast, Wesely et al. (1978) reported estimates of r_i ranging from 300 to 400 s m^{-1} for ozone fluxes to corn canopies in the field. A value of 60 s m^{-1} was estimated for alfalfa (Heath, 1980). Appreciable values for r_i were also estimated by Taylor et al. (1982) for soybean leaves exposed to the relatively high range of ozone concentrations from 0.25 to 0.6 ppmv, using the model depicted in Figure 5.4d. Total ozone flux, F_t, was found, in general, to be linearly related to ozone concentration, C_a, as expected, and after rising during the first hour of exposure, remained constant over the next 3 hours when C_a was less than 0.30 ppmv. At higher ozone concentrations, F_t declined after the initial rise. Flux to the surface, F_e, also increased with C_a, but after its initial rise, it declined at all concentrations. The contribution of F_e to F_t was usually less than 30%. The internal flux, $F_s = F_t - F_e$, tended to rise slowly with time at low concentrations; at high concentrations it rose to still higher initial levels and then tended to decline, so that after 4 hours the values of F_s for all concentrations, C_a, were similar. These changes were accompanied by changes in calculated internal resistance, r_i, as shown in Figure 5.5. For $C_a < 0.30$ ppmv, r_i was less than 100 s m^{-1} and remained steady over four hours. For higher concentrations, r_i increased with both time and concentration, with maxima ranging from 530 to 930 s m^{-1} for different leaf ages. Furthermore, the contribution of r_i to r_t increased with r_t, accounting for 10 to 20% when $100 < r_t < 400$ s m^{-1}, and rising to 25 to 50% when r_t approached 1000 s m^{-1}. There was a tendency for r_i to contribute more to r_t in young leaves than in old ones. In these, as in other studies, r_b and r_s for ozone were calculated from data obtained for water vapor efflux by multiplying by 1.63, the ratio of the diffusivities of water vapor and ozone.

There are many possible explanations for these reported differences in calculated internal resistances. The assumption of identical pathways of water vapor and ozone movement within the leaf (and, hence, the validity of using

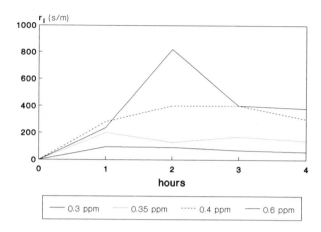

Figure 5.5.   Effect of ambient ozone concentration and duration of exposure on internal resistance, r_i, in soybean leaves (*Glycine max* L., cv Hood). (Adapted from Taylor et al., 1982.)

water vapor flux or resistance measurements for estimating ozone values) has been questioned on the grounds of the relatively low solubility of ozone in water and the consequent likelihood that the ozone diffusion path length is longer and involves the air spaces beyond the substomatal cavities (McLaughlin and Taylor, 1981). Related to this are the features of the anatomy of the leaves of the different species. The soybean leaf has a clearly developed spongy mesophyll tissue in contrast to the typical Kranz anatomy of a C-4 plant such as corn. In addition, although both corn and soybean are amphistomatous, in the former the stomatal frequencies are approximately the same on both surfaces, while in soybean there are approximately twice as many stomata on the abaxial as on the adaxial surface (Taylor et al., 1982). Such differences in stomatal distribution have been incorporated into some models of uptake in which the routes through the abaxial and adaxial surfaces are dealt with as separate parallel pathways (Bennett et al., 1973; Leuning et al., 1979a,b).

Yet another explanation for differences in estimates of r_i may lie in differences in chemical reactions that deplete ozone in the interior air spaces or at the mesophyll cell surfaces. A low value of r_i does not provide any information about the mechanisms involved, only that the interior gas phase concentration is rapidly reduced.

The model depicted in Figure 5.6 illustrates some of the potential component resistances within r_i that could be involved in determining the ultimate fate of ozone within the leaf. At this conceptual level, it is immaterial whether the species involved in the various transfers and reactions is ozone itself or its reaction products, since unless the reaction products are also gases that escape from the liquid phase, they are all directly interrelated through the rate constants of the reactions by which they are formed or utilized. The important feature is that progressive increases in the resistances associated with the individual branch

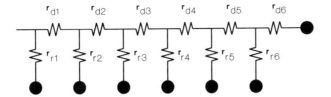

Figure 5.6. Components resistances of r_i. Resistances to physical processes: r_{d1} gas-phase transfer within the intercellular spaces; r_{d2} partition/aqueous-phase diffusion in the cell wall; r_{d3} transfer through the plasmalemma; r_{d4} diffusion in the cytosol; r_{d5} transfer through the membranes of cell organelles (chloroplasts, mitochondria, etc.); r_{d6} diffusion within organelles; Resistances to chemical processes: r_{r1} reactions within the intercellular spaces; r_{r2} reactions within the cell wall matrix; r_{r3} reactions with the plasmalemma; r_{r4} reactions within the cytosol; r_{r5} reactions with organelle membranes; r_{r6} reactions within cell organelles.

reactions along the network, as a result of saturation of sites or depletion of reactants, results in a progressive increase in the overall internal resistance.

Table 5.1 lists some values for the component transfer/diffusion resistances depicted in Figure 5.6 (r_{d1}, r_{d2}, etc.), based on published values for resistances to water and CO_2 movement. The diffusive resistances for the cell wall onwards are based on diffusion in water. However, in the case of ozone, it may also be necessary to consider diffusion within the lipid phase of the cell membranes and the resistances to partitioning between the aqueous and lipid phases.

The related chemical-reaction resistances depicted in Figure 5.6 (r_{r1}, r_{r2}, etc.) are more difficult to estimate in comparable s m^{-1} units. However, recourse can again be made to the methodology developed in modeling CO_2 fixation, the latter stages of which involve liquid phase reactions catalyzed by phosphoenolpyruvate carboxylase in C-4 plants, and ribulose-bis-phosphate carboxylase/oxygenase (rubisco) in both C-3 and C-4 plants, as discussed in Chapter 6.

For example, one such model was developed by Sinclair et al. (1977), based

Table 5.1. Estimates of Some of the Component Resistances to r_i for the Transfer of Ozone, Based Upon Estimates for the Transfer of Water Vapor and CO_2[a]

Component	Resistance (s m^{-1})
r_{d1}: intercellular space	16 –82
r_{d2}: cell wall	103
r_{d3}: plasmalemma	514
r_{d4}: cytosol	10
r_{d5}: chloroplast	< 514

[a] Calculated in proportion to diffusivities, according to $r_1 = r_2(D_2/D_1)$ (Bennett et al., 1973). $D_{H2O} = 2.42 \times 10^{-5}$, $D_{CO2} = 1.51 \times 10^{-5}$, and $D_{ozone} = 1.47 \times 10^{-5}$ m^2.s^{-1} (Leuning et al., 1979a; Nobel, 1983). The notations for individual resistances are as for Figure 5.6.

upon diffusion of CO_2 within a spherical cell in which the chloroplasts are uniformly distributed in the cytosol between the plasmalemma (at radius r_2) and the tonoplast membrane surrounding the cell vacuole (at radius r_1). In C-3 plants in which rubisco acts as both a carboxylase and an oxygenase (leading to significant photorespiration), an approximation for r_1 (for CO_2) that includes both diffusion within the cell and enzymic reaction (but not diffusion through the internal air spaces to the mesophyll cells) is given by

$$\frac{3K_c\left(K_o + [O]\right)}{r_2 V'_c K_o} \cdot \frac{A}{A^{mes}} \qquad (5.4)$$

where K_c and K_o are the Michaelis-Menten constants for CO_2 and O_2, respectively (g m^{-3} units), [O] is the oxygen concentration (g m^{-3} units), r_2 is the cell radius (approximately 5 μm), V'_c is the maximum enzymic velocity for the carboxylase reaction (g m^{-3} s^{-1} units), and A and A^{mes} are as previously defined. In C-3 plants, the term $(K_o + [O])/K_o$ in Equation 5.4 equals approximately 1.7, while for C-4 plants with negligible photorespiration, it is assumed to equal 1.0, leading to values for r_i of 600 and 80 s m^{-1} for C-3 and C-4 plants, respectively.

As Cooke and Rand (1980) point out, these values are based upon numerous assumptions and offer order-of-magnitude approximations rather than statistical estimates. Indeed, since the overall processes are light dependent, they vary with light intensity. Tenhunen et al. (1980), for example, show values for r_i (for CO_2) for wheat increasing from 300 to more than 1500 s m^{-1} as light intensity falls. Any application of such approaches to ozone uptake is, of course, dependent upon knowledge of the specific reactions involved and their rate constants. At the present time, such information is rudimentary; we are still at the stage of speculating about the significance of the various potential reactions that may be involved, many of which are discussed in Chapter 6.

However, Chameides (1989) has recently presented a general model that involves both diffusive and chemical resistances to ozone flux into the leaf and to the plasmalemma. He assumes that, having entered the interior air spaces of the leaf, ozone reaches the plasmalemma by diffusing through the aqueous medium within the cell wall. Its rate of change within the wall is given by the one-dimensional diffusion equation, modified to include depletion by any chemical reactions occurring during the process:

$$\frac{d[O_3]}{dt} = D_w \frac{d^2[O_3]}{dz^2} - k[O_3]$$

where D_w is the aqueous phase diffusivity of ozone, $[O_3]$ is the aqueous phase ozone concentration (M), z is the distance moved within the cell wall, and k is the first order coefficient for loss of ozone in the cell wall solution as a result of chemical reaction (in s^{-1} units).

The *steady state* solutions for the ozone fluxes to the cell wall and plasma-lemma are

$$F_{cw} = \frac{H\,Av}{10^3 \cdot r'_{cw}}\left[X_{cw} - \frac{[O_3]_p}{H\cdot\cosh\,(q)}\right] \tag{5.5}$$

and

$$F_{pl} = \frac{H\,Av}{10^3 \cdot r'_{cw}}\left[\frac{X_{cw}}{\cosh\,(q)} - \frac{[O_3]_p}{H}\right] \tag{5.6}$$

where F_{cw} and F_{pl} are the fluxes (in molecules cm^{-2} s^{-1}) at the interior air space/cell-wall and cell-wall/plasmalemma boundaries, respectively, Av is Avogadro's Number; X_{cw} is the ozone partial pressure at the interior air-space–cell-wall boundary (ppmv); H is the ozone solubility constant (0.01 M atm^{-1}), and $[O_3]_p$ is the ozone concentration. The reactive transfer resistance of ozone within the cell wall, r'_{cw}, which combines both physical and chemical resistances, is given by

$$r'_{cw} = \frac{\tanh\,(q)}{\left(k\cdot D_w\right)^{0.5}}$$

in which

$$q = \left(\frac{k}{D_w}\right)^{0.5}\cdot L$$

where L is the cell-wall thickness (10^{-4} cm). It should be noted that although the concentrations are dimensionless (since they are defined as partial pressures), the particular form of the solutions that incorporates the pressure-dependent term H, results in reactive transfer resistances being expressed in s m^{-1} units.

The fraction (FR) of the ozone reaching the cell wall that reacts during transfer across it is given by:

$$\frac{F_{cw} - F_{pl}}{F_{cw}}$$

Table 5.2. Estimates of Reactive Resistances of Cell Wall Ascorbate Oxidation[a]

Ascorbate concentration (M)	Resistance, r'_{cw} (s m^{-1})
0	1500
10^{-7}	1490
10^{-6}	1400
10^{-5}	905
10^{-4}	316
10^{-3}	100

[a] Based upon rate constant, k = 5 × 10^7 [ascorbate] (s^{-1}), aqueous diffusivity of ozone, D = 2 × 10^{-5} (cm^2 s^{-1}), and cell wall thickness, L = 3 × 10^{-4} cm (based on Chameides 1989).

Combining Equations 5.5, 5.6, and 5.7, and assuming that $[O_3]_p$ is negligibly small, leads to FR = 1 − sech(q).

FR increases steadily with L (for values ranging from 0.1 to 30 μm). FR also increases with k, but the significant increase occurs only over the range $0.1 < k < 10^4$ s^{-1}. With k < 0.1 s^{-1}, virtually all of the ozone reaches the plasmalemma, whereas above 10^4 s^{-1}, it all reacts in the cell wall. Hence, any reaction with a rate constant greater than about 10^3 s^{-1} will represent a significant sink for ozone. As discussed more fully in Chapter 6, ozone is a strong oxidant and is highly reactive towards unsaturated sulfhydryl and various types of ring compounds (Mudd, 1982). It has been suggested that it may be scavenged by reactions in the interior air spaces with metabolic volatiles such as ethylene (Melhorn and Wellburn, 1987) or other olefins. However, the rate constants for such reactions led Chameides (1989) to concur with Tingey and Taylor (1982) in concluding that reactions with olefins in the intercellular air spaces are probably insignificant. On the other hand, he presents evidence for a first-order loss coefficient ranging from 3×10^3 to 6×10^4 s^{-1} for the reaction of ozone with ascorbate. Such values of k are within the range likely to result in significant depletion of ozone, especially since ascorbate levels up to 10^{-3} M have been reported in leaf intercellular fluid (Castillo and Greppin, 1986; Castillo et al., 1987).

Chameides (1989) further argues that, while flux to the leaf is largely dependent upon stomatal resistance, flux to the plasmalemma, F_{pl}, is highly dependent upon the ascorbate content of the mesophyll cell walls and, hence, upon the reactive resistance of ascorbate oxidation by ozone, in the cell walls. Values of r'_{cw} calculated from his data are presented in Table 5.2 and show that, for a typical leaf situation, this resistance can vary from about 100 to 1500 s m^{-1} as the ascorbate content decreases from 10^{-3} M to zero.

Several points emerge from this study. First, the modest decrease in total flux resulting from a decrease in cell-wall ascorbate is consistent with an increase in the branch resistance related to the ozone-ascorbic acid reaction in the cell wall (e.g., r_{r2}; Figure 5.6). Second, although the model can be solved for different stomatal resistances and ascorbate levels, it is a steady-state model. While the results support the idea that cell-wall ascorbate may play an important protective role, no provision is made for the replenishment of ascorbate within the walls. Third, terminating the model at the plasmalemma permits Chameides (1989) to focus on the potential importance of cell-wall ascorbate in accounting for differences in response, but discounts other potentially important scavenging reactions that may occur within the cell, as discussed in Chapter 6. Nevertheless, the approach is a welcome addition to current thinking about ozone uptake, its regulation, and its fate.

In spite of the differences that may exist in r_i among species in different environmental conditions, there is no question that stomatal resistance exercises a major control of ozone uptake, as shown by the daily rise and fall in flux that accompanies the typical daily cycle of stomatal opening and closing, regardless of any changes in ambient ozone levels (Leuning et al., 1979a; Heath, 1980). The dominant role played by stomatal resistance has been a cornerstone to the development of the "big-leaf" models of canopy uptake already referred to. However, as Runeckles and Rosen (1977), Olszyk and Tibbitts (1981), and others have shown, stomatal response is itself capable of being modified by exposure to ozone. While most reports indicate that ozone causes reductions in aperture (Winner et al., 1988), the response may be significantly modified by the leaf's previous history of exposure (Runeckles and Rosen, 1977).

As discussed more fully in Chapter 6, other factors may also influence stomatal response, especially water status and light intensity (Cape and Unsworth, 1988). Since shading within a canopy reduces the average light intensities incident upon lower leaves, and this in turn may reduce stomatal aperture (i.e., increase r_s), some canopy-scale models have dealt with shaded leaves separately as fractions of the total leaf area index (Leuning et al., 1979a,b) and by incorporating the intensity of photosynthetically active incident radiation as a modifier of average canopy resistance (Baldocchi et al., 1987; Baldocchi, 1988).

Uptake at the canopy level has also been modeled by viewing the canopy as consisting of discrete layers of foliage, each with its own characteristic resistances to flux. Unsworth (1981) used this approach to describe the midday ozone fluxes through a tobacco crop. For each layer of the canopy, the incoming flux was defined by the average concentration within it, C_i, the transfer resistances between the canopy layers, r_{ij} (where i and j = i + 1 refer to the canopy layers), and the adaxial and abaxial boundary and stomatal resistances within each layer. Figure 5.7 illustrates the concentrations, fluxes, and resistances predicted using this approach when the ambient concentration, C_a, was 0.96 ppmv (192 μg m^{-3}).

In this model the soil resistances (200 s m^{-1}) were assumed to be constant (Turner et al., 1974). The figure clearly shows the importance of the stomatal resistances. Furthermore, the differences in r_s between the adaxial and abaxial surfaces become progressively more pronounced in the lower layers, with the abaxial resistances progressively decreasing while the adaxial resistances increase. It should also be noted that, although the model indicates that the canopy was the major sink for ozone, almost half of the flux was to the soil.

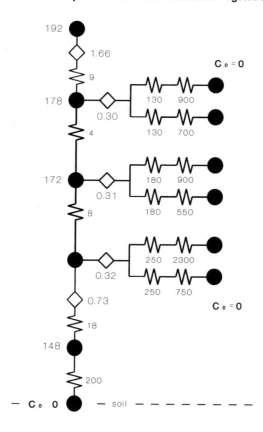

Figure 5.7.   Concentrations, fluxes, and resistances to ozone transfer in a corn *(Zea mays)* canopy. The pairs of horizontal networks are for the fluxes to the upper and lower surfaces, respectively, at each of three levels within the canopy. In each horizontal path, from left to right, the resistances refer to the boundary layer and bulk stomatal resistance. Vertical transfer resistances between the canopy levels are based on eddy diffusivity and spacing between the layers. Concentrations are $\mu g \, m^3$ ($1000 \, \mu g \, m^3 = 0.5$ ppmv), fluxes are $\mu g \, m^{-2} \, s^{-1}$, and resistances are s m^{-1} (adapted from Unsworth, 1981).

Although in many field situations the boundary layer resistances, r_b, have generally been found to be of lesser importance to uptake than stomatal resistance and the transfer to the boundary layer through atmospheric turbulence and mixing (Hicks and Matt, 1988), the situation may be quite different within a controlled environment and other chambers and cuvettes used experimentally. In such situations, inadequate air movement may lead to artificially high r_b values, which in turn lead to the need for misleadingly high "ambient" concentrations of pollutant gas to elicit a given plant response (Ashenden and Mansfield, 1977). Data from such experimentation is therefore of limited use in establishing dose-response relationships.

An alternative approach to the determination of the overall deposition of ozone involves the use of micrometeorological measurements alone, without specific information about the component resistances to ozone flux (Leuning et al., 1979a,b). Measurement of the temperature and humidity gradients above a canopy permits computation of Bowen ratios, B:

$$B = \frac{j \cdot (T_2 - T_1)}{e_2 - e_1}$$

where j is the psychrometer constant (Pa °C^{-1} units), and $(T_2 - T_1)$ and $(e_2 - e_1)$ are measured gradients of temperature (T; °C) and water vapor pressure (e; Pa units), respectively. Since the net radiation at the canopy, R, may be partitioned into sensible heat flux, H, latent heat flux, LE, and soil heat flux, G (all in W m^{-2} units),

$$LE = \frac{R - G}{1 + B}$$

Assuming that the eddy diffusivities of ozone and water vapor are the same above the canopy, the combined flux of ozone to the canopy and the soil is given by

$$F = \frac{f \cdot j}{d \cdot C_p} \cdot LE \cdot \frac{(X_2 - X_1)}{e_2 - e_1}$$

where f is the conversion factor relating the ozone mole fraction, X (m^3 m^{-3} units), to density (kg m^{-3} units) (approximately 2.0 kg m^{-3} at 20°C and 101.3 kPa); d is the density of air (kg m^{-3} units); and C_p is the specific heat of air (J kg^{-1} °C^{-1} units). $(X_2 - X_1)$ and $(e_2 - e_1)$ are measured over the same height interval.

Leuning et al. (1979a) point out that the main shortcoming of this approach is that it does not distinguish between flux to the vegetation and flux to the soil and, hence, cannot be used to explore specific plant uptake-response relationships unless the usually appreciable deposition to the soil is also determined (Macdowall, 1974; Turner et al., 1982). Nevertheless, in spite of this need to be able to partition the major fluxes, the micrometeorological approach has distinct advantages when dealing with the diversity of species found in many natural habitats (Unsworth, 1981). Furthermore, the instrumentation available permits the continuous collection of data, rather than the collection of intermittent diffusive resistance data under field conditions necessitated by minimizing the time during which the selected leaves are subjected to porometry, not to mention the tedium of such work!

Recognition of the roles of environmental conditions in determining uptake has led to a few reports in which such information was used as an alternative to stomatal resistance measurements in order to provide better estimates of dose. Perhaps the best example is the now classical work of Macdowall et al. (1964) and Mukammal (1965) on commercial tobacco exposed to ambient ozone in southern Ontario. They found that the previous day's exposure, expressed as C_a t, was a poor predictor of the severity of the ozone-induced leaf injury that occurred episodically throughout the season. However, the availability of measurements of wind speed, water vapor pressure, and evapotranspiration rate permitted them to apply the mass transfer equation to calculate hourly coefficients of evaporation:

$$A = \frac{E}{U_2\left(e_2 - e_1\right)}$$

where A is the coefficient of evaporation, E is evapotranspiration (determined by lysimeter), U_2 is the wind speed at 340 cm, and $(e_2 - e_1)$ is the water vapor pressure difference between 340 and 180 cm elevations. Use of the product of the mean hourly ambient concentration and the mean hourly coefficient of evaporation, C_a A, as the independent dose variable greatly improved the linear relationship with severity of leaf injury (adjusted $r^2 = 0.81$). A was assumed to be an empirical index reflecting the physiological conditions of the plant (especially the condition of the stomata) coupled to the environmental conditions in the atmosphere and soil at the time.

More recently, Mukammal et al. (1982) applied the concept of utilizing meteorological data as an alternative to direct stomatal information in developing an empirical linear relationship between ozone-induced injury on white beans (*Phaseolus vulgaris*). In this they used a modified exposure term given by

$$D = R \cdot CHU \cdot O \cdot \frac{M}{E}$$

where D is the modified dose, R is a relative rainfall factor to normalize different sites, CHU is the ratio of corn heat units accumulated to the time of assessment of injury to the accumulated corn heat units required for crop maturity, O is a weighted cumulative daily ozone exposure (ppmv-h), E is the weighted cumulative pan evaporation, and M is the minimum weighted pan evaporation observed during the July–August season, also used to normalize E. The weighting function, running backwards in time, used to derive O, E, and M is given by

$$f = e^{-\frac{(x-1)}{k}}$$

where x = 1 is the day before assessment, and k is a time constant found to equal 4 d for the best fit of the data.

Leung et al. (1982) incorporated temperature, rainfall, and relative humidity in developing an economic model of crop losses in Southern California. Adomait et al. (1987) found that the relationship between the crop loss of beans and the natural logarithm of cumulative exposure to ozone, again in southern Ontario, was improved when the logarithms of mean monthly temperature and rainfall were included in multiple linear regression.

While these examples are empirical in nature, they nevertheless show that meteorological data can be used to good effect in bridging the gap between exposure and dose.

5.5 EXPOSURE, UPTAKE, AND DOSE

Resistance models have led to improved understanding of the importance of the different pathways of uptake. However, such information has so far seen little application in response modeling.

Runeckles (1974) introduced the concept of "effective dose," which incorporates uptake, as the basis for establishing dose-response relationships. As a development of this concept, and by analogy with the medical usage of the RAD as the unit of absorbed dosage of ionizing radiation, Fowler and Cape (1982) proposed the dose unit PAD (pollutant absorbed dose or cumulative uptake, $g\,m^{-2}$ units), defined as the product of ambient concentration, time, and stomatal (or canopy) conductance:

$$PAD = C_a^* \cdot \left(\frac{1}{r_s^*} \right) \cdot t$$

or

$$PAD = \sum_{t=1}^{N} C_a \cdot \left(\frac{1}{r_s} \right)$$

where t is the time interval over which n measurements are taken, and C_a^* and r_s^* are mean concentrations and resistances for the period t.

Dose expressed in such terms has the advantage of being rational, since it attempts to define the amount of pollutant *capable* of eliciting a response that is taken up per unit area (of leaf or of ground surface) over a period of time, rather than as the conventional expression of dose as the simple product of the mean ambient concentration and time, $C_a\,T$ (Munn, 1970; Lefohn and Runeckles, 1987).

There is no doubt that changes in stomatal aperture during the day can result in dramatic differences in flux and uptake, and Leuning et al. (1979a) expressed high confidence in the fluxes calculated from ozone concentration and stomatal-resistance measurements. Furthermore, Reich (1987), in his review of ozone effects on photosynthetic rates and growth, justified his computation of ozone uptakes as the products of dose and mean diffusive conductance even when dose-response and conductance data were only available (for a given species) from unrelated experiments, "because the error is likely to be small compared to interspecific variation in leaf conductance which covers a 10-fold range."

Several workers have suggested that flux (g m^{-2} s^{-1}) be used to define dose (Black and Unsworth, 1979; Leuning et al., 1979a; Tingey and Taylor, 1982). The definition of PAD is in keeping with this concept, since it is essentially the accumulated flux over time. However, the use of stomatal resistance, rather than canopy resistance, would appear to be justified only in situations where the other foliar resistances are shown to be either negligible or constant. For example, experiments in which various species were subjected to different types of exposure to ozone (steady-state exposures and exposures in which the concentration peaked at the beginning, in the middle, or at the end of a 7-h daytime exposure period) failed to reveal any simple relationship between degree of foliar injury and PAD, based solely on stomatal resistance, although the magnitude of the response decreased dramatically the later the peak concentration occurred during the exposure period (Bicak, 1978; Runeckles, 1987). Similarly, the studies of Taylor et al. (1982) with soybean showed that r_i could exceed r_s at concentrations greater than 0.3 ppmv and, hence, play the key role in uptake. The relative importance of r_s and r_i are discussed further in Chapter 6. Several other studies, summarized by Tingey and Taylor (1982), have also reported failure to find a clear relationship between overall gas-phase resistance and response (measured as acute injury).

Short-term variations in uptake computed in PAD units will reflect variation in both concentration and canopy (stomatal) resistance. The daily changes in the plant's external atmospheric environment are stochastic and are governed by meteorology, photochemical synthesis, chemical reaction, and transport. In many locations, the overall consequences of photochemical formation of ozone and its scavenging lead to the typical daily pattern of rising concentrations during the daylight hours and a subsequent decline during the night, with minima usually occurring between 0400 and 0700 h.

Stomatal resistance (over which the plant has limited control) similarly follows a general cycle of daytime opening and nighttime closure. Hence, in situations of relatively low ambient ozone concentration in which uptake is largely regulated by stomatal resistance, ozone flux is the resultant of the superimposition of these two cycles. The more closely the cycles are in phase, the greater will be the flux and the greater the uptake.

Cape and Unsworth (1988) have described the simple hypothetical case in which, with the pollutant concentration fluctuating with minima/maxima at 0000/1200, 0300/1500, or 0600/1800 or maintained at a steady level resulting in the same total daily exposure and a constant deposition velocity between 0600 and 1800 h (with zero uptake at other hours), an almost twofold difference in cumulative uptake can be demonstrated between the steady-state or 0600 h minimum scenarios and the 0000 h minimum case. A more realistic but still hypothetical model in which the stomatal aperture also changes cyclically leads to the curves presented in Figure 5.8. The cumulative uptake-curves clearly show the degree to which phase differences dictate the cumulative uptake, although the magnitude of the differences is somewhat less than that demonstrated by Cape and Unsworth (1988). It is important to stress that these examples make no provision for the effects of the pollutant on stomatal aperture per se. Nevertheless, the curves in Figure 5.8 may be used to illustrate several ways in which differences in uptake patterns may dictate plant response. Thus, although the fluxes are greatest overall when the concentration and stomatal conductance cycles are synchronized, the steady-state situation, or one with a much less pronounced early morning minimum concentration than any of the three cycles depicted in Figure 5.8, may result in greater flux values for the first few hours following stomatal opening than those in the synchronous case. Since plants appear to be most sensitive to ozone at the beginning of an exposure period (Bicak, 1978), the occurrence of such situations may have a significant bearing on response. On the other hand, in the synchronized case, the high flux levels are more likely to overwhelm any mechanisms for detoxification. In the case of the concentration wave with a minimum at 0600 h, the maximum flux exceeds that in the steady-state case and occurs later in the day, but the cumulative uptakes are the same. Hence, although a knowledge of uptake is essential in attempting to predict response, information about flux rates and their rates of change may be equally important.

The matter of relating dose and exposure to response is discussed further in Chapter 6, but these examples and the work of Mukammal (1965), Mukammal et al. (1982), Leung et al. (1982), and Adomait et al. (1987), described above, illustrate the importance of uptake in the development of dose-response rather than exposure-response relationships.

5.6 FUTURE DIRECTIONS AND RESEARCH NEEDS

Although models based on electrical resistance networks have provided useful insights into ozone uptake and its regulation, several caveats and limitations attend their use.

At the outset it should be reiterated that the system is dynamic and that the important resistances, r_b, r_s, and r_i, are not fixed or constant. Changes in these resistances occur over time as a result of changes in external conditions, in stomatal aperture, and in internal sorptive and reactive capacity. The time scales for these changes may be widely different, but as illustrated above for stomatal aperture alone, the consequences for uptake can be highly significant. The time scales of other changes, such as those related to season and ontogeny, which may

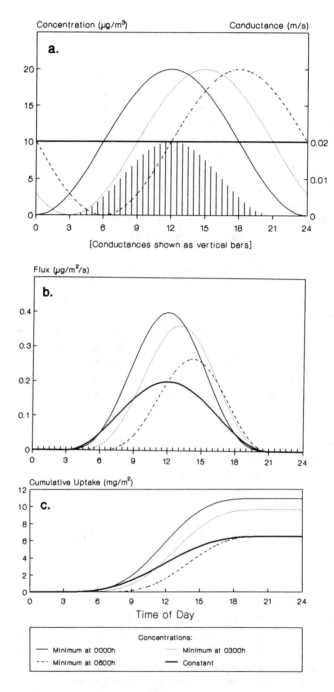

Figure 5.8. Hypothetical changes in (a) ambient ozone concentration and stomatal conductance during a day-night cycle, and their consequences to (b) flux and (c) cumulative uptake.

profoundly influence response, also need to be kept in mind, since they will affect the applicability of any uptake model based upon short-term observations to long-term responses, such as effects on growth and yield.

It should also be stressed that at the reaction-site level, ozone uptake is the *consequence* of concentration or partial pressure differences between the outside of the leaf and its interior, and the chemical reactions that lead to its utilization. Although uptake is an essential prerequisite for response, it is not the driving force. The ultimate fates of any ozone molecules that enter the leaf may be limited by, but are not controlled by, for example, stomatal resistance, but are controlled by the rate constants of the various physical and chemical processes that are available to the ozone molecules and their products. Regardless of the magnitude of the stomatal resistance, the overall uptake and ultimate sinks for ozone will show some dependence on these rate constants.

As has already been pointed out, we know little about the causes of variations in r_i. Estimates of r_i cited above range from zero to more than 1000 s m^{-1}. Furthermore, Taylor et al. (1982) showed that, in soybean, r_i generally increased with ambient concentration. Their work supports the case made by O'Dell et al. (1977), Wesely et al. (1978), and others, for recognition of r_i as a potentially important component of the overall resistance to uptake, as discussed further in Chapter 6. Their data further suggest that within the complex of physicochemical processes that constitutes r_i, there are pathways of ozone depletion that become progressively saturated or blocked. The suggestions of Chameides (1989) are based on such reactions.

Further studies of the factors affecting r_i are obviously needed, both in order to provide a broader base of overall r_i values and their variation and to gain an understanding of the component processes and reactions and their relative importance. Such studies are an important prelude to the development of better models of the overall process of uptake.

The finite element modeling approach developed by Parkhurst (1986) for CO_2, which treats individual stomata as the sources of the gas, offers a means for further investigating the interior physical processes of ozone uptake, particularly if linked to recent biochemical-process models of the type developed for CO_2 assimilation in photosynthesis (e.g., Farquhar and von Caemmerer, 1982; Giersch, 1986), as suggested by Luxmoore (1988).

As an aside it is worth noting the advantage offered by the ready availability of ^{35}S-labeled sources, which has led to the elegant models of SO_2 uptake developed by Pfanz et al. (1987a,b) and Laisk et al. (1988a,b). Although ^{18}O-labeled ozone has been used to estimate the sorption of ozone in the rat respiratory tract (Hatch et al., 1989), comparable methodology appears not yet to have been applied to uptake by plant tissues.

Reference has already been made to the fact that resistances expressed in $m^2 \text{ s mol}^{-1}$ units are independent of atmospheric pressure, whereas those expressed as $s m^{-1}$ are not. The magnitude of the response at any receptor within the plant is a function of the number of molecules of ozone or its products that reach the receptor. It is therefore important to be consistent in the usage of units of concentration and to recognize that resistances expressed in terms of s m^{-1} will vary with pressure, and, hence, with altitude, in exactly the same way that concentrations determined at different altitudes to have the same mole fraction of ozone will have different absolute concentrations when expressed in g m^{-3} units (Lefohn et al., 1990).

Pressure phenomena may also affect uptake as a result of the mass flow caused by pressure differences between different sides or between the exterior and interior of a leaf. Such pressure differences are a feature of air turbulence and result in the movement of leaves induced by wind. Turbulence results in the eddy diffusivities of gases in the air being considerably greater than their molecular diffusivities in still air, and plays an important role in transport to the boundary layer and in reducing boundary layer resistance. However, the influence of such pressure differences on gas movement into and within the leaf is less clear, largely because little information exists as to their magnitude, either between the different sides of a leaf or between the outside and inside air spaces.

Dacey (1987) has questioned the assumption that, within the internal gas spaces of leaves, flow of gases is essentially diffusive, and has reported pressure differences of up to 0.3 kPa, albeit in aquatic plant leaves. Woolley (1961) observed a maximum pressure difference of 0.057 kPa across a leaf normal to a 7 m s^{-1} wind. Shive and Brown (1978), on the other hand, reported that oscillations of cottonwood leaves could reduce the total resistance to gas flow by between 350 and 550 s m^{-1} while r_b was reduced by only 50 to 220 s m^{-1}. Day and Parkinson (1979) questioned the magnitude of the differences observed, but their own calculations suggested that mass flow could account for as much as 46% of transpiration in a 7 m s^{-1} wind, although in most situations the contribution of mass flow would probably be not more than a few percent. This is an area that needs further investigation in order to ascertain the magnitude of the contribution made by mass transfer to ozone uptake, since it is excluded from the resistance models that have been described.

In conclusion, it should be stated that our knowledge of the mechanisms by which ozone is taken into the plant and reaches its sites of reaction continues to expand. For many applications, it may be sufficient to assume that stomatal resistance provides the principal limiting factor to uptake and therefore to "effective dose" and dose response. However, the value of stomatal resistance information is limited unless it is obtained on a continuous basis: a nearly impossible task in field situations because of the magnitude of the work entailed and the need to avoid adverse effects on the plants caused by obtaining the measurements themselves. In addition, reassurance is needed that resistances, usually measured on areas of individual leaves, are representative both of the leaves themselves and of the canopy as a whole. Hence, further developments aimed at providing better estimates of stomatal condition from micrometeorological observations can be anticipated. Finally, there have been sufficient questions raised as to the importance of the overall internal resistance that the next few years will undoubtedly see considerable attention paid to developing better methods for its determination, to increasing the body of knowledge about internal resistance and its variability, and to the analysis of its component processes and reactions and their importance to uptake.

REFERENCES

Adomait, E. J., Ensing J., and Hofstra, G. (1987) A dose-response function for the impact of O_3 on Ontario-grown white bean and an estimate of economic loss. *Can. J. Plant Sci.* **67**, 131–136.

Ashenden, T. W. and Mansfield, T. A. (1977) Influence of wind speed on the sensitivity of ryegrass to SO_2. *J. Exper. Bot.* **28**, 729–735.

Bache, D. H. (1986) On the theory of gaseous transport to plant canopies. *Atmos. Environ.* **20**, 1379–1388.

Baldocchi, D. (1988) A multi-layer model for estimating sulfur dioxide deposition to a deciduous oak forest canopy. *Atmos. Environ.* **22**, 869–884.

Baldocchi, D. B., Hicks, B. B., and Camara, P. (1987) A canopy stomatal resistance model for gaseous deposition to vegetated surfaces. *Atmos. Environ.* **21**, 91–101.

Bennett, J. H. and Hill, A. C. (1973) Absorption of gaseous air pollutants by a standardized plant canopy. *JAPCA* **23**, 203–206.

Bennett, J. H., Hill, A. C., and Gates, D. M. (1973) A model for gaseous pollutant sorption by leaves. *JAPCA* **23**, 957–962.

Bicak, C. J. (1978) Plant response to variable ozone regimes of constant dosage. M.Sc. Thesis, University of British Columbia, Vancouver, B.C.

Black, V. J. and Unsworth, M. H. (1979) Resistance analysis of sulphur dioxide fluxes to *Vicia faba*. *Nature* **282**, 68–69.

Cape, J. N. and Unsworth, M. H. (1988) Deposition, uptake and residence of pollutants. In *Air Pollution and Plant Metabolism* (eds., Schulte-Hostede, S., Darrall, N. M., Blank, L. W., and Wellburn, A. R.). Elsevier, London, 1–18.

Castillo, F. J. and Greppin, H. (1986) Balance between anionic and cationic extracellular peroxidase activities in *Sedum album* leaves after ozone exposure. Analysis by high-performance liquid chromatography. *Physiol. Plant.* **68**, 201–208.

Castillo, F. J., Miller, P. R. and Greppin, H. (1987) Extracellular biochemical markers of photo-chemical oxidant air pollution damage to Norway spruce. *Experientia* **43**, 111–115.

Chamberlain, A. C. (1986) Deposition of gases and particles on vegetation and soils. In *Air Pollutants and Their Effects on the Terrestrial Ecosystem* (eds., Legge, A. H., and Krupa, S. V.). John Wiley & Sons, New York, 189–209.

Chameides, W. L. (1989) The chemistry of ozone deposition to plant leaves: the role of ascorbic acid. *Environ. Sci. Technol.* **23**, 595–600.

Cooke, R. J. and Rand, R. H. (1980) Diffusion resistance models. In *Predicting Photosynthesis for Ecosystem Models* Vol. 1, (eds., Hesketh, J. D. and Jones ,J. W.). CRC Press, Boca Raton, FL., 93–121.

Dacey, J. W. H. (1987) Knudsen-transitional flow and gas pressurization in leaves of *Nelumbo*. *Plant Physiol.* **85**, 199–203.

Day, W. and Parkinson, K. J. (1979) Importance to gas exchange of mass flow of air through leaves. *Plant Physiol.* **64**, 345–346.

Elkiey, T. and Ormrod, D. P. (1981a) Absorption of ozone, sulphur dioxide and nitrogen dioxide by Petunia plants. *Environ. Exp. Bot.* **21**, 63–70.

Elkiey, T. and Ormrod, D. P. (1981b) Sorption of O_3, SO_2, NO_2 or their mixtures by nine *Poa pratensis* cultivars of differing pollutant sensitivity. *Atmos. Environ.* **15**, 1739–1743.

Endress, A. G. and Grunwald, C. (1988) Similarity of proportional yield response in greenhouse- and field-grown soybeans exposed to O_3. *Environ. Pollut.* **53**, 424–425 (poster summary).

Evans, L. S. and Ting, I. P. (1974) Ozone sensitivity of leaves; relationship to leaf water content, gas transfer resistance and anatomical characteristics. *Am. J. Bot.* **61**, 592–597.

Farquhar, G. D. and von Caemmerer, S. (1982) Modeling of photosynthetic response to environmental conditions. In *Physiological Plant Ecology. II. Water Relations and Carbon Assimilation* (eds., Lange, O. L. et al.). Springer-Verlag, Berlin, 549–588.

Farquhar, G. D. and Sharkey, T. D. (1982) Stomatal conductance and photosynthesis. *Ann. Rev. Plant Physiol.* **33**, 317–345.

Fowler, D. (1981) Turbulent transfer of sulphur dioxide to cereals: A case study. In *Plants and their Atmospheric Environment* (eds., Grace, J., Ford, E. D., and Jarvis, P. G.), Blackwell Scientific, Oxford, 139–146.

Fowler, D. and Cape, J. N. (1982) Air pollutants in agriculture and horticulture. In *Effects of Gaseous Air Pollution in Agriculture and Horticulture* (eds., Unsworth, M. H. and Ormrod, D. P.). Butterworth Scientific, London, 3–26.

Fowler, D., Cape, J. N., and Unsworth, M. H. (1989) Deposition of atmospheric pollutants on forests. *Phil. Trans. R. Soc. Lond.* B **324**, 247–265.

Gaastra, P. (1959) Photosynthesis of crop plants. *Mededelingen van de Landbouwhogeschool te Wageningen* **59**, 1–68.

Galbally, I. E. and Roy, C. R. (1980) Destruction of ozone at the earth's surface. *Q J. R. Meterol. Soc.* **106**, 599–620.

Giersch, C. (1986) Oscillatory response of photosynthesis in leaves to environmental perturbations: a mathematical model. *Arch. Biochem. Biophys.* **245**, 263–270.

Graham, S. and Ormrod, D. P. (1989) Sorption of ozone by 'New Yorker' tomato leaves. *Environ. Pollut.* **58**, 213–220.

Hatch, G. E., Wiester, M. J., Overton, J. H., Jr., and Aissa, M. (1989) Respiratory tract dosimetry of ^{18}O-labeled ozone in rats: Implications for a rat-human extrapolation of ozone dose. In *Atmospheric Ozone Research and its Policy Implications* (ed., Schneider, T. et al.), Elsevier, Amsterdam, 553–560.

Heath, R. L. (1980) Initial events in injury to plants by air pollutants. *Ann. Rev. Plant Physiol.* **31**, 395–431.

Hicks, B. B. and Matt, D. R. (1988) Combining biology, chemistry and meteorology in modeling and measuring dry deposition. *J. Atmos. Chem.* **6**, 117–131.

Hicks, B. B., Baldocchi, D. D., Meyers, T. P., Hosker, R. P., Jr., and Matt, D. R. (1987) A preliminary multiple resistance routine for deriving dry deposition velocities from measured quantities. *Water Soil Air Pollut.* **36**, 311–330.

Hosker, R. P., Jr. (1986) Practical application of air pollutant deposition models — current status, data requirements and research needs. In *Air Pollutants and Their Effects on the Terrestrial Ecosystem* (eds., Legge, A. H. and Krupa, S. V.). John Wiley & Sons, New York, 505–567.

Hosker, R. P., Jr. and Lindberg, S. E. (1982) Review: atmospheric deposition and plant assimilation of gases and particles. *Atmos. Environ.* **16**, 889–910.

Jarvis, P. G. (1971) The estimation of resistances to carbon dioxide. In *Plant Photosynthetic Production: Manual of Methods* (eds., Sestak, Z., Catsky, J. and Jarvis, P. G.). Dr. Junk, The Hague, 566–622.

Jeffree, C. E., Read, N. D., Smith, J. A. C., and Dale, J. E. (1987) Water droplets and ice deposits in leaf intercellular spaces: redistribution of water during cryofixation for scanning electron microscopy. *Planta* **172**, 20–37.

Laisk, A., Kull, O. and Moldau, H. (1989) Ozone concentration in leaf intercellular air spaces is close to zero. *Plant Physiol.* **90**, 1163–1167.

Laisk, A., Pfanz H., and Heber, U. (1988a) Sulfur-dioxide fluxes into different cellular compartments of leave photosynthesizing in a polluted atmosphere. II. Consequences of SO_2 uptake as revealed by computer analysis. *Planta* **173**, 241–252.

Laisk, A., Pfanz, H., Schramm, M. J., and Heber, U. (1988b) Sulfur-dioxide fluxes into different cellular compartments of leaves photosynthesizing in a polluted atmosphere. I. Computer analysis. *Planta* **173**, 230–240.

Lefohn, A. S. and Runeckles, V. C. (1987) Establishing standards to protect vegetation — ozone exposure/dose considerations. *Atmos. Environ.* **21**, 561–568.

Lefohn, A. S., Shadwick, D. S., and Mohnen, V. A. (1990) The characterization of ozone concentrations at a select set of high-elevation sites in the eastern United States. *Environ. Pollut.* **67**, 147–178.

Leung, S. K., Reed, W., and Geng, S. (1982) Estimations of ozone damage to selected crops grown in southern California. *JAPCA* **32**, 160–164.

Leuning, R. (1983) Transport of gases into leaves. *Plant Cell Environ.* **6**, 181–194.

Leuning, R., Neumann, H. H., and Thurtell, G. W. (1979a) Ozone uptake by corn (*Zea mays* L.): a general approach. *Agric. Meteorol.* **20**, 115–135.

Leuning, R., Unsworth, M. H., Neumann, H. N., and King, K. M. (1979b) Ozone fluxes to tobacco and soil under field conditions. *Atmos. Environ.* **13**, 1155–1163.

Luxmoore, R. J. (1988) Assessing the mechanisms of crop loss from air pollutants with process models. In *Assessment of Crop Loss from Air Pollutants* (eds., Heck, W. W., Taylor, O. C., and Tingey, D. T.). Elsevier, London, 417–444.

Macdowall, F. D. H. (1974) Importance of soil in the absorption of ozone by a crop. *Can. J. Soil Sci.* **54**, 239–240.

Macdowall, F. D. H., Mukammal, E. I., and Cole, A. F. W. (1964) Direct correlation of air-polluting ozone and tobacco weather-fleck. *Can. J. Plant Sci.* **44**, 410–417

McLaughlin, S. B. and Taylor, G. E. (1981) Relative humidity: important modifier of pollutant uptake by plants. *Science* **211**, 167–169.

Melhorn, H. and Wellburn, A. R. (1987) Stress ethylene formation determines plant sensitivity to ozone. *Nature* **327**, 417–418.

Monteith, J. L. (1981) Coupling of plant to the atmosphere. In *Plants and Their Atmospheric Environment* (eds., Grace, J., Ford, E. D., and Jarvis, P. G.). Blackwell Scientific, Oxford, 1–29.

Mudd, J. B. (1982) Effects of oxidants on metabolic function. In *Effects of Gaseous Air Pollution on Agriculture and Horticulture* (eds., Unsworth, M. H. and Ormrod, D. P.). Butterworth Scientific, London, 189–203.

Mukammal, E. I. (1965) Ozone as a cause of tobacco injury. *Agric. Meteorol.* **2**, 145–165.

Mukammal, E. I., Neumann, H. H., and Hofstra, G. (1982) Ozone injury to white bean (*Phaseolus vulgaris* L.) in southwestern Ontario, Canada: correlation with ozone dose, pan evaporation, plant maturity and rainfall. In *Effects of Gaseous Air Pollution in Agriculture and Horticulture* (eds., Unsworth, M. H. and Ormrod, D. P.). Butterworth Scientific, London, 470–471.

Munn, R. E. (1970) *Biometeorological Methods.* Academic Press, New York, 336.

Nobel, P. S. (1983) *Biophysical Plant Physiology and Ecology.* W. H. Freeman, San Francisco, 608.

O'Dell, R. A., Taheri, M., and Kabel, R. L. (1977) A model for uptake of pollutants by vegetation. *JAPCA* **27**, 1104–1109.

Olszyk, D. M. and Tibbitts, T. W. (1981) Stomatal response and leaf injury of *Pisum sativum* L. with SO_2 and O_3 exposures. II. Influence of moisture stress and time of exposure. *Plant Physiol.* **67**, 545–549.

Parkhurst, D. F. (1986) Internal leaf structure: a three-dimensional perspective. In *On the Economy of Plant Form and Function* (ed., Givnish, T. J.). Cambridge University Press, London, 215–249.

Pfanz, H., Martinoia, E., Lange, O. L., and Heber, U. (1987a) Mesophyll resistances to SO_2 fluxes into leaves. *Plant Physiol.* **85**, 922–927.

Pfanz, H., Martinoia, E., Lange, O. L. and Heber, U. (1987b) Flux of SO_2 into leaf cells and cellular acidification by SO_2. *Plant Physiol.* **85**, 928–933.

Reich, P. B. (1987) Quantifying plant response to ozone: a unifying theory. *Tree Physiol.* **3**, 63–91.

Runeckles, V. C. (1987) Exposure, dose, vegetation response and standards: will they ever be related? *Proc. 80th Annual Meeting of the Air Pollution Control Association,* Paper 87-33.2. Air Pollution Control Association, New York.

Runeckles, V. C. and Rosen, P. M. (1977) Effects of ambient ozone pretreatment on transpiration and susceptibility to ozone injury. *Can. J. Bot.* **55**, 193–197.

Shive, J. B., Jr. and Brown, K. W. (1978) Quaking and gas exchange in leaves of cottonwood (*Populus deltoides* Marsh). *Plant Physiol.* **61**, 331–333.

Sinclair, T. R., Goudriaan, J., and DeWit, C. T. (1977) Mesophyll resistance and CO_2 compensation concentration in leaf photosynthesis models. *Photosynthetica* **11**, 56–66.

Taylor, G. E., Jr. (1978) Plant and leaf resistance to gaseous air pollution stress. *New Phytol.* **80**, 523–534.

Taylor, G. E., Jr., Tingey, D. T., and Ratsch ,H. C. (1982) Ozone flux in *Glycine max* (L.) Merr.: sites of regulation and relationship to leaf injury. *Oecologia* (Berlin) **53**, 179–186.

Taylor, G. E., Jr., Hanson, P. J., and Baldocchi, D. D. (1988) Pollutant deposition to individual leaves and plant canopies: Sites of regulation and relationship to injury. In *Assessment of Crop Loss from Air Pollutants* (eds., Heck, W. W., Taylor, O. C., and Tingey, D.T.). Elsevier, London, 227–257.

Tenhunen, J. D., Hesketh, J. D., and Gates, D. M. (1980) Leaf photosynthesis models. In *Predicting Photosynthesis for Ecosystem Models* (eds., Hesketh, J. D. and Jones, J. W.). CRC Press, Boca Raton, FL, 123–181.

Thorne, L. and Hanson, G. P. (1972) Species differences in rates of vegetal ozone absorption. *Environ. Pollut.* **3**, 303–312.

Tingey, D. T. and Taylor, G. E., Jr. (1982) Variation in plant response to ozone: a conceptual model of physiological events. In *Effects of Gaseous Air Pollution in Agriculture and Horticulture* (eds., Unsworth M. H. and Ormrod D. P.). Butterworth Scientific, London, 113–138.

Turner, N. C., Waggoner, P. E., and Rich, S. (1974) Removal of ozone from the atmosphere by soil and vegetation. *Nature* **250**, 486–489.

Unsworth, M. H. (1981) The exchange of carbon dioxide and air pollutants between vegetation and the atmosphere. In *Plants and their Atmospheric Environment* (eds., Grace J., Ford, E. D., and Jarvis, P. G.). Blackwell Scientific Publications, Oxford, 111–138.

Unsworth, M. H. (1982) Exposure to gaseous pollutants and uptake by plants. In *Effects of Gaseous Air Pollution in Agriculture and Horticulture* (eds., Unsworth, M. H. and Ormrod, D. P.). Butterworth Scientific, London, 43–63.

Wesely, M. L. and Hicks, B. B. (1977) Some factors that affect the deposition rates of sulfur dioxide and similar gases on vegetation. *JAPCA* **27**, 1110–1116.

Wesely, M. L., Eastman, J. A., Stedman, D. H., and Yalvac, E. D. (1982) An eddy-correlation measurement of NO_2 flux to vegetation and comparison to O_3 flux. *Atmos. Environ.* **16**, 815–820.

Winner, W. E., Gillespie, C., Shen, W. S., and Mooney, H. A. (1988) Stomatal responses to SO_2 and O_3. In *Air Pollution and Plant Metabolism* (eds., Schulte-Hostede, S., Darrall, N. M., Blank, L. W., and Wellburn, A. R.). Elsevier, London, 255–271.

Woolley, J. T. (1961) Mechanisms by which wind influences transpiration. *Plant Physiol.* **6**, 112–114.

CHAPTER 6

Crop Responses to Ozone

Victor C. Runeckles, Department of Plant Science, University of British Columbia, Vancouver, B.C., Canada

Boris I. Chevone, Department Plant Pathology and Physiology, Virginia Polytechnic Institute and State University, Blacksburg, VA

6.1 INTRODUCTION

Since first being identified as a significant, phytotoxic, gaseous air pollutant in Southern California half a century ago, ozone has progressively become the major air pollutant in many parts of the world. Although the build-up of tropospheric ozone is typical of urban areas with high automobile densities, its ready transport to nonurban, rural, and pristine areas can result in adverse effects on the growth of crops, forest trees, and other natural vegetation. Such impacts are the consequence of effects on the biochemistry and physiology of the plants subjected to such exposures.

Several general reviews of the effects of air pollutants on vegetation, which include some discussion of the biochemical, metabolic, and physiological consequences of exposure to ozone, have appeared over the past decade, e.g., Darrall (1989) and Roberts (1984), in addition to several multiauthored review volumes (Guderian, 1985; Koziol and Whatley, 1984; Heck et al., 1988; Lee, 1985; Legge and Krupa, 1986; Schulte-Hostede et al., 1988; Treshow, 1984; Unsworth and Ormrod, 1982).

There is no clear dividing line between many of the biochemical and physiological effects of ozone, because the latter have their origins in the chemical reactions of ozone (or its reaction products) with cellular constituents.

The effects may be the result of reactions with proteins and other components of various cell membranes, or may be on soluble enzymes and substrates, resulting in modifications of normal metabolic processes. Although this chapter reviews the effects of ozone on all aspects of the growth of crops, the initial discussion concerns plant metabolism and biochemistry in general, including effects reported on forest tree species, since the mechanisms involved are similar if not identical. Subsequent sections deal solely with physiological and other effects on crop species that lead to modifications of growth and yield (including the interactions of ozone with other abiotic and biotic stress factors, and ecological effects). The impacts of ozone on forest trees are dealt with specifically in Chapter 7.

6.2 EFFECTS OF OZONE ON PLANT BIOCHEMISTRY AND METABOLISM

6.2.1 Reactions of Ozone with Biological Materials

As discussed in Chapter 5, in order to elicit an effect, ozone or its reaction products must first reach the active sites within the plant tissues. Leaves are the primary route of uptake, and the first consequences of such uptake are the reactions directly related to the chemical nature of ozone itself.

Ozone can react with a diverse array of biological compounds and metabolites that are normally present in plant cells (Mudd, 1982; Heath, 1984; Pryor et al., 1984; Giamalva et al., 1985). Historically, ozone toxicity has been attributed to lipid peroxidation and/or ozonolysis of the plasma membrane (Tomlinson and Rich, 1969), followed by increases in cell permeability and subsequent failure of chemiosmoregulatory processes (Chimiklis and Heath, 1975; Sutton and Ting, 1977). However, while this series of events may represent conditions at extreme concentrations (>0.5 ppmv), reviews by Mudd (1982), Heath (1984), and Heath and Castillo (1988) have emphasized the importance of the interactions of ozone with the plant cell, which involves the oxidation of reactive sulfhydryl groups located on membrane and other proteins or the oxidation of other cellular scavenging compounds. They present convincing arguments against an initial direct attack of membrane lipids.

A wide range of cellular constituents can be directly oxidized by ozone *in vitro*, including the nucleic acids, purine and pyrimidine derivatives, several amino acids (especially cysteine, methionine, and tryptophan), many lipids, peptides such as glutathione, and proteins, including enzymes such as glyceraldehyde-3-phosphate dehydrogenase, catalase, peroxidase, papain, ribonuclease, and urease (U.S. EPA, 1986). The rate constants for several *in vitro* reactions are presented in Table 6.1. The most reactive compound is the sulfhydryl-containing tripeptide, glutathione (c-glutamylcysteinylglycine), with a rate constant $>1.0 \times 10^9 \ M^{-1} \ s^{-1}$ at pH 7.0. Cysteine, the sulfhydryl-containing amino acid of glutathione, has a similar rate constant ($>1.0 \times 10^9 \ M^{-1} \ s^{-1}$), which suggests that

Table 6.1. Rate Constants for the Reaction of Ozone with Biological Compounds *in vitro*

Compound	Rate constant $(M^{-1}s^{-1})$	References
Glutathione (pH 7.0)	$>1.0 \times 10^9$	Pryor et al., 1984
Free amino acids		
cysteine (pH 7.0)	$>1.0 \times 10^9$	Pryor et al., 1984
tryptophan (pH 7.0)	7.0×10^6	Pryor et al., 1984
methionine	4.0×10^6	Pryor et al., 1984
proline	4.3×10^6	Pryor et al., 1984
Ascorbic acid (pH 7.0)	5.6×10^7	Giamalva et al., 1985
α-tocopherol (pH 7.0)	7.5×10^5	Giamalva et al., 1985
Simple polyunsaturated fatty acids (PUFA)	$\approx 1.0 \times 10^6$	Giamalva et al., 1985
oleic acid	$9\ 8 \times 10^5$	Giamalva et al., 1985
methyl oleate	$8\ 7 \times 10^5$	Giamalva et al., 1985
linoleic acid	1.0×10^6	Giamalva et al., 1985
2°, 3° Amines	$\approx 1.0 \times 10^2$	Pryor et al., 1984
Water	$\approx 5.0 \times 10^1$	Staehelin and Hoigne, 1982 ,1985
1°, 3° Alcohols	$\approx 5.0 \times 10^0$	Pryor et al., 1984
Carboxylic acids	0.0	Pryor et al., 1984
Amides	0.0	Pryor et al., 1984

other sulfhydryl-rich peptides and proteins are highly susceptible to ozone attack. The reactivity of ascorbate is relatively high ($5.6 \times 10^7\ M^{-1}\ s^{-1}$ at pH 7.0) and is one to two orders of magnitude greater than the reaction rates with simple polyunsaturated fatty acids. Amines, amides, alcohols, and carboxylic acids are among the least reactive compounds.

In vivo rate constants may differ substantially from *in vitro* constants that apply to reactions in solution because of such considerations as the compartmentalization of reactive substrates within the cell, the accessibility of sensitive species within the cell structure, the accessibility of sensitive sites within the physical architecture of a biomolecule, and physicochemical properties such as the degree of hydrophobicity of the interacting compound (Saran et al., 1988). In addition, the probability of ozone reacting with a specific chemical entity depends upon both the rate constant of the reaction involved and the cellular concentrations of the reactants. Thus, within the chloroplast (where the concentration of glutathione is 1 to 5 mM and of ascorbate is 25 to 40 mM), the difference in reaction rates with ozone *in situ* may be closer to 1-fold to 3.5-fold, rather than an order of magnitude or more.

Ozone reacts with the sulfhydryl group of glutathione (GSH) to yield the oxidized form (GSSG), according to Equation 6.1 (Mudd, 1982). The GSSG can then be reduced by NADPH-dependent glutathione reductase (Equation 6.2).

$$2GSH + O_3 \rightarrow GSSG + H_2O + O_2 \qquad (6.1)$$

$$GSSG + NADPH + H^+ \rightarrow 2GSH + NADP^+ \qquad (6.2)$$

In the chloroplast, glutathione is maintained predominantly in the reduced state (95%) under light conditions and is thought to react with oxygen species that are generated during photosynthesis (Salin, 1988).

Mudd (1982), Heath (1984, 1987a), Heath and Castillo (1988), and Wellburn (1988) have emphasized the important distinctions between ozonolysis and lipid peroxidation, in their reviews of the reactions of ozone with unsaturated carbon bonds in cell membrane lipids. Ozonolysis of carbon double bonds in the presence of water (Criegee, 1975) proceeds initially by the formation of a primary ozonide (Equation 6.3). The ozonide is then cleaved to form a carbonyl and a carbonyloxide (Equation 6.4). The carbonyloxide reacts with water to form a second carbonyl and hydrogen peroxide (Equation 6.5).

$$R - HC = CH - R' + O_3 \rightarrow R\underset{\underset{H}{|}}{\overset{\overset{O}{\diagup \diagdown}}{\underset{}{C}}}\underset{\underset{H}{|}}{\overset{O \quad O}{C}}-R' \qquad (6.3)$$

$$\text{ozonide} \rightarrow R - HC = O + R'HC = O^+ - O^- \qquad (6.4)$$

$$R'HC = O^+ - O^- + H_2O \rightarrow R'HC = O + H_2O_2 \qquad (6.5)$$

Ozone may also react indirectly with lipid components of the membrane through the formation of free radicals and subsequent lipid peroxidation (Halliwell, 1982). Both ozonolysis and peroxidation of polyunsaturated fatty acids produce malondialdehyde, a product commonly assayed to indicate nonspecific "lipid peroxidation".

The reaction between ozone and ascorbate presumably occurs by attack at the double bond between carbons 2 and 3. If the process is similar to lipid ozonolysis, then one of the oxidation products would also be hydrogen peroxide. At the present time little information is available on the direct interaction of ozone with ascorbate.

Free sulfhydryl groups on enzymes are potentially highly susceptible to oxidation by ozone (Equation 6.1; Mudd, 1982; Heath, 1987a). If disulfide bond formation is the extent of the oxidation, as in plasma membrane ATPases (Heath, 1987a), the process is easily reversed by the addition of a reducing agent such as dithiothreitol. The formation of such disulfide bonds within the cell could be temporary and reversible when the oxidative stress is relaxed (Equation 6.2). As Mudd (1982) has pointed out, the impact of reactions of ozone with sulfhydryl groups on membrane-bound or cytosolic enzymes will be correlated with the functional relationship of the sulfhydryl group to the enzyme's catalytic activity.

The localization of the different reactions with ozone will influence the terminal effect of ozone on the plant cell. Glutathione and ascorbate may be considered among the primary reactants and scavengers of ozone because of their high rate constants for oxidation, molecular mobility, high cytoplasmic (and chloroplastic) concentrations, their generalized dispersion throughout the cell, and, in the case of ascorbate, its occurrence within the cell wall, as discussed further in Section 6.2.4.2.

6.2.2 Ozone and the Production of Oxyradicals

The dissolution of ozone in water can produce active decomposition products that include superoxide, peroxyl, and hydroxyl radicals (Hoigne and Bader, 1975; Staehelin and Hoigne, 1982, 1985). Decomposition is favored in alkaline solutions (pH > 8.0), but is relatively slow at physiological pH. Heath (1987a) calculated an ozone rate loss of 0.015% min^{-1} in a saturated aqueous solution at pH 7.0. Under these conditions, the steady-state concentration of the superoxide anion radical (O_2^-), one of the initial products formed, would be only 8.75×10^{-15} M. Indeed, Grimes et al. (1983) showed by electron paramagnetic resonance (EPR) spectrometry and the use of spin traps in $vitro$ that, although ozone could dissolve in water to yield H_2O_2, O_2^-, and OH, only hydroxyl radicals could be detected under physiological conditions, especially in the presence of phenolics such as caffeic or ferulic acids.

The interactions of ozone with substrates that yield hydrogen peroxide can lead to increased levels of O_2^- and to the subsequent generation of the highly reactive hydroxyl radical. Hydrogen peroxide itself combines with superoxide to produce the ˙OH radical through the Haber-Weiss reaction (Equation 6.6; Haber and Weiss, 1934).

$$O_2^- + H_2O_2 \rightarrow {}^{\cdot}OH + OH^- + O_2 \qquad (6.6)$$

The rate of this reaction is considerably enhanced by metal catalysts as a result of the reactions shown in Equations 6.7 and 6.8 (Saran et al., 1988).

$$Me^{n+} + O_2^- \rightarrow Me^{(n-1)+} + O_2 \qquad (6.7)$$

$$Me^{(n-1)+} + H_2O_2 \rightarrow Me^{n+} + {}^{\cdot}OH + OH^- \qquad (6.8)$$

The potential importance of systems that result in the production of ˙OH radicals needs to be viewed in the context of the extreme reactivity of such radicals and their concomitant brief lifetimes (1.2 ns) and diffusion path lengths (3.5 nm). In particular, the diffusion path length is comparable to such intracellular dimensions as the thickness of phospholipid bilayers (5.0 nm), membrane globular proteins (7.5 nm), and the mean water thickness around biomolecules in general (3 nm). These considerations led Saran et al. (1988) to conclude that such radicals could only react with specific target biomolecules in the immediate proximity of the site of their production. In contrast, the less reactive superoxide anion (with an approximate lifetime of 1 ms) could travel several molecular distances to reach a specific target site.

The production and scavenging of the various forms of active oxygen during normal photosynthesis have been reviewed by Asada and Takahashi (1987). The superoxide anion is naturally generated by the single electron transfer from reduced ferredoxin to oxygen (Equation 6.9), and is then dismutated to hydrogen peroxide by superoxide dismutase according to the reaction shown in Equation 6.10 (Elstner, 1982; Halliwell, 1982). The reduction of oxygen to hydrogen peroxide by electrons generated

$$Fe_{red} + O_2 \rightarrow Fe_{ox} + O_2^- \qquad (6.9)$$

$$2O_2^- + 2H^+ \rightarrow H_2O_2 + O_2 \qquad (6.10)$$

from the photolysis of water in the Mehler reaction (Mehler, 1951) occurs continuously in the illuminated chloroplast. Electron transfer to oxygen is increased under conditions that limit the normal electron flow through $NADP^+$ to carbon dioxide fixation (Robinson, 1988).

The production of the superoxide anion and hydroxyl radicals thus presents an attractive hypothesis for the initial effects of ozone that lead to cellular perturbations (Tingey and Taylor, 1982; Bennett et al., 1984; Fong, 1985; Heath, 1987a; Alscher and Amthor, 1988). However, it should be pointed out that its attractiveness comes from its theoretical plausibility rather than from experimental verification *in vivo*.

Although Grimes et al. (1983) were unable to detect it in aqueous solutions of ozone under physiological conditions, most investigations involving oxyradicals have focused on the possible role of the superoxide anion, since there are numerous O_2^--generating systems within the plant whose activities could be enhanced independently of any direct O_2^--production from ozone *per se* (Fong, 1985).

However, the evidence for the involvement of O_2^- in ozone phytotoxicity is both indirect and conflicting. Lee and Bennett (1982) inferred a role for superoxide based on observations of the effects of ozone on the levels of superoxide dismutase (SOD) in bean leaves. Decleire et al. (1984) reported increased levels of SOD in spinach leaves, and Castillo et al. (1987) found increases in both extracellular and total "SOD-like activity" in Norway spruce needles exposed to ozone. McKersie et al. (1982a) and Becker et al. (1989), on the other hand, found no relationship between endogenous SOD levels and susceptibility to ozone injury in ranges of cultivars of bean or white clover, respectively. Chanway and Runeckles (1984a) could only detect ozone-induced increases in SOD levels in bean leaves following the onset of distinct cellular injury. In spinach leaves, Sakaki et al. (1983) observed decreased SOD activity, with no change in total protein content. Matters and Scandalios (1987) found no significant effects of ozone exposures (lasting up to 4 d) on corn leaf SOD levels.

Lee and Bennett (1982) also reported that SOD and catalase levels were increased in the leaves of bean plants treated with N-[2-(2-oxo-1-imidazolidinyl)ether]-N-phenylyurea ("Ethylene diurea", EDU), an antiozonant that has been found to be an effective protectant against visible foliar injury on all plant species tested since its introduction (Carnahan et al., 1978). EDU has also been reported to cause increases in SOD in human gingival fibroblast cell cultures and in heart, liver, and lung tissue of rats, following interperitoneal administration (Stevens et al., 1988).

However, Chanway and Runeckles (1984b) were unable to demonstrate any SOD enhancement in bean leaves exposed to ozone and were forced to conclude that EDU-induced tolerance is based on mechanisms other than those related to SOD activity.

In the context of a potential role for SOD in determining sensitivity to ozone, it should be noted that Tepperman and Dunsmuir (1990) recently reported that genetically transformed tobacco plants carrying a chimeric Petunia nuclear gene encoding chloroplast SOD showed no resistance to the light-activated herbicide paraquat in spite of up to 50-fold increases in SOD activity. Similarly, two- to four-fold increases in SOD in transformed tomato plants failed to result in increased tolerance to conditions known to cause enhanced O_2^- levels, such as high, photoinhibitory light intensities.

In a preliminary study, Rowlands et al. (1970) observed several changes in the free radical signals observed by EPR spectrometry in pieces of *Phaseolus* and *Glycine* leaves taken from plants after treatment with ozone, but they could not identify any signals other than those associated with Photosystems I and II and the Mn^{2+} ion that are typically seen in photosynthesizing cells. The effects of ozone on such signals is discussed in Section 6.2.7.

Vaartnou (1988), however, has detected the EPR signal of the superoxide anion *in situ* in intact leaves of grass plants (*Poa pratensis* and *Lolium perenne*) treated with ozone, but only at relatively high concentrations or after prolonged exposures. Presumably, with lesser treatments, SOD and other scavengers are able to prevent the build-up of O_2^- concentrations to above the detection limit. Attempts to detect O_2^- and other radicals *in situ* by means of their adducts with spin labels such as tiron (1,2-dihydroxybenzene-3,5-disulphonate) or DMPO (5,5-dimethyl-pyrroline-N-oxide) have so far been unsuccessful because of the toxicity of the spin labels (Runeckles, unpublished).

Cassab and Varner (1988) have speculated that histidine residues within cell-wall protein may be attacked by oxygen radicals (Rivett, 1986), which would result in changes in cell-wall pH and the interrelationships of ionized wall constituents such as proteins, pectins, and Ca^{2+}. Oxyradicals have also been implicated in the chemical modification of DNA. In the case of hydrogen peroxide toxicity to *Escherichia coli*, which is largely attributable to DNA damage resulting from the reaction of H_2O_2 with DNA-bound iron in the presence of a source of reducing equivalents, oxygen itself may work together with SOD and catalase to form a scavenging system that converts other oxyradicals to O_2^-, which is then destroyed (Imlay and Linn, 1988). Among the many possible reactions of oxyradicals with DNA, hydroxyl radicals can lead to the hydroxylation of guanine bases to form the 8-hydroxy-derivative, as discussed in Section 6.2.10, below.

6.2.3   The Extent of Ozone Penetration into the Cell

The chemical reactivity of ozone and its capacity to form oxyradicals inevitably leads to questions as to whether the effects of exposure are attributable to molecular ozone per se, and which reactions are most likely to occur and where, in the pathway of ozone from the ambient air to the interior of the leaf, as discussed in general in Chapter 5.

The observation that the ozone concentration in the intercellular air space of sunflower leaves is close to zero, even when the leaves are exposed to ozone concentrations as high as 1 ppmv (Laisk et al. 1989), suggests that the apoplastic space and the plasmalemma are major sinks for ozone, at least during the first few minutes of exposure. Potential effects of ozone on the plasmalemma that relate to permeability and membrane function are discussed more fully in Section 6.3. However, the suggestion of Laisk et al. (1989) that the combination of apoplast and plasmalemma reactions prevents the penetration of ozone per se into the deeper layers of the cell has important ramifications with regard to the *in vivo* relevance of many of the known reactions of ozone with biomolecules.

Although numerous effects of ozone on intracellular enzymes and other constituents have been demonstrated *in vivo*, and many of them appear to parallel observations of effects observed on *in vitro* systems, we have no direct evidence that the *in vivo* reactions are caused by ozone itself.

There are, however, several lines of indirect evidence. There have been many reports of ultrastructural changes to internal membranes, such as indentations of chloroplast envelope, the swelling of chloroplast thylakoids, Golgi body cisternae, the endoplasmic reticulum, the nuclear envelope, the shrinkage of mitochondrial cristae, and the appearance of crystalline bodies within the chloroplast (Thomson et al., 1974; Miyake et al., 1984; and others). Many of these effects of ozone might be interpreted as indicating disturbed osmotic relationships within the cell, caused by changes in the plasmalemma, since they resulted from exposures to relatively high ozone concentrations (0.4 to 0.5 ppmv) which led to the appearance of visible symptoms of injury. However, the rapidity of the appearance of many of these changes, especially the swelling of the thylakoid, which can be detected before any visible injury is apparent, suggests that unreacted ozone penetrates beyond the plasmalemma (Miyake et al., 1984; Guderian et al., 1985). Similar conclusions may be drawn from many of the observations of effects on *in vivo* chloroplast functioning discussed below in Section 6.2.7.

Castillo and Greppin (1988) found that the decrease in total ascorbate in *Sedum* leaves exposed to ozone could not be accounted for by the decrease in total apoplastic ascorbate, and concluded, therefore, that there was either ozone-induced destruction of intracellular ascorbate or impaired synthesis. However, neither explanation is dependent on direct reactions of ozone itself.

Indirect support for the penetration of ozone may come from the possibility of its reacting to form oxyradicals that, because of their generally high reactivities and consequent short diffusion path lengths, would have to be formed close to the sites at which they react, as discussed in Section 6.2.2.

Even where the evidence for intracellular effects of molecular ozone appears to be convincing, the caveat that the ozone levels used in many studies have been considerably higher than any experienced in even the most heavily polluted ambient air must still be borne in mind.

If ozone enters the cell, its cytoplasmic reactions with glutathione or ascorbate are probably less important than those occurring elsewhere. Reduced glutathione can be regenerated by NADH-dependent glutathione reductase (Equation 6.2), and the hydrogen peroxide produced should be removed quite rapidly by catalase. On the other hand, the interactions of ozone with chloroplast components may result in the formation of a number of species that are highly toxic to chloroplast function. Ascorbate and glutathione are, in fact, important components of a protective system (see Section 6.2.6) that prevents the accumulation of endogenous, reactive, toxic oxygen species, such as the superoxide and hydroxyl radicals and hydrogen peroxide, within the chloroplast (Halliwell, 1982; Robinson, 1988; Salin, 1988).

6.2.4 Leaf Surface and Apoplastic Reactions of Ozone

The preceding sections have focused on possible reactions of ozone with metabolites, enzymes, and membrane constituents, for several of which there is experimental support. However, their importance *in vivo* has to be judged in the context of ozone uptake from the ambient air. Hence, before turning to the potential importance of metabolic effects within the cell and its organelles, it is appropriate to review the reactions that may occur at the leaf surface and within the intercellular air spaces and mesophyll cell walls, i.e., the apoplastic space outside the plasmalemma.

6.2.4.1 Reactions at the Leaf Surface

The surface of the typical leaf comprises the cuticle and, depending upon the species, various types of trichomes or leaf hairs, glands, and wax deposits of various types and thicknesses. The waxes consist largely of long-chain alkanes, aldehydes, ketones, alcohols, fatty acids and their esters, and cyclic triterpenoids (Baker, 1982), laid down in a matrix of ester-linked hydroxy-fatty acids and polysaccharide fibrils (Holloway, 1982). While these compounds are generally much less reactive with ozone than compounds with unsaturated bonds (Table 6.1), oxidation and cleavage can nevertheless occur, leading to changes in composition and the physical properties of the leaf surface. In particular, the introduction of covalently bound oxygen results in greater polarity and renders the surface less hydrophobic (Kersteins and Lendzian, 1989). In fact, although little appears to be known about the specific chemistry involved (Trimble et al., 1982), increased wettability has been found to result from long-term exposures of Norway spruce to ozone (Barnes et al., 1990). However, while such changes may have far reaching physiological consequences, such as the increased leaching of cations and other solutes, Kersteins and Lendzian (1989) found no significant effects on water permeance of the cuticles of several species.

6.2.4.2 Intercellular Reactions

The leaf apoplast comprises the intercellular air spaces, the aqueous medium on and within the walls, and the wall constituents outside the plasmalemma of the mesophyll cells.

Ozone may react with olefins, such as ethylene, and terpenoids, released into the intercellular spaces. Indeed, Melhorn and Wellburn (1987) suggested that the foliar injury attributed to ozone is the result of ozonolysis of stress-induced ethylene, leading to the formation of free radicals such as the superoxide anion, since injury and rate of ethylene production were found to be directly correlated. At levels of ozone insufficient to cause acute injury or necrosis, Taylor et al. (1988) similarly observed that the reductions in CO_2 assimilation rates and stomatal conductance caused by ozone were diminished or eliminated when stress-ethylene production was metabolically inhibited. However, this hypothesis has been questioned both by Chameides (1989), who has argued on thermodynamic grounds that a causal relationship between ethylene production and ozone injury is unlikely, and by Zwoch et al. (1990) as a result of their observations on ozone-treated sunflower leaves, which revealed increased rates of ethylene release after 10 or more days of exposure, without any visible injury occurring.

Tingey and Taylor (1982) concluded that there was little likelihood of intercellular terpenoids acting as effective scavengers of gas-phase ozone, but again, this does not preclude a role for traces of the products of such reactions.

Reactions of ozone with hydroxyl ions in the aqueous phase (with a rate constant of $=50 \ M^{-1}s^{-1}$, Table 6.1) or with major cell-wall constituents, such as cellulose, lignin, and pectins, are considered by Chameides (1989) to have insignificant effects on ozone flux to the plasmalemma. However, this does not preclude the possibility that trace reaction products may play a role in cell responses to ozone.

On the other hand, there is increasing evidence for ozone-induced changes in apoplastic constituents, especially ascorbate. Castillo and Greppin (1988) reported that although ozone treatment of *Sedum* leaves caused a decrease in apoplastic ascorbate itself, the level of ascorbate *plus* dehydroascorbate increased. Furthermore, the evidence indicated a continuous supply of ascorbate to the apoplast. The activity of extracellular peroxidases was also found to increase (Castillo and Greppin, 1986; Castillo et al., 1984). This has been confirmed for bean leaves (Peters et al., 1989) in which both extracellular cationic (ascorbate-dependent) and anionic (guaiacol-dependent) isoforms were found to increase, and for spruce needles (Castillo et al., 1987). Extracellular diamine oxidase, on the other hand, was found to decrease bean leaves. These changes led Peters et al. (1989) to suggest that the increased level of ascorbate peroxidase activity might result from ozone-induced increases in the levels of phenolic compounds (Howell, 1970), since they found that caffeic acid increased extracellular ascorbate peroxidase activity. However, as discussed below in Section 6.2.12, increased phenolic levels are thought to result from polyphenoloxidase action following ozone-induced breakdown of membrane integrity and compartmentalization. Such breakdown also affects the balance of Ca^{2+} ions across the plasmalemma (Gaspar et al., 1985). Since Ca^{2+} is a peroxidase activator, both mechanisms may be involved and, in either case, imply an initial effect of ozone on the plasmalemma.

Peters et al. (1989) also suggested that decreased diamine oxidase activity would result in decreased polyamine breakdown. Hence, the combination of increased peroxidase activity and decreased diamine oxidase activity could reduce the impact of ozone by increasing the scavenging of potentially harmful hydrogen peroxide, on the one hand, and by increasing di- and polyamine levels, on the other. Polyamines stabilize membranes and buffer cells against changes in ionic composition (Smith, 1985), and exogenous polyamines have been shown to protect leaves from ozone-induced visible injury (Ormrod and Beckerson, 1986).

The increased levels of extracellular ascorbate could act as direct sinks for ozone, as suggested by Chameides (1989). However, in order to function as an enhanced detoxification system for ozone, the increased activity of ascorbate peroxidase requires that ozone is converted to hydrogen peroxide either directly or via the superoxide anion. Indirect evidence for the involvement of superoxide comes from the work of Castillo et al. (1987), who studied the changes in extracellular enzymes in the needles of Norway spruce exposed for up to 30 d to daily concentrations of ozone up to 160 ppbv. They observed that, although the extracellular superoxide dismutase-like activity amounted to only about 1% of the total needle activity, it increased markedly following ozone exposure. Such increases imply the increased production of superoxide in the intercellular space of exposed needles. They also imply concomitantly increased removal of protons (Equation 6.10), with possible effects on the pH of the wall space.

Independent of SOD-catalyzed dismutation, superoxide may also react with hydrogen peroxide in the cell-wall fluid to yield hydroxyl radicals (Equation 6.6). Comparable reactions leading to superoxide anion, hydrogen peroxide, and hydroxyl radical may also occur within the cell envelope. The various possible reactions are incorporated in the scheme depicted in Figure 6.1, based in part on the cycle of reactions involving ascorbate proposed by Castillo and Greppin (1988) that is discussed further in Section 6.2.6.

As a result of these observations, it becomes apparent that, although many of the effects of ozone observed in leaves involve intracellular processes such as photosynthesis and respiration, the actions of ozone on nonstructural constituents of the cell wall outside the plasmalemma are important indirect determinants of the magnitudes of these internal effects.

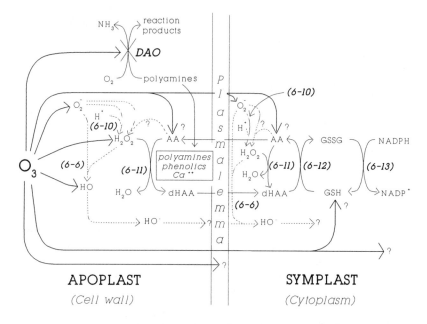

Figure 6.1.   Potential reactions of ozone (heavy lines) and some oxyradicals (broken lines) within the apoplast and symplast, involving ascorbate, diamine oxidase, superoxide dismutase, and ascorbate-dependent peroxidases. Polyamines, phenolics, and Ca^{2+} ions are peroxidase activators. Numbers refer to the Haber-Weiss reaction (6-6), and the coupled reactions involving superoxide dismutase (6-10), ascorbate peroxidase (6-11), dehydroascorbate reductase (6-12), and glutathione reductase (6-13) described in Sections 6.2.5 and 6.2.6. AA = ascorbate; dHAA = dehydroascorbate; DAO = diamine oxidase; GSH = glutathione; GSSG = oxidized glutathione (based in part on Castillo and Greppin, 1988).

6.2.5   Potential Intracellular Ozone-Scavenging Systems

In addition to the ability of ascorbate to act as a sink for ozone prior to its reaching the plasmalemma, both ascorbate and glutathione have the potential to scavenge ozone within the cell. There are numerous reports of changes in ascorbate and glutathione levels during and following exposure to ozone. In *Sedum* leaves (Castillo and Greppin, 1986, 1988), total ascorbate and glutathione levels decreased during exposure to ozone, but rapidly increased immediately following the exposure. Total ascorbate was observed to decrease in spinach and sensitive cultivars of soybean and snapbean within 1 or 2 h after the initiation of ozone fumigation (Lee et al., 1984; Tanaka et al., 1985). But in tolerant soybean and snapbean genotypes, ozone exposure increased ascorbate concentrations within 1 h. The activities of both ascorbate peroxidase and glutathione reductase also increased after ozone exposure in bean and spinach (Guri, 1983; Tanaka et al., 1988).

In seedlings of several species of Pinus, ascorbate concentrations doubled after 11 weeks of treatment with low (0.05 ppmv) or moderate (0.15 ppmv) ozone concentrations (Barnes, 1972), the effect being more pronounced in primary than in secondary needles. Melhorn et al. (1986) reported small increases in the glutathione and ascorbate contents of the needles of spruce trees exposed for 2 years to very low ozone concentrations (0.037 ppmv). Ascorbate concentrations in the needles of an oxidant-tolerant white pine genotype have been observed to remain consistently higher throughout the growing season and into the winter months than in those of an oxidant-sensitive genotype (Chevone et al., 1989).

It is apparent that additional research concerning ozone effects on ascorbate and glutathione metabolism is needed to characterize the specific roles that these antioxidants play in mediating the toxicity of ozone to plant cells. Furthermore, in light of the known compartmentalization of the potential scavengers of ozone and its reaction products, it is important to recognize the limited usefulness of gross tissue analyses for compounds such as ascorbate and glutathione, or enzymes such as SOD.

6.2.6 Free-Radical-Scavenging Systems

Oxyradical formation is an inevitable consequence of life in an oxygen-rich environment (Halliwell, 1974). As a consequence, numerous scavenging systems have evolved in order to minimize their harmful effects.

In the chloroplast the transfer of electrons from ferredoxin to oxygen and the subsequent formation of hydrogen peroxide via superoxide and superoxide dismutase necessitates a metabolic pathway to prevent the accumulation of peroxide. Since catalase is not present in the chloroplast, an alternative system of enzymes and metabolites that can reduce hydrogen peroxide is present (Halliwell, 1982; Alscher and Amthor, 1988). The initial reaction between reduced ascorbate (AA) and hydrogen peroxide is catalyzed by ascorbate peroxidase, as shown in Equation 6.11 (Nakano and Asada, 1980, 1981; Asada and Badger, 1984), with the monodehydroascorbate radical as an intermediate (Hossain et al., 1984). The monodehydroascorbate radical can dismutate to dehydroascorbate (dHAA) or be reduced to ascorbate by NADPH-dependent monodehydroascorbate reductase. The oxidized dehydroascorbate is reduced back to ascorbate (Equation 6.12) by dehydroascorbate reductase, using two molecules of glutathione (Nakano and Asada, 1981; Jablonski and Anderson, 1982).

$$H_2O_2 + AA \rightarrow dHAA + 2H_2O \qquad (6.11)$$

$$dHAA + 2GSH \rightarrow AA + GSSG \qquad (6.12)$$

The oxidized glutathione (GSSG) is then reduced to GSH by NADPH-dependent glutathione reductase, as shown in Equation 6.13 (Schaedle and Bassham, 1977).

$$GSSG + NADPH + H^+ \rightarrow 2GSH + NADP^+ \qquad (6.13)$$

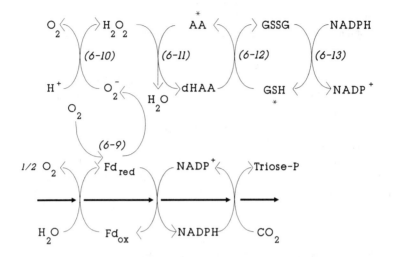

Figure 6.2.   Normal and alternate electron flow in the chloroplast. Heavy straight arrows indicate normal electron flow from water through ferredoxin (Fd) to triose phosphate of the Calvin cycle. Pseudocyclic electron flow is initiated by the single electron transfer from reduced ferredoxin (Fd$_{red}$) to oxygen. Numbers refer to equations in the text. The enzymes involved are superoxide dismutase (6-10), ascorbate peroxidase (6-11), dehydroascorbate reductase (6-12), and glutathione reductase (6-13). Asterisks indicate potential direct interactions with ozone.

The major components of the pseudocyclic electron transport system of the chloroplast (Egneus et al., 1975) are shown in Figure 6.2. Normal electron flow arises from the photolytic cleavage of water associated with photosystem II. The electrons are ultimately accepted by oxidized ferredoxin, transferred to NADP$^+$, and utilized in carbon dioxide reduction. However, ferredoxin can donate a single electron to oxygen, forming the superoxide anion. A subsequent series of reactions involves the production of hydrogen peroxide (through the action of superoxide dismutase) and its reduction to water through ascorbate, glutathione, and NADPH. The NADPH is then regenerated by normal electron flow through ferredoxin.

Hydrogen peroxide is a pivotal metabolite in this alternate pathway of electron flow and, if not removed, will damage chloroplast function. At concentrations as low as 2 to 3 mM, hydrogen peroxide (>50%) substantially inhibits the light-activated enzymes of the Calvin cycle, presumably by interacting with the reduced sulfhydryl groups (Kaiser, 1976; Tanaka et al., 1982). In addition, the reaction of the superoxide anion with hydrogen peroxide produces the hydroxyl radical ($\cdot$OH), which is a potent initiator of lipid peroxidation. In this respect, it may not be surprising that α-tocopherol, which has been found to be predominantly associated with chloroplast membranes, is an efficient scavenger of lipid radicals (Finckh and Kunert, 1985; Kunert and Ederer, 1985).

6.2.7   Ozone and Chloroplast Metabolism

As a result of the various reactions and other effects of ozone, the potential for disruption of chloroplast metabolism is considerable, whether by direct or indirect action. Because of the high reaction rate of ozone and oxyradicals with sulfhydryl groups, the enzymes of the reductive pentose cycle are particularly vulnerable. Oxidation of their SH groups would require electrons from NADPH and ferredoxin (thioredoxin) for subsequent reduction, and these two constituents are critical for the maintenance of carbon dioxide fixation and the prevention of increases in hydrogen peroxide.

The direct reaction of ozone with ascorbate in the chloroplast is also potentially a highly damaging aspect of ozone toxicity. If one of the reaction products is hydrogen peroxide, not only is ascorbate removed as the peroxidase substrate (Equation 6.11) but additionally, hydrogen peroxide is generated, placing a twofold metabolic demand on the remaining ascorbate pool. If the chloroplast hydrogen peroxide concentration increases above 1 to 2 mM, then a cascading production of toxic oxygen species would ensue, with the eventual peroxidation of chloroplast membranes.

In spite of the potential for these various direct or indirect reactions of ozone to damage the functioning chloroplast, there is little direct experimental evidence to indicate which of the possible disruptive reactions predominates or even occurs. There have been relatively few investigations that have attempted to elucidate effects on specific reactions within the photosystems and electron transport chains *in vivo*. In contrast, the adverse effects of ozone on the overall photosynthetic process, as revealed by gas exchange measurements, are well established and are discussed below in Section 6.4.2.

One *in vitro* approach that has been used is based on the ability of isolated chloroplasts to undertake the Hill reaction: the photolysis of water (with the release of oxygen) when supplied with appropriate exogenous electron acceptors, as shown in Equation 6.14.

$$2H_2O + 4A \rightarrow O_2 + 4AH \qquad (6.14)$$

The electron transport pathway depicted in Figure 6.2 is part of the more complex photosynthetic system depicted in summary form in Figure 6.3. The different energy levels involved in the intermediate reactions of photosystems I and II make it possible to supply exogenous electron acceptors or electron donors that can act as sinks or sources of electrons at different locations within the overall scheme.

Dichlorophenol-indophenol (DCIP) is such an acceptor. Its photoreduction by water to DCIPH$_2$ in illuminated chloroplasts, together with the release of oxygen, indicates that photosystem II is functioning, while photoreduction of NADP by DCIPH$_2$ indicates the functioning of photosystem I. The decreased

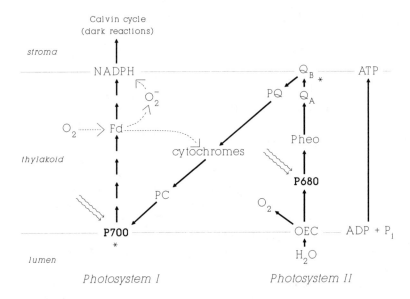

Figure 6.3. General scheme of electron transport photosystems I and II. Several intermediates are omitted for clarity. Pseudocyclic electron flow (see Figure 6.2) is shown by the broken lines. OEC = oxygen evolving complex; P680, P700 = active chlorophyll a; PC = plastocyanin; Pheo = pheophytin; PQ = plastoquinone; Q_A, Q_B = iron-protein-bound plastoquinones. Asterisks indicate sites of electron paramagnetic resonance signals affected by ozone (see Figure 6.5). (Based on Govindjee et al., 1985; Haehnel, 1984, and Andreasson and Vanngard, 1989).

Hill reaction activity observed by Koiwai and Kisaki (1976) and others in chloroplasts isolated from several species treated with ozone was confirmed for spinach chloroplasts by Sugahara et al. (1984), who analyzed the H_2O-DCIP and $DCIPH_2$-NADP photoreductions separately. However, inhibition of either photosystem was only observed when the 4-h exposures to ozone were at the unrealistically high concentration of 0.5 ppmv; at 0.1 ppmv, no impairment of either photosystem was observed.

An *in vivo* approach that is gaining in popularity involves the measurement of chlorophyll fluorescent transients (CFTs). The forms of chlorophyll associated with photosystems I and II (P700 and P680, respectively, Figure 6.3) are activated by photons absorbed by their associated antenna pigment systems. Some of the light energy is passed to the electron transport systems associated with each photosystem, some is lost as heat, and some is reemitted as fluorescence. Since these processes are competitive, a change in one results in complementary changes in the others. Most available instrumentation and methodology detect changes in the predominant fluorescence arising from the chlorophyll *a* of photosystem II, although measurements attributable to photosystem I are possible (Kyle et al., 1983). The method has been the subject of several recent reviews, including those of Schreiber and Bilger (1987) and Bolhar-Nordenkampf et al. (1989).

Relative fluorescence

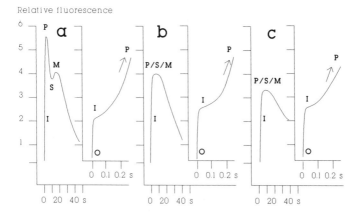

Figure 6.4. Chlorophyll fluorescence transients of bean leaves. (a) Controls; (b) and (c) measurements made 20 h after exposures to 0.3 or 0.5 ppmv, respectively. (Modified from Schreiber et al., 1978.) The O-I rise occurs as the plastoquinone pool becomes reduced, and is typical of dark-adapted tissues in which electron-transport capacity on photosystem II acceptor side has been deactivated in darkness. Subsequent declines through S and M result from photochemical fluorescence quenching in photosystem II reaction centers, and nonphotochemical quenching due to other pathways of deexcitation, including transfer to photosystem I and heat loss (see text).

Schreiber et al. (1978) observed that the major effect of ozone on CFTs in bean leaves was in reducing the magnitude of the I-P rise, as shown in Figure 6.4. They interpreted this change, together with the relative stability of the O-I rise, as indicating that the effect was attributable to a decrease in water-splitting activity, leading to increased oxidation of the plastoquinone pool (Figure 6.3), rather than to a direct effect on the reaction center of photosystem II. These observations were subsequently confirmed by Shimazaki (1988) using spinach leaves.

Barnes et al. (1988) noted an initial increase in the O-I rise in pea leaves exposed to 0.075 ppmv ozone and an increase in I-P, which they attributed to impaired plastoquinone reoxidation. The subsequent decline was attributed to reduced electron flow from the water-splitting component of photosystem II, which resulted in a shift in the plastoquinone equilibrium to the reducing side.

Several recent investigations, such as those of Cape et al. (1988), Davison et al. (1988), and Rowland-Bamford et al. (1989), have tended to use CFT measurements as general "early-warning" indicators of response to ozone rather than for the analysis of specific effects on the photosynthetic process.

The limited investigations carried out on photosynthetic free radicals and paramagnetic species using EPR spectrometry have shown that exposure of soybean leaves to 0.05 ppmv ozone for several hours results in enhancement of the typical six-peak Mn^{2+} signal (Rowlands et al. 1970). Such studies cannot differentiate among the different pools of manganous ion within the cell, but it should be noted that Mn^{2+} plays an important role in the "oxygen clock" within the oxygen-evolving complex of photosystem II (Yocum, 1987). Although the functional manganese in chloroplast membranes was originally thought to be

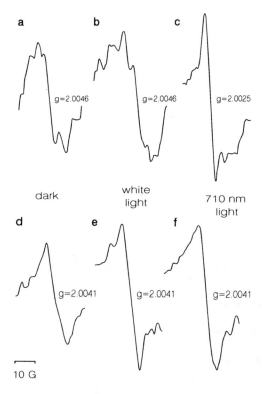

Figure 6.5. Electron paramagnetic resonance (EPR) signals from entire attached *Poa pratensis* leaves. (a), (b), and (c) = controls, in darkness, white light, or 710 nm light, respectively. (d), (e), and (f) = comparable signals after exposure to 0.1 ppmv ozone for 3 h (from Vaartnou,1988).

nonparamagnetic, work at cryogenic temperatures has revealed a multiline EPR signal associated with photosystem II (Ghanotakis and Yocum, 1990).

EPR signals attributed to oxidized chlorophyll *a* in the reaction center of photosystem I (signal I) and to a tyrosine radical in photosystem II (signal II)(Debus et al., 1988) can be demonstrated in isolated chloroplasts and in leaf tissue at room temperature (Figure 6.5). Signal II is observable in healthy tissue in darkness and in white (or 650 nm) light, while signal I (superimposed on signal II) can only be detected on illumination with far-red light (700 to 730 nm). Vaartnou (1988) found that exposure of an attached *Poa pratensis* leaf to 0.1 ppmv ozone within the EPR spectrometer cavity revealed progressive changes in these signals, as shown in Figure 6.5. After 3 h, a signal with identical characteristics in each light regime (g-value 2.0041; peak-to-peak width 10 gauss) replaced signals I and II. The disappearance of signal I (g-value 2.0025; width 7.5 to 9 gauss) indicates an effect of ozone or its reaction products on the oxidation of chlorophyll *a* in the reaction center of photosystem I. The elimination of signal II (g-value 2.0046; width 18 to 20 gauss) indicates disruption of the electron flow through photosystem II, possibly in a manner analogous to the action of herbicides such as diuron and atrazine (Vermaas et al., 1987). Although EPR spectrometry alone cannot determine the identity of the new radical, its transience is indicated by its disappearance within 20 min of ending the exposure to ozone with the reappearance of typical signals I and II. Nevertheless, the nature of the changes indicates a relatively rapid and severe perturbation of the electron flows associated with both photosystems.

The spatial localization of the different reactive sites on and within the thylakoid membrane results in photophosphorylation by the establishment of a proton gradient across the membrane. Protons pass into the lumen, making it more acidic; the consumption of protons in NADPH formation makes the stromal side less acidic. The energy represented by this imbalance is used to produce adenosine triphosphate (ATP), the principle energy "storehouse" used by the plant in biosynthesis and regulation.

Analyses of leaf tissues from plants exposed to ozone have led to reports of both increases and decreases in total ATP, as noted by Wellburn (1984). However, such analyses fail to differentiate among the different pools of ATP within the cell. Because of the dependence of chloroplast ATP formation on the pH gradient across the thylakoid, decreases in the latter should result in reduced ATP formation. Using the fluorescence of 9-aminoacridine as a probe, Robinson and Wellburn (1983) found that pulses of ozone supplied to suspensions of oat leaf thylakoid membrane preparations resulted in progressive decreases in the pH gradient, suggesting a decrease in ATP synthesis, in agreement with Heath's (1980) reinterpretation of earlier findings (Coulson and Heath, 1974).

Effects on chloroplast ATP levels will influence the dark reactions of the Calvin cycle occurring in the stroma. The key enzymes involved in the initial capture of CO_2 are ribulose-1,5-*bis*phosphate carboxylase/oxygenase (rubisco), and phosphenolpyruvate (PEP) carboxylase. Rubisco is ubiquitous, but PEP carboxylase is an important CO_2 acceptor in C-4 plants in which the overall process of photosynthesis involves reactions in both mesophyll and bundle sheath cells, as discussed in Chapter 5. Effects on these and other enzymes involved in the complex system of reactions of the Calvin cycle that ultimately lead to the production of glucose or its polymeric form, starch, may be caused by direct reaction with ozone or its products. Alternatively, the reactions that they catalyze may be impaired as a result of reduction of the ATP supply.

Although there have been numerous reports of the effects of other pollutants on enzymes involved with CO_2 fixation, little is known about the effects of ozone. Enzymes that require cysteine or tryptophan for their biological activity would be expected to be highly susceptible; this is borne out by the high *in vitro* susceptibility of glyceraldehyde-3-phosphate dehydrogenase (Mudd, 1982). Pell and Pearson (1983) reported reduced levels of rubisco in alfalfa leaves after ozone treatment. Lehnherr et al. (1987) also observed decreases in both initial activity and activity after activation with CO_2 and Mg^{2+} on a leaf-area basis, but no effect on activated rubisco activity on a chlorophyll basis (Lehnherr et al., 1988). Associated measurements of RuBP, ATP/ADP ratios, and triosephosphate/ phosphoglyceric acid ratios led the authors to suggest that decreased rubisco activity, rather than reduced recycling of RuBP, was responsible for the con-comitant decrease in photosynthetic CO_2 assimilation observed. Ozone may cause the activation of glucose-6-phosphate dehydrogenase, since H_2O_2 has been shown to activate the enzyme in pea chloroplasts (Brennan and Anderson, 1980). By analogy, Koziol et al. (1988) suggested that phosphofructokinase might also be activated. Such effects would lead to disruption of normal carbohydrate metabolism in the chloroplast.

Effects of ozone on chloroplast functioning may also include effects on nitrogen metabolism. The enzyme catalyzing the conversion of nitrite to ammonium, nitrite reductase (NiR), is located within the chloroplast in close association with photosystem I. Cytoplasmic nitrate reductase (NaR) supplies NiR with nitrite, a highly toxic intermediate, the reduction of which constitutes a sink for photosynthetic-reducing power that competes with the needs of the Calvin cycle. Ozone reduces NiR activity in corn and soybean leaves (Leffler and Cherry, 1974). The superoxide anion radical also inactivates NiR presumably by attacking its molybdenum center (Aryan and Wallace, 1985; Mikami and Ida, 1986).

Chloroplast DNA can be modified by ozone or its reaction products. Increased levels of 8-hydroxy-2'-deoxyguanosine (8-OHdG) have been detected in digests of chloroplast DNA from the leaves of ozone-injured bean and pea plants (Floyd et al., 1989). Since no 8-OHdG could be detected in digests of calf thymus DNA subjected to ozone, which revealed other evidence of reaction, the effect on chloroplast DNA appears to involve hydroxyl radicals resulting from the action of ozone on the chloroplast per se, rather than from the direct reaction of ozone with the DNA present. Since the presence of 8-OHdG can result in the misreading of DNA templates (Kuchino et al., 1987), its formation in chloroplast DNA may have implications both to protein synthesis and DNA replication.

Ozone induces changes in chloroplast lipids. General glycolipid formation in spinach was found to be reduced (Mudd et al., 1971), and Sakaki et al. (1985) observed that ozone decreased the levels of both mono- and digalactosyl lipids (which are located exclusively in thylakoid membranes).

The most widely observed chemical change induced by ozone in the chloroplast is the destruction of chlorophyll. The literature abounds with reports of decreases in leaf chlorophyll content following a wide range of ozone treatments (Guderian et al., 1985). Most reports indicate that chlorophyll *b* is more sensitive than chlorophyll *a*. Carotenoids appear to be less sensitive to ozone than the chlorophylls, which accounts for the yellowing, rather than bleaching, of leaf tissue in which chlorophyll loss becomes appreciable.

Since chloroplast reactions provide the precursors of many other plant pigments, phenolic compounds (such as the flavonoids), alkaloids, and other secondary compounds (Luckner, 1980), ozone may also affect their biosynthesis.

Finally, it is worth repeating that metabolic processes within the chloroplast may also be affected by ozone in nonspecific ways via effects on the integrity of the envelope and thylakoid membranes that perturb the chemiosmotic status and the delicate balance of proton and other ionic flows.

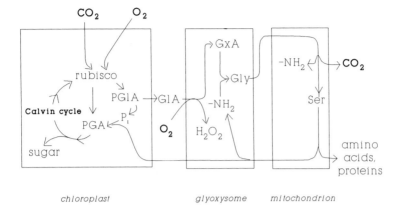

Figure 6.6.   Pathways of carbon flow in photorespiration. GlA = glycolate; GxA = glyoxylate; Gly = glycine; PGA = phosphoglycerate; PGlA = phosphoglycolate; Ser = serine; rubisco = ribulose-1,5-*bis*phosphate carboxylase/oxygenase.

6.2.8   Effects of Ozone on Photorespiration and Respiration

Photorespiration involves reactions within the chloroplast, the cytoplasmic glyoxysomes, and the mitochondria and results from the action of the oxygenase function of rubisco. In the process of photorespiration, depicted in Figure 6.6, a two-carbon fragment of ribulose-1,5-*bis*phosphate is oxidized to glycolic acid, which undergoes further reactions leading to the release of CO_2. Although this results in an overall reduction in the net efficiency of CO_2 utilization, it has been argued that photorespiration may offer a protective device to permit the dissipation of light energy when the amounts of CO_2 are limiting (Wellburn, 1988). In such conditions the formation of injurious oxyradicals would be favored, but the "recycling" of carbon provided by photorespiration provides a means for maintaining higher internal CO_2 levels that would minimize these effects. It should be noted in passing that the oxgenase function of rubisco has been suggested to involve the formation of the superoxide anion, which is retained within the "cage" of the catalytic site on the enzyme (Lorimer, 1981).

As shown in Figure 6.6, the pathway involves two amino acids (glycine and serine) and, hence, has close links to nitrogen metabolism. Although effects of other pollutants on photorespiration have been reported (Hallgren, 1984), there appear to have been few studies of specific effects of ozone, probably because of the difficulties in making quantitative measurements (Miller, 1987). Increased photorespiration is only one of several explanations for the increased pool sizes of both glycine and serine observed by Ito et al. (1985a) in bean plants.

Normal respiration is primarily a function of the mitochondria. As in the case of ozone effects on chloroplast ultrastructure, changes to mitochondrial structure have been reported. For example, Lee (1967) reported mitochondrial swelling when mitochondrial preparations from tobacco leaves were treated with ozone *in vitro*, while Miyake et al. (1984) observed *in situ* shrinkage of mitochondrial

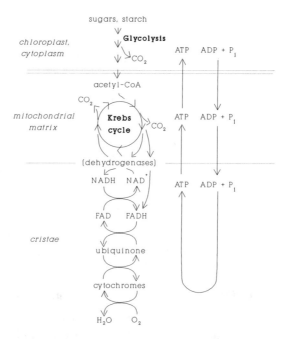

Figure 6.7.   Pathways of carbon and electron flow in respiration, involving cytoplasmic reactions (glycolysis) and reactions in the mitochondria. FAD (FADH) = flavin adenine dinucleotide; NAD (NADH) = nicotinamide adenine dinucleotide.

cristae. Since the enzyme systems responsible for terminal oxidation and oxidative phosphorylation are located on the cristae, it is therefore not surprising that these activities have been reported to be influenced by ozone. However, as Black (1984) has pointed out, it is difficult to differentiate between direct effects on the respiratory processes and changes due to leaf injury. The difficulty is compounded by the conflicting evidence that has been reported of effects on the overall process, as revealed by measurements of O_2 and CO_2 gas exchange, as discussed below in Section 6.4.2.

The overall processes of respiration are summarized in Figure 6.7. The reactions of the Krebs cycle that include two decarboxylations occur in the intramitochondrial matrix and require the supply of acetyl-CoA by the processes of glycolysis occurring in the chloroplast and cytoplasm. The progressive oxidation of the Krebs cycle intermediates, involving the reduction of the cofactors NAD and FAD, leads to a flow of electrons to the ultimate acceptor, oxygen. This electron transport occurs within the infolded internal mitochondrial membrane that forms the cristae. The oxidative phosphorylation system has some features in common with the chloroplast thylakoid, since here, too, the spatial distribution of the individual reaction centers involved results in the establishment of a proton (pH) gradient across the membrane between the matrix and intracristal spaces, which ultimately leads to the production and release of ATP into the matrix.

Lee (1967) observed a rapid decrease in oxidative phosphorylation in tobacco mitochondria from leaves subjected to high (1 ppmv) ozone exposures. Macdowall (1965) also noted a decrease in cytochrome oxidase activity and an uncoupling of oxygen uptake and ATP formation, suggesting either that perturbation of the cristal membrane was uncoupling the electron and proton flows or that there was a shift to the widely distributed nonphosphorylating cyanide-resistant pathway (Dizengremel and Citerne, 1988). Earlier evidence supporting the latter view comes from the observation that a high concentration of ozone (0.5 ppmv) increased glucose-6-phosphate dehydrogenase and decreased glyceraldehyde-3-phosphate dehydrogenase activities in soybean (Tingey et al., 1975).

It should be noted that indirect inhibitory effects of ozone on the respiration of roots have also been reported (Hofstra et al., 1981), although this probably reflects reduced supply of assimilates (Miller, 1987). In a study of nonphotosynthesizing tissues, Anderson and Taylor (1973) observed that the respiration of tobacco callus culture tissue was increased by ozone.

6.2.9 Effects of Ozone on Carbohydrate Metabolism

Since carbohydrates are the principal products and substrates of photosynthesis and respiration, respectively, it is to be expected that the effects of ozone on these processes will influence other aspects of carbohydrate metabolism and the sizes of the pools of various carbohydrates that occur throughout the plant, independent of any direct effects that ozone may have on the enzymes that regulate individual pools *in situ*. As Koziol et al. (1988) point out in their recent review: "... metabolism is not a mosaic composed of interesting bits and pieces but rather a continuum of interrelated events." The key roles that carbohydrates play in the carbon and energy economies of the plant, therefore, also lead to effects on almost all other aspects of metabolism and growth.

Reference has already been made to the importance of rubisco and its carboxylase and oxygenase functions and the fact that some of the enzymes within the chloroplast are light modulated, notably glucose-6-phosphate dehydrogenase; light inactivation of this enzyme serves to prevent starch catabolism. Effects of ozone on such enzymes will therefore exert major influences on carbohydrate translocation.

Most investigations of carbohydrate metabolism have focused on the effects on pool sizes of the free sugars and storage polysaccharides, such as starch, rather than on the metabolically active phosphorylated intermediates. Nevertheless, pool sizes provide useful information about overall metabolic status. Among the reported effects of ozone on carbohydrate levels, a common finding has been a general decrease in free sugars and storage polysaccharides in roots (Jensen, 1981; McCool and Menge, 1983; Ito et al. 1985b), although the levels of individual sugars, such as fructose and sucrose, have been found to increase (Ito et al. 1985b). In shoots the responses are more varied: increases have been observed in pine needles (Barnes, 1972; Tingey et al., 1976) and lima bean (*Phaseolus lunatus*) seeds (Meredith et al., 1986); increases and decreases in bean (*P. vulgaris*) (Ito et al., 1985b); and decreases in soybean (Tingey et al., 1973b).

These varied responses probably reflect the combination of effects on biosynthetic and catabolic processes influenced by source-sink relationships, translocation, and partitioning (see Section 6.4.3).

6.2.10 Effects of Ozone on Nitrogen Metabolism

The close linkages of nitrite and nitrate reductases with the chloroplast, and of glycine and serine with the photorespiratory pathway, have already been alluded to. There have been several reports of ozone causing decreases in nitrate reductase activity (Tingey et al., 1973a; Purvis, 1978; Flagler et al., 1987). The reductions appeared to be caused by a decreased supply of precursors rather than by any direct effects on the enzyme.

In several species (particularly many important leguminous crops), root symbioses with bacteria (e.g., *Rhizobium*) and actinomycetes that are capable of fixing atmospheric nitrogen play an important part in overall nitrogen metabolism.

There are several reports of decreased nodulation of the roots of legumes caused by exposure to ozone, and more importantly in the present context, of reduced nitrogen-fixing activity (Jones et al., 1985; Flagler et al., 1987). However, Letchworth and Blum (1977) found no reductions in N fixation in ladino clover. Changes in nodulation and nodule activity are, of course, secondary effects, since N fixation is highly dependent upon the supply of photosynthate from the shoot, but their occurrence illustrates another means by which ozone can disturb nitrogen metabolism.

Reported effects of ozone on total nitrogen content are extremely variable, ranging from significant decreases to significant increases (Rowland et al., 1988).

Heath (1984) has presented a summary of the relationships between the major amino acid "families" and key intermediates of glycolysis, the pentose shunt, and the Krebs cycle. As a result of these relationships, effects of ozone may be manifested in changes in the pools of amino acids present in the plant. He concluded that, although little information existed about the effects of ozone on the levels of certain amino acids, there was a general pattern of increased pool sizes.

Polyamines derived from methionine and amino acids such as arginine have been reported to possess antiozone properties (Ormrod and Beckerson 1986) and can function as radical scavengers (Drolet et al., 1986). Ozone-induced increases in polyamine concentrations (Borland and Rowland; in Heath, 1988) may result from diamine oxidase inactivation by hydrogen peroxide (Mondovi et al., 1967; Peters et al., 1989).

The nucleic acids are important nitrogen-containing constituents, but apart from the study of Floyd et al. (1989) on chloroplast DNA, there is little known about the effects of ozone on their levels and functioning. Queiroz (1988) has reviewed the ramifications of changes that may affect gene expression and post-translational effects on enzymes. He has also raised the possibility of "pollutant protein" formation, analogous to the "shock" proteins synthesized by plant tissues in response to a range of environmental stresses (heat, salinity, drought, certain heavy metals), and the potential importance of protein phosphorylation/ dephosphorylation as a post-translational regulator of enzyme activities.

6.2.11 Effects of Ozone on Organic Acids and Lipids

Many organic acids are intermediates in the processes of respiration (Krebs cycle), photorespiration, and amino acid metabolism, while others are involved in lipid metabolism. In some tissues appreciable amounts of the Krebs cycle intermediates, such as malic or citric acids, accumulate. Acetyl-CoA derived from glycolysis is the main starting point for the biosynthesis of the fatty acids that occur in lipids.

Because of the key role of lipids in membrane structure and function and the potential for their direct chemical reactions with ozone, much early work was directed towards relating plant injury to lipid decomposition and the formation of malondialdehyde (MDA), as discussed above in Section 6.2.1.

An emphasis on lipid analysis persists in more recent studies; there appear to be no reports of investigations of the effects of ozone on the specific enzyme systems involved. Various lipid and fatty acid fractions have been shown to undergo changes as a result of exposure to ozone. Crude fat in tall fescue was decreased (Flagler and Youngner, 1985), and lipid oxidation in spinach was inferred from increased MDA levels (Sakaki et al., 1983). A decrease in wheat leaf phospholipids was accompanied by an increase in free fatty acids (Mackay et al., 1987). However, during exposures lasting up to 8 h, ozone had no effect on the glycolipids and unsaturated fatty acids of either bean or morning glory (*Pharbitis nil*) (Nouchi and Toyama, 1988). But in morning glory, phospholipids were initially decreased, and significant decreases in glyco- and sulfolipids were observed after 24 h. An increase in sterols was observed in treated soybean leaves (Grunwald and Endress, 1985).

Few studies have been undertaken on storage lipids in seeds, but Grunwald and Endress (1988) found that ozone increased the oil content of soybean seeds largely as a result of increased amounts of linoleic and stearic acids.

6.2.12 Effects of Ozone on Secondary Metabolism

The secondary metabolites of plants constitute an immense and diverse array of different chemical types: phenolics, flavonoids, alkaloids, terpenoids, betalains, glucosinolates, to name but a few. The group includes pigments, volatiles (including ethylene), endogenous plant growth regulators, and many com- pounds that appear to play roles in plant-insect relations. Because of the diversity of their chemical structures, which may contain nitrogen and sulfur, their biosyntheses follow many different pathways, but a feature of many secondary constituents is that they accumulate in mature or senescing tissues in which their synthesis outstrips their breakdown.

The effects of ozone on the levels of phenolic compounds have received the most attention, possibly because of their known involvement with enzymatic browning reactions and the frequent observation of pigmentation as a visual symptom of ozone injury. Early observations were that ozone caused significant increases in the levels of anthocyanin pigment in *Rumex* leaves (Koukol and Dugger, 1967) and of caffeic acid in bean leaves (Howell, 1970). In a subsequent paper (Howell and Kremer, 1973) it was shown that initiation of a brown-pigment complex containing protein, sugar, and an *o*-diphenol occurred in bean leaves within 2 h of the start of exposure to 0.16 ppmv ozone, long before the appearance of any clearly visible injury symptoms.

Keen and Taylor (1975) observed that 2-h exposures of soybean plants to ozone concentrations, ranging from 0.4 to 0.8 ppmv, led to injury and progressive increases in the levels of the isoflavonoids, coumestrol, daidzein, and sojagol. However, Skarby and Pell (1979) and Hurwitz et al. (1979) found no evidence of coumestrol in alfalfa leaves, but observed that "at least seven fluorescent compounds accumulated," one of which was identified as 4',7-dihydroxy flavone, a compound undetectable in untreated foliage.

In each case, the conclusion was drawn that the increases in phenolics resulted from the activity of polyphenoloxidases acting upon precursors released as a result of membrane breakdown. In some instances, the changes were detectable without the subsequent appearance of visible injury symptoms.

6.3 EFFECTS OF OZONE ON THE PHYSIOLOGY OF PLANT CELLS AND TISSUES

6.3.1 Membrane Reactions

An important feature of plant cell structure is compartmentalization. Not only does the cell contain distinct organelles, such as the chloroplasts and mitochondria, and distinct structures, such as the vacuole, but the organelles themselves are also compartmentalized. Membranes separate the compartments, and other membranes, such as those of the endoplasmic reticulum and the Golgi bodies, provide a further degree of organization within the ground cytoplasm.

One consequence of the existence of these extensive membrane structures, already alluded to in Section 6.2.1, is that it becomes necessary to think of many of the chemical reactions that occur in the cell in terms of semisolid membrane-surface reactions rather than in terms of classical solution kinetics. Since many cellular enzymes are associated with or bound to membranes, similar considerations apply to the reactions that they catalyze.

An important physical feature of membranes is their semipermeability to many solutes, and this property is frequently accompanied *in vivo* by mechanisms that permit energy-requiring reactions to occur (for example the active transport of solutes from one compartment to another) against a concentration gradient. Conversely, solute concentration gradients across membranes may be harnessed to convert their energy to chemically useful forms, such as is seen in the generation of ATP in the chloroplast and mitochondrial membranes.

The plasmalemma that surrounds the cell is particularly vulnerable; it separates the apoplast and symplast and is the first membrane likely to be encountered by ozone following its entry into the leaf. Hence, the reactions of ozone with its protein and lipid components, such as those discussed in Section 6.2.1, will have far-reaching consequences for the plasmalemma's physicochemical functions. These in turn will affect intracellular activities and may lead to effects at the tissue level.

In three recent reviews, Heath (1987a,b) and Heath and Castillo (1988) have described some of the experimental evidence for ozone-induced modifications to the plasmalemma and its functioning. Heath (1987a) also makes the important distinction between chronic responses, which may be reversible, and acute responses, which are not. The initial acute response is the appearance of waterlogging within the palisade mesophyll tissues of exposed leaves, indicating a breakdown of membrane function and leakage into the apoplastic space.

Increases in plasmalemma permeability have been observed both with isolated algal cells (*Chlorella sorokiniana*; Heath and Frederick, 1979) using $^{86}Rb^+$ as a tracer for potassium fluxes, and in leaf tissues (Evans and Ting, 1974). The release of a nonionic compound such as deoxyglucose occurs at somewhat less than half the rate of $^{86}Rb^+$ in *Chlorella* cells, indicating that both depolarization and permeability of the membrane are affected (Schwab and Komor, 1978). If the ozone treatment is prolonged or at a high enough level, these changes are irreversible.

Leakage of ions may affect the pH gradient across a membrane. Conversely, any change in bulk pH caused by reactions of ozone outside the cell can influence the normal transport of ions and metabolites. For example, both influx and efflux of $^{86}Rb^+$ are increased by increasing pH (Heath, 1987a), while uptake of deoxyglucose is decreased (Schwab and Komor, 1978). Heath (1987a) has calculated that for *Chlorella*, with a wall space of 3.38×10^{-12} cm^3 and a buffer capacity of 20, a pH change from 6.5 to 9.5 would require the release of only 2 $\times 10^{-17}$ equivalents of hydroxyl ion, a situation that is well within the range of possibilities, based on the reactions of ozone during its dissolution in water.

Ozone can also alter the physical condition of membranes. Pauls and Thompson (1981) observed that treatment of microsome membranes increased the temperature of the transition from liquid-crystalline to gel-liquid phase. There was a concomitant decrease in sterol and unsaturated fatty acid contents in relation to total phospholipids, a situation similar to that occurring during senescence.

In the few studies conducted directly on membrane proteins and enzymes, Dominy and Heath (1985) and Finchk and Heath (in Heath, 1987a) found that ozone inactivated the Mg^{2+}-dependent and K^+-stimulated plasmalemma ATPases that are thought to be associated with the ion pumps on the membrane, presumably by oxidizing sulfhydryl groups.

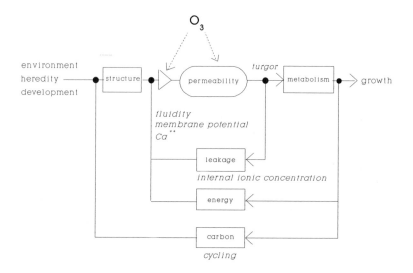

Figure 6.8. Control diagram for membrane function(adapted from Heath and Castillo, 1988).

6.3.2 Permeability and Leakage

Heath and Castillo (1988) have applied the principles of control theory (Gaines, 1958) to plant membrane function. They have viewed the membrane as a "controlled system" with its associated "control elements" and "feedback elements," as depicted in Figure 6.8. The system can be disturbed directly or through changes to the set-point. In the latter case, if the set-point is close to the conditions favoring maximum throughput, increased input may "saturate" the system, rendering it unable to respond coherently.

In their membrane model, active transport is the controlled system, and leakage is feedback. The set-point is defined by heredity, stage of development, and environmental factors. The output (leading to growth) consists of ion and metabolite transport and metabolism and includes changes in cell turgor and water relations. Such a view facilitates differentiating between responses to brief exposures to high ozone levels that lead to disruption of the system and its control elements directly, and responses to prolonged, low-level exposures that merely change the set-point and may lead to acclimation or hardening and increased tolerance.

Heath and Castillo (1988) stress the potential importance of membrane disturbances that affect the transport of K^+ and Ca^{2+} ions, because of their important regulatory roles in a wide range of cellular functions. They point out that the cell expends considerable energy in maintaining low levels of free Ca^{2+} in the cytoplasm ($< 10^{-6} M$) vs. higher levels outside the plasmalemma (in the cell wall) and within organelles such as the mitochondria and the vacuole. Studies with isolated membrane fraction vesicles from bean leaves (mixed plasmalemma, endoplasmic reticulum, and tonoplast fractions) showed that ozone pretreatment increased the ATP-dependent uptake of $^{45}Ca^{2+}$ significantly. However, the vesicles also showed a much greater passive release (leakage) of Ca^{2+} into the medium than those from control plants, indicating a breakdown in the ability of the membranes to maintain the normal Ca^{2+} balance.

They therefore suggest that ozone may cause an increase in free Ca^{2+} in the cytoplasm. Minor perturbation would probably initiate restorative processes, such as stimulation of the Ca^{2+} pumps to reestablish Ca^{2+} balances. However, if excessive free Ca^{2+} levels occur, the recovery processes would be overwhelmed, leading to irreversible and fatal changes. They further speculate that Ca^{2+} imbalance may disturb the normal functioning of scavenger systems for ozone and its products. For example, peroxidases are activated by Ca^{2+}, and as discussed in Section 6.2.4.2, Castillo et al. (1984) observed ozone-induced increases in extracellular peroxidase activity. Castillo and Greppin (1986) subsequently reported that the anionic and cationic peroxidase isoforms behaved differently. Initially the anionic form (using syringaldazine as an electron donor) was unaffected, while activity of the cationic form (with ascorbate as donor) more than doubled. After 24 h, the increased activity of the cationic form was only 25%, but that of the anionic form was almost threefold. Although the evidence is still somewhat fragmented, Heath and Castillo (1988) have suggested that the increases in extracellular peroxidases are the consequence of greater secretion from the cell caused by an increase in cytoplasmic Ca^{2+} levels, which is, in turn, the result of the cell membranes' inability to maintain normal Ca^{2+} balances.

The potential vulnerability of the histidine residues in cell wall proteins to oxyradical attack (Cassab and Varner, 1990) also has implications with regard to pH-mediated changes in Ca^{2+} levels in the cell wall.

Although Heath and Castillo (1988) suggest a crucial role for Ca^{2+} ion balances because of their potential for regulating apoplast-symplast relations and potential effects on apoplastic scavenging, other ions and metabolites may also be involved. There is certainly abundant evidence at the tissue level for the ozone-induced breakdown of general membrane integrity in leaf cells, as revealed by an increased ability to leach a range of ionic and nonionic solutes from treated leaves (Barnes et al., 1988; Beckerson and Hofstra, 1980; Dijak and Ormrod, 1982; McKersie et al., 1982b; and others).

Thus, although changes induced by ozone may involve specific chemical reactions that exert direct effects on metabolic processes, it is important to recognize that many effects may be mediated by changes in membrane structure and function that profoundly influence the physiology of the affected cells, the tissues in which they occur, and, hence, the plant as a whole. Plant responses that result in visual changes in foliar characteristics are secondary processes of ozone toxicity, which appear after initial defense mechanisms are overrun. Cellular biochemical and physiological alterations occur without such visible injury symptoms appearing, and these modifications affect critical metabolic functions capable of limiting oxidative stress and ozone toxicity both directly and via more complex physiological interactions within the cell. It is the integrated cellular system that confers and determines plant sensitivity to ozone.

6.4 EFFECTS OF OZONE ON PLANT PHYSIOLOGY AND GROWTH

6.4.1 Stomatal Physiology

The importance of stomatal aperture in the flux of ozone into the leaf has been discussed in Chapter 5. Under field conditions, 80% or more of the total resistance to ozone flux resides in the boundary layer and stomatal resistances. Reich (1987) has suggested that since leaf gas-phase resistance controls uptake nearly exclusively at typical ambient ozone levels (0.08 to 0.15 ppmv), differences in sensitivity among species could be explained by differences in leaf resistance. However, there are several lines of evidence (discussed in Chapter 5) that indicate that this is, at best, a first approximation. While stomatal resistance is unquestionably important, mesophyll reactions related to internal resistance and metabolic processes appear to be equally important factors, in several species, for regulating flux and for determining toxicity and plant sensitivity (Bicak, 1978; Elkiey et al., 1979; Coyne and Bingham, 1982).

Stomatal action is controlled by several interacting factors: CO_2 concentration in the substomatal cavity; leaf water status; functional integrity of the subsidiary cells of the epidermis; and fluxes of ions (especially K^+) and water, regulated by the phytohormones, abscisic acid and indolylacetic acid (Mansfield and Freer-Smith, 1984). Hence, although stomatal closure may appear to provide a mechanism for minimizing uptake and the subsequent adverse effects of ozone (i.e., a typical "stress-avoidance" mechanism) (Levitt, 1972), such action is itself stressful, since it limits CO_2 uptake, although it also reduces transpirational water loss (Winner et al., 1988). At the same time, the interacting factors make it difficult to differentiate between direct effects of ozone on the guard, or subsidiary, cells and indirect responses such as those resulting from effects on photosynthesis that influence internal CO_2 levels (Winner et al. 1988). However, Sheng and Chevone (1988) have reported that changes on soybean leaf stomatal resistance occurred subsequent to detectable decreases in net photosynthetic gas exchange. Numerous other investigators have failed to provide any evidence to support a direct effect of ozone on guard cell function. There is, therefore, general support for the concept that ozone flux into a leaf is controlled by the interactive effects of photosynthetic rate and internal CO_2 concentration on stomatal resistance.

The numerous reports of effects of ozone on stomatal response to ozone include the early suggestion by Engle and Gabelman (1966) that the difference in susceptibility of two onion cultivars to acute injury was explained by stomatal closure in the tolerant cultivar, as a result of rapid loss of turgor of the guard cells. A similar situation has been reported for bean (*Phaseolus*) cultivars (Butler and

Tibbetts, 1979). Darrall (1989) has reviewed the subject of stomatal response and noted that, at ozone levels less than 0.2 ppmv, stomatal opening was reported in two species, closure in four, with a further species showing no effect. At higher concentrations, stomatal closure was the usual response.

However, the pattern of stomatal behavior is not solely influenced by the current ozone level, since the sensitivity of the stomatal response varies with stage of leaf development and its previous history of exposure (Runeckles and Rosen, 1977).

The water status of the leaf and the ambient relative humidity also exercise control over stomatal response to ozone. Stressed bean (Rich and Turner, 1972) and soybean (Reich et al., 1985) plants were found to exhibit a more rapid stomatal closure than unstressed plants, although Temple (1986) observed no such effects in field-grown cotton. The corollary is, of course, that since the stomata provide the principal route for the exchange of all gases with the ambient air, direct or indirect effects of ozone on stomatal aperture will affect transpiration and, hence, water-use efficiency. At high relative humidities there is less ozone-induced stomatal closure than at lower humidities (Elkiey and Ormrod, 1979; Rich and Turner, 1972).

Stomatal response is also influenced by the levels of endogenous abscisic acid (ABA). Kondo and Sugahara (1984) observed that, in species with high ABA levels (peanut, tomato), 0.5 ppmv ozone caused rapid stomatal closure, whereas in species with low ABA levels (broad bean, corn, radish, spinach), closure only commenced after a distinct lag period. Adedipe et al. (1973) had earlier shown that application of exogenous ABA to tomato leaves resulted in stomatal closure and reduced ozone-induced injury.

6.4.2 Gas Exchange Functions

6.4.2.1 Photosynthetic Gas Exchange

Most studies of the effects of ozone on photosynthesis have measured rates of net photosynthesis, P_N, the resultant of photosynthetic carbon assimilation and photorespiration. Few studies have determined the effects on gross (true) photosynthetic carbon assimilation, P_G, based on such measurements as the uptake of ^{13}C- or ^{14}C-labeled CO_2. Even fewer have determined effects on P_G in terms of O_2 production.

The general finding has been that ozone inhibits photosynthetic gas exchange, although the degree of inhibition is highly dependent upon the concentration of ozone, the duration of the exposure, and the species or cultivar tested. Although much early work used atypically high ozone concentrations, the recent reviews by Miller (1987, 1988) and Darrall (1989) emphasize work undertaken using exposure regimes that are more appropriate for many regions of the world. A sampling of recent results using a variety of techniques is presented in Figure 6.9. In viewing such a composite, it is important to note that the results were obtained using different facilities and experimental protocols for providing exposures to ozone, ranging from controlled environment chambers to open-top field cham-

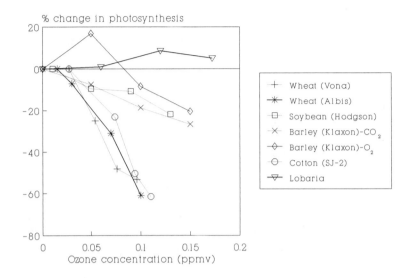

Figure 6.9. Changes in photosynthetic rates resulting from exposure to ozone. Heavy lines: gross photosynthesis, P_G (^{14}C-uptake); solid line: O_2-evolution; dotted lines: net photosynthesis, P_N. Experimental details: "Vona" wheat, P_N, measured 1200–1500 h daily from anthesis to maturity, ozone expressed as daily 7-h seasonal means (Amundson et al., 1987); "Albis" wheat, P_G, measured at anthesis, ozone expressed as daily 8-h seasonal means (Lehnherr et al., 1988); "Hodgson" soybean, P_N, measured at midday after 4-weeks exposure, ozone expressed as daily 7-h seasonal means (Reich et al., 1986); "Klaxon" barley, P_N and O_2-evolution, measured after 12-d exposure, expressed as daily 7-h seasonal means (Rowland-Bamford et al., 1989); "SJ-2" cotton, P_N, measured on youngest fully expanded leaves after 3 months' exposure, expressed as daily 12-h seasonal means (Temple et al., 1988b); *Lobaria pulmonaria*, P_G, measured after 4-h exposure, expressed as 4-h means (Sigal and Johnston, 1986).

bers, with different types of diurnal exposure profiles. Furthermore, normalization of the ozone exposures could only be achieved using 1-h or various seasonal daily average concentrations in spite of the fact that such averages have been recognized as providing poor summarizations of the actual dosages received by foliage (Lefohn et al., 1989; Tingey et al., 1989).

The responses depicted in Figure 6.9 clearly show ozone-induced decreases in the photosynthetic rates of the crop species, regardless of the type of measurement. On the other hand, the algal symbiont of the lichen *Lobaria* appears to be unaffected by the ozone concentrations used (the apparent increase in P_G shown in Figure 6.9 is not statistically significant). Stomatal response is not a factor to be considered in this case, nor in the case of fronds of duckweed, *Lemna*, in which the stomata are open and nonfunctional. Forberg et al. (1987) observed no impairment of net photosynthetic CO_2 uptake by fronds of *Lemna* until the exposure to ozone exceeded about 0.25 ppmv for 1 h. The apparent stimulation of O_2 evolution in "Klaxon" barley at low ozone exposure should also be noted.

From the earliest studies it was recognized that the effects of ozone on CO_2 exchange rates could be modified by numerous environmental variables (light, humidity, ambient carbon dioxide concentration, other gaseous pollutants) and other biological factors (leaf age, plant water status, respiration rate) and that it was necessary to distinguish between the effects of short-duration exposures to high ozone levels and to prolonged or repeated exposures to lower levels (Miller, 1987; Reich, 1987; Darrall, 1989). In particular, at low exposure levels, reductions in CO_2 exchange rates are usually only observed during the exposure to ozone, with almost full recovery afterwards. At higher levels, (e.g., greater than 0.2 ppmv for 4 h in the case of broad bean, *Vicia faba* [Black et al., 1982]), lack of subsequent recovery indicated permanent damage to the photosynthetic system. However, repeated daily (7.5 h) exposures of soybean to 0.067 ppmv ozone were found to have progressively less inhibitory effect on whole plant P_N, indicating the development of tolerance (Le Sueur-Brymer and Ormrod, 1984).

The characteristics of the exposure regime are important determinants of the magnitude of the photosynthetic response to ozone. Although the ozone concentration and the duration of the exposure are significant components of the fumigation profile, interest at present is focused on aspects of the exposure profile that relate to temporal and dynamic fluctuations of the ozone concentration. These features include the frequency and extent of peak concentrations, the rate of the concentration increase and decline, and the duration of respite (recovery) periods between peak exposures.

The length of the exposure period, from a biological perspective, has two primary functional phases: (1) short term or diurnal; and (2) long term or seasonal. A short term fumigation typifies the peak segment of a summertime diurnal ozone cycle, where concentrations are increased for 2 to 4 h (or more) from lower, nocturnal levels. The maximum concentrations generally occur from midday to early afternoon (1200 to 1400 h; see Chapter 5). Plant response to this type of exposure profile should favor metabolic processes that utilize endogenous metabolite or enzyme pools for ozone detoxification or repair of damaged chloroplast function. While *de novo* synthesis of substrates or proteins may occur during the later stages of short-term exposures, this response is probably more typical of longer-term, lower-concentration exposure regimes. Under growing-season exposure profiles, plant response may involve various metabolic processes that enhance the photosynthetic capacity of the plant to accommodate oxidative stress.

Jones (1985) has reviewed current techniques to calculate stomatal and nonstomatal limitations to photosynthesis and has presented various methods, with their advantages and shortcomings, for partitioning contributions to a change in the photosynthetic rate. Kropff (1987) selected path-dependent analysis to investigate the initial effect of sulfur dioxide on photosynthesis and stomatal regulation in *Vicia faba* L. He determined that mesophyll resistance to CO_2 increased first, and stomatal resistance subsequently changed to accommodate reduced CO_2 assimilation.

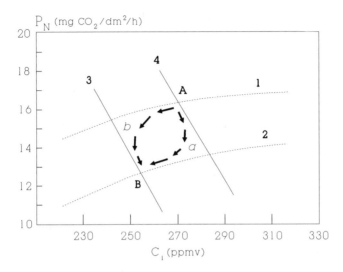

Figure 6.10. Path-dependent changes in net photosynthesis (P_N) and intercellular CO_2 concentration (C_i) when initial P_N limitation results from (a) increased mesophyll resistance, or (b) increased stomatal resistance.

Path-dependent analysis may be applied to ozone exposure studies and can be useful in determining initial photosynthetic limitations. Figure 6.10 represents hypothetical portions of two photosynthetic response curves at light saturation, as C_i, the internal partial pressure of CO_2, increases. Curve 1 describes photosynthesis in unpolluted air, whereas curve 2 defines a reduced steady-state photosynthetic rate resulting from exposure to ozone. At point **A**, the photosynthetic rate is determined by the slopes of the supply function (line 4), which is the gas-phase resistance to CO_2, and the demand function (a line tangential to the P_N curve at point **A**), which is the mesophyll resistance to CO_2 (Raschke, 1979). At point **B**, the photosynthetic rate is again determined by the new supply function (line 3) and the new demand function.

As photosynthesis decreases from **A** to **B**, a time course analysis of the relationship between P_N and C_i can be used to analyze the limitations to the rate change. If mesophyll resistance increases first, (e.g., as a result of the oxidation of sulfhydryl residues on rubisco), P_N will decrease because of reduced CO_2 fixation. Since the stomata will not be affected directly by ozone in this instance, C_i will increase as P_N decreases, following path *a* in Figure 6.10. The initial rise in C_i will then result in an increase in stomatal resistance through the feedback loop between P_N and C_i. As P_N falls to level **B**, the CO_2 supply and demand functions will change until they attain the equilibrium values at point **B**. In this path the P_N limitation is controlled entirely by mesophyll resistance to CO_2, the stomatal limitation being a passive consequence of the relationship between P_N and C_i.

Ozone, however, may directly affect the stomata first, causing an immediate increase in gas-phase resistance to CO_2. Under these conditions, the net CO_2 fixation rate remains unchanged initially and therefore, C_i will first decrease, and only then will P_N decline. The path followed in this situation will be as shown in b in Figure 6.10, and the P_N limitation will be controlled by the stomatal resistance.

Ozone may affect both stomatal and mesophyll functions simultaneously, in which case the path between points **A** and **B** will be more direct than either a or b. Situations in which the impairment of gross photosynthesis is not accounted for by reduced g_s and increased C_i have been described by Lehnherr et al. (1988) and others. The range of possibilities is shown by the studies of Furukawa et al. (1984), who found that ozone-induced reductions in P_N and transpiration rate, **E** (reflecting stomatal closure), in sunflower and two poplar species (hybrids) were linearly related and comparable in magnitude, while in a third poplar hybrid, P_N was reduced with little decrease in **E**. In this last case, stomatal conductance was found to be unchanged by ozone and the limitations on P_N were attributable almost exclusively to increased mesophyll resistance.

Path-dependent analysis may also be used to study controlling points in the recovery of photosynthesis during stress release after exposure. The P_N operating point will rise from **B** to **A** (Figure 6.10) along path a or b, depending upon the primary limitation that is initially removed. If mesophyll resistance is decreased first, C_i will initially decrease as P_N rises, and the path followed will be the reverse of b. Alternatively, if stomatal resistance decreases first, then the P_N–C_i path will be the reverse of a.

The accuracy of C_i curves for determining limitations to P_N has been questioned recently in light of observations that stomata do not close uniformly under such conditions as drought stress, diminished irradiance, or exogenously applied abscisic acid (Laisk, 1983; Downton et al., 1988a,b). This "patchiness" of stomatal apertures may lead to significant errors in estimating mesophyll resistance, especially at low, mean leaf conductances when apertures may vary over a wide range. If diffusional resistance between adjacent substomatal cavities is low, then the magnitude of the error lessens, since C_i will be more uniform within the leaf. At present, patchy stomatal closure resulting from ozone exposure has not been demonstrated, although it may be a contributor to the observed patterns of ozone-induced leaf pigmentation or necrosis. Nevertheless, this variability in stomatal behavior must be taken into account in future studies of ozone limitations to P_N.

Other types of analysis can be used to assign initial limitations to P_N under ozone exposure (Jones, 1985). The method used is not as critical, however, as the information that can be obtained from such studies. The primary tolerance to ozone of a plant species, or genotype within a species, may reside in either the mesophyll or stomatal response. A knowledge of the limitation to P_N would be most valuable in directing physiological and biochemical investigations that are designed to further understand the pollutant mode of action, detoxification mechanisms, and the basis for differential genotypic sensitivity.

In a few investigations, the effects of ozone on the light- or CO_2-dependence of photosynthetic gas exchange have been studied. Light-response information permits the estimation of the quantum yield of photosynthesis (ϕ; mol CO_2 absorbed or O_2 evolved/Einstein). Rowland-Bamford et al. (1989) observed that ϕ decreased linearly with increasing ozone concentration in barley *during* exposure. On the other hand, Atkinson et al. (1988) reported that *after* 21 daily exposures, ϕ of radish leaves in ozone-free air was found to increase with ozone treatment. The former result was interpreted as indicating that, during exposure, the quantum requirement probably increased as a result of impairment of the photosynthetic electron chain. The latter observation was interpreted as reflecting the fact that the extended ozone exposures had resulted in the development of thinner leaves. Nonlinearity of the relationship between the degree of ozone-induced inhibition of P_N and light intensity has been observed, with the greatest effects appearing at low and at high intensities. Darrall (1989) has suggested that this reflects the limited energy available for repair processes at low intensities, on the one hand, and the increased endogenous production of potentially damaging oxyradicals at high intensities, in combination with those derived from ozone, on the other.

In suspension cultures of isolated soybean leaf mesophyll cells, Omielan and Pell (1988) found that ozone reduced P_G more than cell viability, indicating that reduced photosynthesis occurred in the absence of any constraints imposed by stomatal action and suggesting that cellular injury was not dependent upon active photosynthesis.

There appear to have been no studies directly related to the effects of ozone on photorespiratory CO_2 production, in spite of the importance of the process in C-3 plants. However, the observations that exposure to ozone leads to increases in glycine and serine pool sizes in bean (Ito et al., 1985a) and white clover (Johnson, 1984; cited by Miller, 1987) may be interpreted as indicating a stimulation of photorespiration.

In summary, it is apparent that although there are many independent observations to the effect that ozone causes transient or permanent inhibitions of photosynthetic rates, the diversity of responses reported for different species (and cultivars), environmental conditions, and experimental procedures for conducting the exposures to ozone has yet to lead to a clear picture of the nature of the inhibition and its mechanism.

This results, in part, from our inability to relate short-term responses (which are usually reversible provided that the level of ozone stress is not excessive) to long-term, chronic responses that may include some degree of adaptation, and, in part, from the interaction between effects on P_G, photorespiration, and stomatal function (discussed in Section 6.4.2.2, below). It is nevertheless clear that in chronic exposure situations, impaired photosynthetic ability is closely related to accelerated senescence (including decreases in chlorophyll content) and decreased overall growth (Miller, 1987).

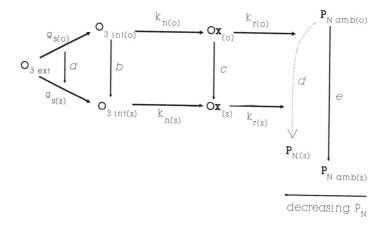

Figure 6.11.   Schematic model representing interactive processes involved in an ozone-induced steady-state inhibition of net photosynthesis. $O_{3\,ext}$ = ambient ozone concentration external to the leaf; $O_{3\,int}$ = intercellular leaf ozone concentration; Ox = ozone or metabolites that inhibit P_N; $P_{N\,amb}$ = net photosynthetic rate in ambient air containing no ozone; g_s = stomatal conductance to ozone; k_n = net accumulation rate of ozone or toxic metabolites; k_r = inhibition rate of Ox on P_N; T = time; (o) subscripts represent initial events in ozone fumigation; (s) subscripts represent steady-state events.

6.4.2.2 Interaction of Photosynthesis and Stomatal Conductance

There are several reports that indicate that reductions in P_N cannot be accounted for solely by stomatal closure and its consequences (Furukawa et al., 1984; Reich and Lassoie, 1984; Reich and Amundson, 1985; Reich et al., 1985; Lehnherr et al., 1988). Nevertheless, the development and maintenance of a reduced steady-state photosynthetic rate observed during short-term ozone exposures (for example, barley, Rowland-Bamford et al., 1989; eastern white pine, Yang et al., 1983; and, less markedly, soybean, Le Sueur-Brymer and Ormrod, 1984) suggests the attainment of an equilibrium condition between ozone influx and P_N inhibition.

Figure 6.11 represents a model system that summarizes the physiological and biochemical processes that are likely to be important in the development of such an equilibrium state. The initial entry of ozone and the subsequent interference with carbon assimilation are depicted by the two series of solid arrows. In the upper sequence of events, ozone influx to the intercellular spaces ($O_{3\,int(O)}$) is controlled by the initial stomatal conductance [$g_{s(O)}$] and the external ozone concentration ($O_{3\,ext}$), as discussed in Chapter 5.

Within the leaf tissue, ozone molecules can undergo a variety of physico-chemical interactions with cellular components, as discussed in Section 6.2. Some of the reactions result in the production of less damaging, or nontoxic, metabolites (e.g., water or molecular oxygen), whereas other reactions may generate chemical entities as toxic as ozone or more so (e.g., hydrogen peroxide, hydroxyl radical).

The mesophyll flux of ozone and toxic metabolites $(Ox_{(o)})$ is therefore controlled by a number of reactions that lead to either an accumulation or removal of chemical species that can interfere with photosynthesis. The net rate of accumulation of $Ox_{(o)}$ ($k_{n(o)}$) is the resultant of those processes producing toxic substances and those involved in detoxification.

As $Ox_{(o)}$ accumulates, the rate of photosynthesis in unpolluted air $(P_{N\,amb(o)})$ declines along curve d. The characteristics of $P_{N\,amb}$ reduction depend upon the mechanism(s) involved (e.g., rubisco inactivation, hydrogen peroxide inhibition of Calvin cycle sulfhydryl enzymes, thylakoid membrane disruption, and impaired electron flow) and the net rate of the inhibition process(es) $(k_{r(o)})$. The rate of inhibition also depends upon the cell's (and its organelles') capacity for repair of biochemical and cytological lesions.

The decrease in CO_2 fixation results in an increase in C_i and a subsequent change in stomatal opening through feedback. Stomatal conductance will therefore gradually decrease (arrow a) to a new steady-state level $(g_{s(s)})$. The intercellular ozone flux will also diminish (arrow b) from $O_{3\,int(o)}$ to the equilibrium rate, $O_{3\,int(s)}$. Assuming that $k_{n(o)}$ remains unchanged, the mesophyll flux of toxic compounds will then decrease (arrow c) from $Ox_{(o)}$ to $Ox_{(s)}$. The decreased flux $(Ox_{(s)})$ will establish a new net inhibition rate $(k_{r(s)})$, and net photosynthesis will attain an equilibrium rate at $P_{N\,amb(s)}$, determined by both $g_{s(s)}$ and $k_{r(s)}$.

The development of a reduced, steady-state $P_{N\,amb(s)}$ rate, caused by ozone exposure, can therefore be viewed as a series of interactive processes whereby ozone influx is regulated by the inhibition of net photosynthesis, the increase in C_i, and the subsequent decrease in g_s.

The model is useful in providing an explanation for differences in tolerance and response. For example, in spite of the importance of leaf (stomatal) conductance, the model provides a mechanism whereby mesophyll processes can exercise a major role in regulating ozone influx and in controlling cytoplasmic sensitivity to ozone, depending upon the magnitude of the internal flux of ozone and its products. Recall that Taylor et al. (1982) found that mesophyll resistance to ozone is minimal at low ozone concentrations and low internal flux rates, but increases at higher flux rates, suggesting a finite capacity of the mesophyll to remove (metabolize) toxic materials. This capacity is genotype dependent and can change with leaf age and probably with other biochemical and physiological conditions.

Differences in the tolerance of net photosynthesis to ozone can be viewed as changes in the component functions depicted in Figure 6.11. A tolerant genotype may possess a relatively low $g_{s(s)}$, resulting in low intercellular flux of ozone $(O_{3\,int(s)})$. In addition, high endogenous concentrations of ascorbate (Lee and Bennett, 1982), glutathione (Guri, 1983), or catalase (Lee and Bennett, 1982), which are all scavengers of ozone or toxic oxygen species (Mudd, 1982; Salin, 1988), may favor a low accumulation rate (k_n) of intermediate toxic metabolites (Ox). The concentration of Ox may never rise sufficiently to surpass the maintenance/repair capacity within the chloroplast, and net photosynthesis will remain unaffected by ozone exposure.

A further consideration is that ozone entry into the cell should be an energy-consuming process, even if photosynthesis is not impaired. Energy expenditure may occur, for example, as increased dark respiration or utilization of NADPH to regenerate compounds that scavenge ozone or other toxic oxygen species. It may be that short-term impairment of net photosynthesis represents an acute response to ozone, in that the metabolic capacity of the cytoplasm and chloroplast is exceeded before physiological and biochemical processes can react to limit ozone uptake and increase the capacity of the endogenous scavenging/detoxification mechanisms. And it must again be emphasized that the model presented in Figure 6.11 does not differentiate between effects of ozone or its products on the individual processes of gross photosynthesis and photorespiration, each of which is a contributor to net photosynthetic CO_2 fixation.

Another approach to differentiating between direct effects on P_N and those resulting from stomatal changes utilizes the measurement of the discrimination against the uptake of $^{13}CO_2$ relative to $^{12}CO_2$. The ability of rubisco to fix $^{12}CO_2$ preferentially results in negative $\delta^{13}C$ values (typically –2.2 to –4.0%) in the photosynthetic products of C-3 plants (Troughton, 1979). The discrimination decreases with decreasing C_i, such as would occur with decreased g_s in the absence of reduced P_G. Greitner and Winner (1988) reported small but significant increases in $\delta^{13}C$ (i.e., to less negative values) in radish (cv Cherry Belle) and soybean (cv Williams) plants exposed to 0.12 ppmv ozone. They interpreted their findings as indicating that ozone treatment resulted in decreased C_i as a result of stomatal closure. Similar findings have been reported for a range of species by Martin et al. (1988) as a result of exposures to mixtures of ozone and SO_2, and ozone, SO_2, and NO_2.

Assumed close relationships of P_N and stomatal conductance formed the basis of the models of leaf response developed by Schut (1985). These models of short-term leaf response to ozone were based on the published studies of Hill and Littlefield (1969), Larsen and Heck (1976), Elkiey and Ormrod (1979), and Black et al. (1982). Schut assumed that stomatal regulation maintained C_i constant; that ozone uptake, CO_2 uptake, and stomatal conductance were proportional, and that the inhibition of P_N was proportional to ozone uptake. The introduction of a threshold concentration of ozone below which no effect on P_N could be detected (Black et al., 1982), and the incorporation of "repair" functions (either constant or P_N dependent), led to the development of models that simulated the published responses of oats, field bean, Pinto bean, and Petunia with some success, including the prediction of acute injury.

6.4.2.3 Effects on Respiratory Gas Exchange

Early investigations of the effects of ozone on dark respiration led to conflicting reports, with both stimulations and inhibitions of CO_2 production in darkness. For example, although there is general agreement that, if exposure to ozone is sufficient to result in visible injury, respiration increases. Macdowall (1965) reported an initial decrease in respiration and oxidative phosphorylation in tobacco. On the other hand, Dugger and Ting (1970; citrus), Barnes (1972; pine), Pell and Brennan (1973; bean), Reich (1983; poplar), and others have consistently reported increases, although Reich et al. (1986) could detect no effects of prolonged exposures of soybean to ozone levels ranging from 0.05 to 0.13 ppmv on dark respiration. In an apparently unique study with nonphotosynthesizing tissues, Anderson and Taylor (1973) observed that the respiration of tobacco callus culture tissue was increased by ozone. The reported decrease in respiratory CO_2 production by the roots of ozone-treated bean plants (referred to in Section 6.2.8) is considered by Miller (1987) to be a consequence of reduced photosynthesis and translocation from the aerial parts of the plant.

However, Lehnherr et al. (1987) have reported that the dark respiration of flag leaves removed from wheat plants exposed to ozone for 18 or more days in open-top field chambers decreased significantly with increasing average ozone concentration.

Therefore, as in the case of photosynthetic rates, there appear to be differences between short-term and long-term responses. And similar to the ozone-induced changes in photosynthesis, the effects of short-term, low-level exposures on respiration rates are reversible. For example, Myhre et al. (1988) observed that although exposure of oats (*Avena sativa* L. cv Titus) to 0.13 ppmv for 2 h caused a doubling of the rate of CO_2 release, determined on immediately placing the plants in darkness, the stimulation steadily declined over the ensuing 6 h.

Effects on dark respiration are the consequences of direct or indirect effects on the mitochondrion and the enzymes of glycolysis, as discussed in Section 6.2.8, together with feedback effects related to the respiratory process's key role in the provision of energy for maintenance and repair. This latter feature has been recently studied in relation to plant growth and is discussed below in Section 6.4.3.

6.4.2.4 Effects on Transpiration

Ozone-induced changes in stomatal diffusive resistance will affect the efflux of water vapor as well as the exchange of other gases. Stomatal closure has the immediate effect of decreasing the transpiration rate, **E**, as has been shown in several studies: for example, Kondo and Sugahara (1984)

Stomatal closure induced by ozone may thus serve to conserve water as well as reduce ozone stress, by reducing ozone flux. Conversely, there is considerable evidence indicating that drought-stressed plants are less susceptible to ozone than are well-watered plants, because of their increased diffusive resistance, regardless of the mechanism responsible for stomatal closure.

Where **E** is reduced by stomatal closure, CO_2 flux is also reduced. However, the magnitudes of the fluxes of H_2O and CO_2 on a molar basis differ by at least two orders of magnitude, and the flux of H_2O changes proportionally more than that of CO_2 with changes in $\mathbf{g}_s$ (Nobel, 1983). Hence, ozone-induced decreases in $\mathbf{g}_s$ may result in increased water-use efficiency (WUE; usually expressed as g CO_2 per kg H_2O), as observed by Greitner and Winner (1988) in soybean leaves. In such cases, although WUE may increase and therefore appear to confer benefit in situations in which water supply is limited, concomitant reductions in CO_2 assimilation resulting from decreased flux of CO_2 will inevitably lead to reduced growth. In contrast, Reich et al. (1985) reported the opposite response in well-watered soybean, i.e., a decrease in WUE as a result of exposure to ozone. They argued that this resulted from the ozone-induced reduction of $\mathbf{P}_N$ exceeding the reduction in $\mathbf{E}$. Inspection of other data from the same experiment (Reich et al., 1986) reveals that the percent reduction of $\mathbf{P}_N$ (162 ppmv^{-1}) was considerably less than the average reduction of diffusive conductance (228 ppmv^{-1}), indicating that $\mathbf{P}_N$ was inhibited independently of any reduction that might have been caused by reduced CO_2 flux. Atkinson et al. (1988) found decreased WUE in individual radish leaves.

Temple and Benoit (1988) reported that exposure of alfalfa to ozone in the field resulted in a reduction of season-long WUE (determined as grams harvested of dry matter per millimeter of water used) averaging 12%. Concurrent measurements of $\mathbf{P}_N$ indicated inhibition, but the effect of the ensuing stomatal closure in preferentially reducing $\mathbf{E}$ was more than offset by the accelerated senescence and abscission of leaves that reduced biomass.

Since WUE is itself influenced by water availability, further experimentation is needed before a clear understanding emerges.

6.4.3 Effects on Growth

Plant growth results from the integration and coordination of the processes of photosynthesis, respiration, translocation, and biosynthesis, controlled by the plant's genetic composition and modified by environmental factors. Photosynthetic carbon gain cannot be equated directly with increase in biomass, because of the needs for maintenance respiration and the supply of respiratory energy for synthesis and turnover of constituents.

Growth can be readily determined as a gain in weight or size of the plant as a whole or of any of its parts, e.g., fruits, seeds, or tubers. Much of the information about the effects of ozone on growth is confined to above-ground biomass. In some cases, this is simply a reflection of the difficulties in recovering the whole root system from the soil, but in many cases the intent has been specifically to determine effects on the harvested yields of different crops, rather than on total biomass, as in the extensive series of field studies undertaken in the U.S. National Crop Loss Assessment Network (NCLAN) program (summarized in Heck et al., 1988). These and other studies have provided abundant evidence that exposure to increased ozone levels usually results in decreased growth and yield. The relationships between overall growth and yield (or their reductions) and ozone exposure are discussed in Section 6.4.4.

6.4.3.1 Effects on Translocation and Assimilate Partitioning

A reduction in growth is to be expected if exposure to ozone results in decreased photosynthesis, but the relative impacts on the growth of the different parts of the plant are the result of differential effects on the translocation of assimilates from the leaves, i.e., their distribution or partitioning. As a result, short-term estimates of assimilation can rarely be used to predict long-term effects on productivity, although some exceptions have been noted (Reich and Amundson, 1985).

Since growth is a long-term process, various types of multiplier effects are to be expected. For example, changes in the partitioning between the leaves and the rest of the plant frequently reduce both the size and longevity of individual leaves and thereby influence the plant's long-term capacity for photosynthetic carbon gain (Mooney and Winner, 1988).

A frequently observed whole-plant response to gaseous air pollutants in general is that root growth is reduced more than shoot growth. Ozone is no exception, and there are numerous reports of ozone-induced increased shoot/root (S/R) dry matter ratios. The lists compiled by Miller (1987) and Cooley and Manning (1987) together cover 26 examples of such S/R reductions taken from studies with 21 different crops; only with peppers (*Capsicum annuum*; Bennett et al., 1979), millet (*Panicum miliaceum*; Agrawal et al., 1983), and peanut (*Arachis hypogaea*; Heagle et al., 1983) was no effect or a decreased ratio observed, while in bean (*Phaseolus vulgaris*; Ito et al., 1985b) and annual ryegrass (*Lolium multiflorum*; Bennett and Runeckles, 1977a), effects were only observed at the highest exposure levels used.

In spite of its implied involvement in the differential growth responses of individual plant parts, few studies have been conducted on the effects of ozone on translocation per se. Sucrose is the major metabolite translocated via the phloem, following "phloem loading," a process that requires ATP and involves the cotransport of sucrose and K^+ ions (Giaquinta, 1983). It also requires functional integrity of the cell membranes of the sieve tubes and companion cells and involves "carrier" proteins that contain essential SH groups (and, hence, are potentially vulnerable to ozone) on the plasmalemma surface.

Even if the process of phloem loading is unimpaired, translocation of sugars and other metabolites is also dependent upon their availability, i.e., their distribution within the various cellular compartments, since there is evidence for the existence of transportable and nontransportable pools within the cell (Koziol et al., 1988).

One approach to determining the effects of ozone on translocation has used isotopically labeled CO_2 and analysis of the distribution patterns of labeled assimilates. Thus, increases in the retention of assimilates by the leaves of ozone-treated bean plants, ranging from 24 to 57%, were reported by McLaughlin and McConathy (1983), based on ^{14}C studies, and by Okano et al. (1984), who studied effects on the partitioning of ^{13}C-labelled assimilates. The greatest retention was in the primary leaves and therefore occurred at the expense of downward translocation to the roots. In contrast, export from the trifoliate leaves (which tend to supply the upward translocation stream) was increased, although this was balanced by decreased CO_2 assimilation. Such experiments provide some explanation for the observed enhancement of S/R ratios by ozone, although they fail to elucidate the mechanisms involved.

Of equal importance to phloem loading is the unloading into sink tissues that must also occur in translocation and may involve passage into and out of the apoplast or within the symplast via the plasmodesmata (Wyse, 1985). In the former case, expenditure of energy is needed for the transfers across the plasmalemma, again involving sucrose-proton carrier proteins. If sucrose is stored in sink tissues, a further membrane transfer is needed across the tonoplast into the vacuole (Miller, 1988). Mass flow-down concentration gradients may also occur. However, there appear to have been no studies of the effects of ozone on these processes.

One consequence of reduced translocation to the root is the reduction in the supply of metabolites and energy to heterotrophic microorganisms. These include both symbionts, such as *Rhizobium* and mycorrhizae, and soil microorganisms in the rhizosphere. The limited information on the effects of ozone on mycorrhizae has been reviewed by McCool (1988).

Although ectomycorrhizae typify many forest tree species, endomycorrhizal associations are widespread throughout the plant kingdom and serve to increase the net absorptive area of the root systems of many crop and native species, thereby enhancing nutrient uptake. McCool et al. (1979) found that, although brief exposures to ozone did not reduce the growth of nonmycorrhizal citrus seedlings, both the growth of mycorrhizal seedlings and spore production by the infecting fungus, *Glomus fasciculatus*, were reduced. Longer exposures resulted in reduced rates of mycorrhizal infection. The spread of infection by the same fungus on tomato plants was also slowed by ozone, which countered the beneficial effects of infection on growth (McCool et al., 1982). *G. geosporum* infection of soybean reduced the yield loss due to ozone from 48 to 25% (Brewer and Heagle, 1983).

Shafer (1988) investigated the effects of ozone on the rhizosphere bacteria associated with sorghum. An initial stimulation was followed by an inhibition of bacterial population densities with increasing ozone exposure. Although such effects are undoubtedly mediated by effects on the supply to and exudation from the roots, their potential impact on overall plant growth of annual as well as perennial species is considerable and calls for much more extensive investigation.

Throughout growth, partitioning (and hence the potential impact of ozone) varies. The onset of reproductive growth marks the start of a new phase in plant partitioning. Cooley and Manning (1987) have summarized numerous studies with soybean, which clearly show that, while ozone reduces overall growth, it causes enhanced partitioning to leaves during early growth, and although seed set is reduced, those seeds that develop are the preferred sink. In tomato, fruit development is less adversely affected by ozone than the growth of vegetative organs (Oshima et al., 1975). In other cases such as peppers (Bennett et al., 1979) and cotton (Oshima et al., 1979), ozone severely inhibits reproductive growth.

6.4.3.2 The Analysis of the Effects of Ozone on Growth Dynamics

Mention has already been made of the fact that many studies of the effects of ozone have been concerned with harvest yields. However, the methods of plant growth analysis (Radford, 1967; Causton and Venus, 1981; Hunt, 1982; Chappelka and Chevone, 1989) provide the tools for elucidating effects on the dynamics of the shifts in partitioning that influence growth as a whole and lead to the differential growth of different plant parts.

Haas (1970) appears to have been the first to apply some of these techniques in his investigation of the relationship between ozone-induced foliar injury (bronzing) and overall growth of bean plants (*Phaseolus*) in the field. He was able to relate the onset of bronzing injury to situations in which the crop growth rate on a leaf area basis ($d\mathbf{L}/dt$, where $\mathbf{L}$ is total leaf area) had dropped to zero. He also observed that decreases over time in the unit leaf rate, $\mathbf{E}$, defined as $(1/\mathbf{L}) \cdot (d\mathbf{W}/dt)$, where $\mathbf{W}$ is shoot dry weight, were greatest in plants that were ultimately most severely injured; $\mathbf{E}$ provides an estimate of $\mathbf{P_N}$ effeciency. However, the study did not attempt to relate ozone exposure to growth dynamics.

Bennett and Runeckles (1977a) examined such relationships in their studies of the long-term effects of ozone on the growth of crimson clover (*Trifolium incarnatum*) and annual ryegrass (*Lolium multiflorum*). They observed that root growth was most severely impaired (increased S/R ratios). Leaves maintained their priority as sinks for assimilates, as shown by unaffected or slightly stimulated leaf weight ratios (LWR; $\mathbf{W_L}/\mathbf{W}$, where $\mathbf{W_L}$ is total leaf dry weight). In both species, specific leaf area (SLA; $\mathbf{L}/\mathbf{W_L}$), and leaf area ratio (LAR; $\mathbf{L}/\mathbf{W}$) were reduced, indicating smaller, heavier leaves. However, there was no effect on overall relative growth rate, $\mathbf{R}$, defined as $(1/\mathbf{W}) \cdot (d\mathbf{W}/dt)$.

LAR and R are linked with $\mathbf{E}$ by the relationship:

$$\mathbf{R} = \mathbf{E} \cdot \mathbf{LAR}$$

In spite of the ozone-induced decrease in LAR with no change in $\mathbf{R}$, no significant increase in $\mathbf{E}$ was observed.

It is important to distinguish between $\mathbf{E}$ and $\mathbf{P_N}$. The former relates to the gain in total dry matter (often of the shoot alone) as a function of leaf area, whereas $\mathbf{P_N}$ relates to leaf weight gain due to carbon assimilation alone. Bennett and Runeckles (1977) did not measure net photosynthesis, but studies with soybean (Reich et al., 1986) and radish (Atkinson et al., 1988) have shown some agreement between the reductions caused by ozone to both parameters. However, as Haas (1970) has pointed out, in the later stages of growth, $\mathbf{E}$ becomes a complex function of growth and leaf abscission and is therefore significantly influenced by the senescence of leaves and leaf area duration (LAD, defined as

Table 6.2. Effects of Exposure to Ozone on Dynamic Growth Parameters for Several Crop Species

Crop	R	LAR	E	LWR	SLA	LAD	Ref[a]
Alfalfa (Medicago sativa)	↓[b]	nr[c]	↓[d]	nr	nr	nr	1
Bean (Phaseolus vulgaris)	↑/↓	nr	nr	nr	nr	nr	2
Clover (Trifolium incarnatum)	↓	↓		≈	↓	nr	3
Cotton (Gossypium hirsutum)	≈	↑/↓	↓/≈	≈/↓	↑	↓	4
	≈	↑	↓	nr	nr	nr	5
Radish (Raphanus sativus)	↓	↑	↓	nr	nr	nr	6
	↑/↓	↑/↓	≈/↓/≈	nr	nr	nr	7
Ryegrass (Lolium multiflorum)	≈	↓	≈	≈	↓	nr	3
Soybean (Glycine max)	↓	≈	↓	nr	nr	nr	8
	↓/≈	≈	↓/≈	↑/≈	↓/≈	↑/↓	9
	nr	nr	↓	nr	nr	nr	10
	nr	nr	nr	↑	nr	nr	11

[a] References: (1) Cooley and Manning, 1988; (2) Blum and Heck, 1980; (3) Bennett and Runeckles, 1977a; (4) Miller et al., 1989; (5) Oshima et al., 1979; (6) Atkinson et al., 1988; (7) Walmsley et al., 1980; (8) Damicone, et al., 1987; (9) Endress and Grunwald, 1985; (10) Reich et al., 1986; (11) Unsworth et al., 1984.

[b] ↓, ↑, ≈ : decrease, increase, or no change relative to controls over time.

[c] nr: Not Reported.

[d] Except cv Saranac after cutting.

the integral of **L** over time). With clover and ryegrass (Bennett and Runeckles, 1977), it appears that the effects observed reflected changes in partitioning rather than changes in P_N.

Table 6.2 presents a summary of the reported effects of ozone on several dynamic growth parameters of a variety of crop species, together, in some cases, with indications of their change over time. The most consistent observation is decreased **E**, although there are several reports indicating no effect of ozone, and Cooley and Manning (1988) reported that **E** was increased by ozone in Saranac alfalfa after cutting, which they attributed to acclimation. In most cases, **R** is also decreased or remains unaffected by ozone, although slight increases were noted with parsley and radish towards the end of the experiments (Table 6.2). No consistent pattern emerges with regard to changes in LAR, LWR, SLA, and LAD. What is clear, however, is that the effects of ozone on these parameters change over time as a result of developmental changes, leading to modifications of source-sink relationships.

Many of the parameters can themselves be partitioned to reveal responses of individual components: leaves, stems, fruits, buds, etc. While this has most frequently been done with the relative growth rate to determine effects on the growth of individual parts such as the root (Endress and Grunwald, 1985), Walmsley et al. (1980) partitioned **E** in order to show that the partitioning of assimilates to radish hypocotyls was persistently reduced by ozone, whereas partitioning to the leaves was significantly enhanced during the middle of the growth period. As the authors state: "This implies that the accelerated senescence of the earlier leaves induced physiological changes in the plant which caused more of the available assimilate to be used for the production of new leaf tissue." Their study is also of interest because they observed that under relatively severe and continuing ozone stress, increased leaf senescence was more than offset by the accelerated development of new leaves; ozone caused a shift in the occurrence of the maximum LAR from 13 to 21 d after planting. Furthermore, the new leaves were acclimated to ozone, but the acclimation was not related to stomatal function.

A recent application of relative growth rate is its use in partitioning between maintenance and growth respiration. Amthor (1988) and Amthor and Cumming (1988) partitioned bean leaf blade respiration according to the relation

$$r = g \cdot \mathbf{R} + m$$

where r is leaf specific respiration rate (milligrams CO_2 per g dry weight/ per day), g is the growth coefficient (milligrams CO_2 per g dry weight), and m is the maintenance coefficient (milligrams CO_2 per g dry weight per day). The growth coefficient, g, is a measure of efficiency (not rate) with which assimilates are used in biosynthesis, while m is the respiratory cost of maintaining existing biomass.

In both growth chamber (Amthor and Cumming, 1988) and field studies (Amthor, 1988), exposure to ozone caused a significant decrease in $\mathbf{R}$, with no effect on the growth coefficient, but a significant increase in maintenance respiration. This indicates an altered carbon budget in the plant, with increased diversion of assimilates to provide energy for the processes of maintenance and repair.

Accelerated senescence and necrosis of leaves are plant responses to various environmental stresses, including air pollutants. Runeckles (1982) used growth analysis concepts in developing the relative death rate, $\mathbf{R}_d$, that defines the dynamics of senescence as a function of the healthy tissue present, according to the relationship

$$\mathbf{R}_d = \frac{1}{\mathbf{M}_l} \cdot \frac{d\mathbf{M}_d}{dt}$$

where $\mathbf{M}_l$ and $\mathbf{M}_d$ are measures of the living and dead (or senescing) tissues or organs, respectively. The computation of instantaneous values for $\mathbf{R}_d$, based on curve fitting (Hunt, 1982), can reveal how the senescence response varies over time. Runeckles (1982) demonstrated its applicability to the effects of SO_2 on senescence of *Agropyron smithii*, but the method has yet to be used to analyze such responses to ozone.

In spite of their potential for providing insights into the effects of ozone on plant growth over time, plant growth analysis methods have seen only limited use, which may in part explain some of the variability noted in Table 6.2. Another probable contributor to the variability is the shift in methodology from the "classical," arithmetic approach, dependent upon determining average changes over discrete intervals between harvests, and the "functional" approach using frequent harvests and curve fitting to permit the determination of instantaneous estimates of the derived functions (Chappelka and Chevone, 1989).

A limitation to the use of these techniques has been their dependence on the selection of appropriate mathematical functions to accommodate extended data sets in which the relationships among dry weights and leaf areas vary irregularly over time. Such irregularities may be the result of normal developmental changes or of changes in environmental factors, including episodic exposure to a pollutant. Continuous descriptions of irregular data may be resolved by fitting polynomials to the data after partitioning or segmentation into discrete intervals, or by the use of splined regressions (Hunt, 1982). Lieth and Reynolds (1986) developed an alternative approach to the analysis of such data (obtained from episodic exposure of bean plants to ozone), based on the Richards function (Richards, 1959), which can be expressed as

$$\frac{1}{W} \cdot \frac{dW}{dt} = \frac{\kappa}{v}\left[1 - \left(\frac{W}{A}\right)^{v}\right] = f(W) \tag{6.15}$$

where A is the asymptote of W, the dry weight; and parameters v and κ determine the lower asymptote and shape of the growth curve. Function $f(W)$, representing the relative growth rate, $\mathbf{R}$, is continuous (and, hence, inappropriate for episodic growth responses). Lieth and Reynolds (1986) therefore modified Equation 6.15 to incorporate a change in $\mathbf{R}$ occurring at $t = t_{ev}$, as shown in Equation 6.16:

$$\frac{1}{W} \cdot \frac{dW}{dt} = f(W)g(t) \tag{6.16}$$

where $g(t) = 1$ when there is no perturbation, i.e., $t < t_{ev}$, and $g(t) < 1$, when $t > t_{ev}$. The solution to Equation 6.16 is a modified Richards function (MRF):

$$W = A\left[1 \pm e^{\left(c - \kappa \int g(t)dt\right)}\right]^{-\frac{1}{v}}$$

Lieth and Reynolds (1986) selected a form of the function $g(t)$ that incorporated the degree of recovery after $t = t_{ev}$ (a concept similar to that used by Schut, 1985), such that

$$g(t) = b - (b - a)e^{-c\left(t - t_{ev}\right)}$$

for $t \geq t_{ev}$, where a is the percent reduction of **R** immediately following $t = t_{ev}$, b is the percent recovery of **R** relative to the value in the absence of exposure; and c is the "rate of recovery."

They showed that the MRF provided a good fit to experimental data obtained with single 3-h exposures to 0.15, 0.30, 0.45, or 0.60 ppmv ozone administered 15 d after planting. Although the method could be used for repetitive exposures, the computations are extremely complex, but they found that an acceptable simulation of the data from exposures repeated at 2- to 5-d intervals could be obtained, based on adjusted values for a and c obtained from single exposures.

Other approaches to the analysis of effects of environmental variables on long-term growth and its dynamics include demographic analysis (Bazzaz and Harper, 1977), sequential yield component analysis (Eaton and Kyte, 1978), and their combination with plant growth analysis (Jolliffe et al., 1982; Jolliffe and Courtney, 1984). However, none of these appears to have been used in ozone studies, although Banwart et al. (1988) used both growth and yield component analyses in their study of the effects of simulated acid rain on corn.

6.4.3.3 Effects of Ozone on Reproduction

The reproductive yield of many crops is reduced by ozone stress. In some cases the reduction appears to be the result of changes in partitioning, but there are also reports of reduced floral initiation and development, and adverse effects on the success of fertilization, for example, in wheat (Heagle et al., 1979b), corn (*Zea mays*; Heagle et al., 1972, 1979a), and soybean (Heagle and Letchworth, 1982; Kress and Miller, 1983; Reich and Amundson, 1984; Endress and Grunwald, 1985). Kress and Miller (1983), for example, reported decreases in filled pods per plant and seeds per filled pods with increased ozone stress.

Adverse effects on floral initiation and development in several species were reported by Feder (1970). Feder (1968), Harrison and Feder (1974), and others have shown that ozone may severely inhibit pollen germination and pollen tube growth.

Hence, in the case of crops grown for their fruit or seed, ozone may exert adverse effects independent of its impact on the vegetative parts and their functioning.

6.4.4 Exposure-Yield Response Relationships

From the earliest studies of the effects of air pollution on plants, there has been an ongoing interest in defining "dose-response" relationships for all pollutants. The elucidation of quantitative relationships between exposure and crop yields is essential for purposes of estimating the magnitude of the adverse effects of tropospheric, ground-level ozone pollution on agricultural production, and as an important contribution to the process of establishing standards of air quality.

In defining a relationship between exposure to a gaseous pollutant, such as ozone, and a plant response, two points need to be emphasized:

1. "Exposure" and "dose" are not synonymous (Lefohn and Runeckles, 1987). The former merely defines the condition of the ambient air in which a plant is growing. The latter concerns the amount of pollutant (or its products) that reaches the reactive sites within the plant tissue that are adversely affected; and
2. Exposure in field conditions constitutes a sequence of different instantaneous concentrations that are a consequence of the fluid nature of the atmosphere and the influence of meteorological factors, particularly wind and turbulence.

The distinction between exposure and dose has been reviewed in Chapter 5, while various approaches to the consolidation of long-term variations in instantaneous or short-term ozone concentrations into meaningful exposure indices have been discussed in Chapter 3.

Mathematical relationships between cause and effect may be mechanistic (based on an understanding of the processes involved), empirical (based on statistical regressions relating cause and effect), or phenomenological (mechanistic-empirical "hybrids") (Luxmoore, 1988). Mechanistic and phenomenological models are "process" models. The models of Schut (1985) and Lieth and Reynolds (1986) discussed previously are examples of mechanistic and phenomenological models, respectively. Krupa and Kickert (1987) have reviewed other examples of process and empirical models of plant response to gaseous air pollutants, but few of these relate to agricultural crops and even fewer to the effects of ozone.

In addition to the selection of an appropriate measure of ozone response (the "effect"), the development of a response relationship involves two other elements: an appropriate exposure index (the "cause") and an appropriate mathematical form for the response function (Lefohn and Runeckles, 1987; Tingey et al., 1989).

With regard to crop loss in contrast to process models, few attempts have been made to define true dose-response relationships. Approximations of ozone flux rates (based on average ozone concentrations and estimates of leaf or canopy diffusive resistances, as discussed in Chapter 5) have been found to lead to statistically improved empirical relationships (Macdowall et al., 1964; Mukammal, 1965; Reich, 1987), but the majority of the relationships that have been described are exposure-response models.

The selection among different exposure indices and mathematical functions has followed an evolutionary course. As discussed in Chapter 3, the early adoption of various types of averages as appropriate summarizations of long-term ozone exposures has been supplanted by the use of cumulative indices that better reflect the potential for uptake and response over extended periods. Furthermore, the nonlinear potential for ozone impact, frequently referred to as "the importance of peak concentrations," has led to the development of exposure indices that recognize response thresholds (with weightings of 0 or 1, below and above the threshold, respectively) or provide a continuous weighting of the concentration, such as the sigmoidal or logistic weighting function first suggested by Lefohn and Runeckles (1987).

Much of the work on developing improved exposure indices has come from the National Crop Loss Assessment Network (NCLAN) program (Heck et al., 1984a,b). In particular, Lee et al. (1988, 1989) compared a large number of indices of different types, varying in complexity from single hourly maxima, through various averages and censored (i.e., threshold based) or weighted summations, to indices incorporating mathematical functions to adjust for changes in plant sensitivity over time.

Many of these indices (their advantages and their shortcomings) have been reviewed by Hogsett et al. (1988). In spite of their deficiencies, the season-long averages of the daily 7- or 12-h average concentrations were used to develop yield-loss models for the major crops grown in the U.S. (Heagle et al., 1988). The 7- (0900 to 1600 h ST) or 12-h (0900 to 2100 h ST) periods used for computing the daily averages were the periods during which ambient ozone was supplemented in the different NCLAN studies.

Simple linear regressions may be used to define yield responses, but suffer from their inability to reflect the curvilinear nature of the responses observed in most crops. Early NCLAN studies therefore explored the use of plateau-linear and polynomial functions, but the flexibility of the Weibull function led to its ultimate adoption, used in the form

$$Y = \alpha \cdot e^{-\left(\frac{X}{\omega}\right)^{\lambda}}$$

where Y is yield, X is an exposure index, and α, ω, and λ are parameters controlling the shape of the function (Rawlings and Cure, 1985; Rawlings et al., 1988). The important features of the function are that yield is asymptotic to zero and declines with increasing exposure so that $Y = 0.37\alpha$ when $X = \omega$.

The concurrent development of "better" exposure indices and the evaluation of different functional forms have led to the view that the combination of the Weibull function and indices such as the SUM06 (the accumulation of daily 1-h average ozone concentrations equal to or greater than 0.06 ppmv throughout the growing season) or a sigmoidally weighted index provides the most generally useful, simple model for predicting the impact of ozone pollution on crop yields.

A possible shortcoming of this use of the Weibull model is that it is a monotonic-decreasing function, since the literature contains reports of stimulations of various growth measures caused by exposures to low ozone levels (Bennett et al., 1974). Such increases were observed in 109 of 262 NCLAN studies, although Rawlings et al. (1988) found no statistical support for their significance. However, Runeckles and Wright (1988) proposed a gamma-function model of comparable flexibility to the Weibull model to overcome this limitation. Their 3-parameter model has the form

$$Y = \alpha(X+1)^{\gamma} \cdot e^{-\beta X}$$

where Y is yield, and X is exposure index. Parameter α is the yield at zero exposure. The function does not require that increases in yield occur at low exposures, but can respond to their presence. A comparison of the performance of the gamma and Weibull models, using several NCLAN data sets, revealed that the gamma model consistently provided statistically better fits to the data.

The reader is referred to Heagle et al. (1988), Hogsett et al. (1988), Lee et al. (1988, 1989), Lefohn et al. (1989), and Krupa and Kickert (1987) for more detailed discussion of the various indices and models that have been reported. However, mention should be made of several attempts at developing models that recognize the episodic nature of long-term exposures to ozone. The first of these was the mixed, multivariate, polynomial Fourier regression model developed by Nosal (1983). This involves the numbers of episodes of exposure above a threshold, the integral of their concentrations over time, and the highest peak concentration in the overall exposure period. As Krupa and Kickert (1987) point out, the model incorporates the concept of episodicity and fluctuations in concentration over time, rather than attempting to define a single index, although it does not attempt to accommodate temporal variations in plant sensitivity.

Transfer function models involving multivariate time series have been described by Younglove et al. (1988) for relating ozone exposure to stomatal aperture in order to derive improved estimates of flux and dose. Krupa and Nosal (1989) have used spectral coherence analysis in order to identify time-dependent changes in exposure indices that are related to corresponding time-dependent changes in the height growth of alfalfa.

A process model of the influence of drought stress on yield response of soybean to ozone is discussed in Section 6.5.1.3.

6.5   EFFECTS OF OZONE COUPLED WITH OTHER ENVIRONMENTAL FACTORS

The effects of ozone on plants are functions of both the severity of the stress induced by ozone itself, and a wide range of factors that influence susceptibility or modify the plant's metabolism and physiology. Some of these factors are inherent, including those dictated by genetic composition and developmental stage at which the stress is imposed, while others relate to abiotic and biotic components of the plant's environment.

6.5.1 Abiotic Factors

6.5.1.1 Other Pollutants

There are numerous reports of the effects of ozone, in concert with other pollutants such as sulfur dioxide, SO_2, oxides of nitrogen (NO and NO_2), and acid rain, on plant growth and metabolism. However, Lefohn and Tingey (1984) and Lefohn et al. (1987) found that simple co-occurrences of exposure to elevated levels of ozone and SO_2 in ambient air in urban and rural areas were less frequent than sequential or combined sequential/concurrent exposures. As a result, much of the work on mixtures of ozone with other gaseous pollutants, which has been reviewed by Lefohn and Ormrod (1984) and more recently by Mansfield and McCune (1988), is of little more than academic interest because of the atypical exposure levels used.

The effects of mixtures that have been reported include enhancements and synergisms as well as antagonisms, covering responses ranging from the effects on photosynthesis and metabolism, to growth and yield. The situation is admirably summarized by Mansfield and McCune (1988): "... the picture is very complex, and it is not possible to reach many useful conclusions about the magnitude of effects on particular crops ... The reactions of different species ... are highly individualistic. Even within a species there is not always agreement about the type of response expected." The reasons lie in the diverse array of exposure conditions, concentrations of pollutants, durations of exposures to either or both, and choice of species or cultivar.

However, where exposure conditions have been realistic in terms of their likelihood of occurrence in ambient air, certain responses are clear. For example, ozone and SO_2, at sufficiently high levels, act synergistically in causing acute injury (Menser and Heggestad, 1966). The symptoms resembled those of ozone rather than those of SO_2.

With regard to the effects of mixtures on growth and yield, the problem of unmanageable experimental designs resulting from the numerous possible combinations has been minimized by the use of response surfaces rather than full factorial designs (Ormrod et al., 1984). Tingey et al. (1973b) related growth responses of soybean to ozone/SO_2 mixtures to the combined "dose" of the two pollutants, expressed as the weighted sum of the concentrations of each. Furthermore, the responses, which included increased S/R ratio, were synergistic, since there were no effects of exposure to SO_2 at the same concentration as that in the mixture.

With regard to mechanism, effects of mixtures of ozone and SO_2 on stomatal aperture may be involved in some cases, since exposure to SO_2 causes stomatal opening in many species and, hence, could lead to increased uptake of both pollutant gases (Mansfield and Freer-Smith, 1984). However, Elkiey and Ormrod (1974), Ashmore and Onal (1984), and Chevone and Yang (1985) found that the effects on stomatal conductance could not account for responses to petunia, barley, and soybean, respectively; and Beckerson and Hofstra (1979) observed that ozone reversed SO_2-induced stomatal openings in bean leaves.

The possibility exists for a second (or third) pollutant to modify the metabolic consequences of ozone exposure, and vice versa. However, although there has been a systematic study of the effects of SO_2 and NO_2 in this regard, there have been few studies involving ozone. Black et al. (1982) observed synergistic inhibition of photosynthesis in broad bean leaves exposed for 4 h to 0.04 ppmv SO_2 and ozone at concentrations ranging from 0.06 to 0.15 ppmv. At higher ozone levels the effects were simply additive. Synergism between ozone and SO_2 was also observed by Chevone and Yang (1985) in a study of the effects of various combinations of simultaneous and sequential exposures to the two gases on photosynthesis in soybean. Neither pollutant alone affected P_N, but their mixture resulted in a 68% decrease after 2 h. Okano et al. (1984) suggested that the synergistic inhibition of bean leaf photosynthesis by ozone and NO_2 is related to the reduction in nitrite reductase activity induced by ozone.

At the molecular level the potential for ozone to generate oxyradicals has been discussed in Section 6.2.2, but it should be noted that bisulphite and sulphite, the products of SO_2 dissolution, can also generate superoxide during oxidation (Peiser and Yang, 1985), thereby adding to the burden of radicals presented to the scavenging mechanisms present.

Considerable effort has been made, over the past decade, to determine the effects of ozone in combination with exposures to "acid rain" or "acid fog." While most interest has focused on the effects on forest tree species, dealt with in Chapter 7, studies have also been conducted with agricultural crop species. In the case of acid rain, in which treatments involved the intermittent application of simulated acid rain (SAR) of different acidities, the bulk of the findings have indicated that, where adverse effects of ozone or low-pH SAR have been observed, they were additive, with no evidence of interaction. These studies have involved crops such as soybean (Norby et al., 1985; Johnston and Shriner, 1986; Takemoto et al., 1987), radish (Johnston et al., 1986), and alfalfa (Rebbeck and Brennan, 1984). However, in a study of Beeson soybean, Troiano et al. (1983) observed that treatment with low-pH SAR, in the absence of ambient ozone, resulted in yield increases that were not found in the presence of ozone.

Similarly, extensive studies in California on alfalfa (Temple et al., 1987; Takemoto et al., 1988a), peppers (Takemoto et al., 1988b), strawberry (Takemoto et al., 1989), and several other crops (Takemoto et al., 1988c) showed that exposure to ozone and intermittent acid fog at pH values as low as 1.68 resulted in additive responses.

In spite of the more prevalent situation in which plants are exposed to sequences of gaseous pollutants, there have been few studies of the effects of sequences. The study of the effects of ozone and SO_2 on soybean photosynthesis (Chevone and Yang, 1985) included the sequences 0.2 ppmv ozone (1 h) followed by 0.2 ppmv ozone plus 0.7 ppmv SO_2, and 0.7 ppmv SO_2 (1 h) followed by 0.2 ppmv ozone plus 0.7 ppmv SO_2. After 2 h the former resulted in a 33% decrease, and the latter a 56% decrease.

Two recent studies have involved sequences of NO_2 and ozone. Runeckles and Palmer (1987) exposed bean, mint, radish, and wheat plants daily to the sequence 0.09 ppmv NO_2 (0900 to 1200 h) followed by 0.09 ppmv ozone (1200 to 1800 h). With radish and wheat, the sequence acted synergistically, reducing shoot growth more than ozone alone (NO_2 alone increased shoot growth). On the other hand, preexposure to NO_2 reduced the negative impact of ozone on bean growth. Growth of mint was not significantly affected by any pollutant exposure. The effects on above-ground biomass were mirrored by effects on root or hypocotyl growth.

Goodyear and Ormrod (1988) subjected tomato plants to two daytime sequences: NO_2–ozone, or ozone–NO_2; to the sequence: nighttime NO_2–daytime ozone; and to concurrent exposure. Each individual exposure lasted 1 h (0.08 ppmv ozone, 0.21 ppmv NO_2). Although in one experiment the combination resulted in significant reductions in leaf area relative to NO_2 alone; and in the fresh weights of leaves and stems relative to the controls or NO_2 alone; no significant reductions were observed in a second experiment. However, significant reductions in leaf and stem fresh weights relative to the controls were found with the ozone–NO_2 sequence. The authors comment that the discrepancy between the two results with concurrent exposures may have been due to the fact that the exposures occurred at different times of day. It also appears that the single 1-h exposure was insufficient to elicit significant effects on dry-matter production.

It is apparent that considerably more experimentation is needed before the importance of sequential exposures involving ozone can be meaningfully assessed.

Not all air pollutants elicit their effects through foliar uptake. In the case of the heavy metals, uptake from the soil is the major route of entry into the plant. The few studies of ozone-heavy metal interactions indicated that Cd and Ni, at levels that do not induce adverse effects per se, caused increased sensitivity to ozone in peas (Ormrod, 1977) and cress and lettuce (Czuba and Ormrod, 1974), but at phytotoxic levels of the heavy metals, sensitivity to ozone was decreased.

6.5.1.2 Plant Nutrition

It has long been known that plant response to ozone can be influenced by nutritional status. The nonspecific term "influenced" is appropriate because of the diversity of effects that have been reported. Little has changed to provide any clear pattern of response since the topic was reviewed by Heck et al. (1977), Cowling and Koziol (1982), and Guderian et al. (1985).

Reports exist that indicate that sensitivity (in terms of acute injury) increases with both increasing and decreasing levels of general soil fertility and that adverse effects of ozone on growth may be ameliorated or enhanced. Thus, reports of the influence of nitrogen and potassium are contradictory (Guderian et al., 1985).

Nevertheless, the bulk of the evidence suggests that optimal nitrogen minimizes, and nitrogen deficiency increases, foliar injury. With potassium, Leone (1976) observed that K-deficient tomato plants were less susceptible to ozone, whereas Dunning et al. (1974) found the reverse with pinto bean and soybean.

Leone and Brennan (1970) found that phosphorus deficiency reduced ozone-induced injury, although Ormrod et al. (1973) found that reductions in radish growth were independent of either N or P levels.

Nutrient balance may be more important than the levels of individual nutrients, since Brewer et al. (1961) noted several interactive effects of varying P and K levels on spinach and mangel. Heagle (1979) observed that susceptibility of soybean was least when plants were grown at low or at high N-P-K levels; plants grown at intermediate levels exhibited the greatest foliar injury.

In the case of sulfur nutrition, in an isolated study Adedipe et al. (1972) found that high S fertility conferred resistance on bean plants, possibly due to increased sulfhydryl levels. However, no studies appear to have been conducted on Ca or Mg nutrition, in spite of the key roles of these elements in cellular regulation and as essential components of the cell wall and of the chloroplast.

Similarly, little is known of the influence of micronutrient nutrition on response to ozone. Czuba and Ormrod (1974) observed that applications of Zn caused increased ozone phytotoxicity in cress and lettuce, and McIlveen et al. (1975) observed a similar response in bean.

Cowling and Koziol (1982) felt that there was sufficient evidence to suggest that, in spite of the discrepancies among the various findings, the interactive effects of ozone and fertility were probably caused by different levels of fertility shifting the plant's soluble carbohydrate pool away from the optimum. They also pointed out that fertility may influence response to ozone indirectly. For example, in field situations, adequate nutrition results in denser canopies that will affect uptake. In addition, they suggested that, within plant populations, adequate nutrition would lead to compensatory effects favoring the growth of less-sensitive individual genotypes.

6.5.1.3 Drought Stress and the Effects of Salinity

Drought stress is frequently an important modifier of plant response to ozone. Susceptibility has generally been found to decrease inversely with the level of stress, as reviewed by Heagle et al. (1988). Their review covered NCLAN field studies involving four crops; alfalfa, barley, cotton, and soybean, and a fescue-clover mixture (Heagle et al., 1989).

The single barley study (Temple et al., 1985a) and one cotton study (Temple et al., 1988b) showed no significant effect of ozone regardless of the level of drought stress. Three soybean experiments (Heagle et al., 1987; Heggestad et al., 1985, 1988; Irving et al., 1988), two cotton studies (Temple et al., 1985b; Heagle et al., 1988b), and the alfalfa study (Temple et al., 1988a) revealed significant ozone/drought interactions, although in the cases of studies conducted in more than one year, considerable year-to-year variability was noted.

Where interactions occurred, growth was usually more severely reduced by ozone in plots with adequate water supply, but in some study years, there was no significant influence of drought stress on yield reductions due to ozone. Greenhouse studies with potted plants have also shown the reduced impact of ozone on plants provided with limited water supply (Amundson et al., 1986; Reich et al., 1985; Tingey and Hogsett, 1985; Moser et al., 1988). In their soybean study, Heggestad et al. (1985) found that the interaction between drought stress and ozone was concentration dependent; exposures to ozone levels whose daily mean values did not exceed 0.07 ppmv acted synergistically with water deficit in reducing yield, but at higher exposure levels, drought reduced the effect of ozone. They accounted for this complex interaction by observing that, although yield decreased with exposure, regardless of water status, enhanced root growth on the 0.2 to 0.4 m root zone was greatest with plants exposed to low ozone levels (Heggestad et al., 1988). They attributed the differences among the various field and pot studies to changes in the ability of the root system to compensate for the limited availability of soil moisture. Such differences reflect differences in soil type and structure and in irrigation methods and regimes, all of which affect the type and extent of root development.

Several studies examined the effects on components of growth and yield, and Moser et al. (1988) developed ozone-response curves for various components, to further demonstrate the importance of the time of occurrence of drought stress during plant development on the nature and magnitude of the interaction.

King (1987) has described a model for predicting the effect of drought stress on crop losses due to ozone, based on seasonal transpiration and transpiration efficiency estimates for some NCLAN soybean data sets. Kobayashi et al. (1990) have described a mechanistic model of soybean growth, which also incorporates drought stress, based on Sinclair's (1986) model of water and nitrogen limitations to growth.

Where drought stress has been observed to reduce the adverse effect of ozone on yield, the weight of the evidence indicates that this is related to stomatal closure. Reich et al. (1985) and others observed increased stomatal resistance in droughted plants. Tingey and Hogsett (1985) found that treatment of stressed bean plants with fusicoccin (which causes stomatal opening) rendered such plants as susceptible to ozone as well watered plants, indicating that metabolic changes were unlikely to be involved.

However, although Rich and Turner (1972) had observed that ozone accelerated stomatal closure on drought-stressed beans, this has not been confirmed in all recent field studies. Temple (1986) found no effect of drought stress on the rate of ozone-induced stomatal closure in cotton and concluded that the effect of ozone on stomatal resistance was the consequence of impaired photosynthesis, a conclusion also reached by Reich et al. (1985) from their soybean study.

Flagler et al. (1987) observed that, although both ozone and drought stress reduced nitrogen fixation in soybean nodules, the effect of drought stress was considerably greater.

Reduced availability of water can also be the result of soil salinity. The few studies of ozone-salinity interactions have indicated that salinity tended to reduce the adverse effects of ozone, in keeping with the general findings with regard to drought stress (Maas et al., 1973; Bytnerowicz and Taylor, 1983).

6.5.1.4 Other Abiotic Factors

Because of the potential for ozone to influence many features of plant metabolism and physiology, it is to be expected that many physical factors in the plant's environment that influence or regulate these features will interact with the effects of ozone, including effects on overall growth. Darrall (1989) has reviewed the effects of light intensity and CO_2 levels and pointed out that, although it is under high light intensities that the potential for the endogenous production of harmful oxyradicals is greatest, and they may combine with ozone-derived radicals to overcome scavenging capacity, at low intensities, overall CO_2 assimilation may limit the availability of energy for maintenance and repair.

Ambient humidity can play a role in determining ozone uptake. Although stomata are influenced by both ozone and humidity, McLaughlin and Taylor (1981) found that the decreased uptake observed at low RH was the result of increased internal (metabolic) resistance rather than a direct effect on stomatal aperture. Conversely, there are several reports of increased susceptibility to ozone, both short and long term, resulting from increased humidity (Guderian et al., 1985).

The influence of light quality, photoperiod, and temperature have also been reviewed by Guderian et al. (1985). Much of the information covered is of academic interest with little relevance to ambient conditions. However, Barnes et al. (1988) have studied the interactions of ozone with factors related to the onset of winter, i.e., chilling, freezing, and winter desiccation. Exposure to ozone increased freezing injury of pea plants, probably because of increased solute leakage. Cultivar differences were accounted for by differences in the rates of their stomatal closure, which affected ozone flux. Although performed on a summer annual species, they point out the potential importance of ozone as a modifier of the growth and survival of overwintering annual species and woody and other perennials, including forest tree species (discussed in Chapter 7).

6.5.2 Biotic Factors

While the successful growth of plants is dependent upon their optimal utilization of the physical and chemical resources available to them, various biotic and ecological factors also play important roles: the incidence and severity of disease and pest infestations, the establishment of symbiotic relationships, and the competition for resources. These relationships may involve individuals or populations of different species, each of which may be differentially affected by ozone. A table of the important features of such interactions as they relate more particularly to forest trees is presented in Chapter 7.

6.5.2.1 Diseases

The effects of ozone on the interaction of host plants with symbiotic microorganisms has been discussed in Sections 6.2.10 and 6.4.3.1. Plant-pathogen relationships may also be affected by ozone as a result of effects on the host plant, on the pathogen, or on both. Such effects may lead to stimulations or inhibitions of disease incidence or severity (Heagle, 1982; Dowding, 1988).

Dowding (1988) has stressed the importance of viewing such effects of ozone in the context of the complex interactions among the host, the pathogen, and the physical, chemical, and other biotic components of the environment. Depending on the pathogen, of critical importance to an effect of ozone on the establishment of disease may be the coincidence of the timing of exposure and the infective period.

In the case of many fungal pathogens, potential effects of exposure of spores appear to be minimal, for example, while being distributed in the air. They are most vulnerable following deposition (frequently accompanied by rainfall or dew formation), since their carbohydrate energy reserves are rapidly depleted on germination. The diverse interactions with ozone on the leaf surface have been discussed by Dowding (1988) and include effects on cuticular chemistry and surface properties, exuded materials, and stomatal response. The growth and development of the pathogen on the host may be inhibited directly by ozone, since ozone toxicity has been observed in axenic culture (Krause and Weidensaul, 1978). Alternatively, important ozone-induced changes in the host include changes in carbohydrate, nitrogen, and phosphorus metabolism; changes in phytoalexin production; and changes in cork deposition, all of which may have profound effects on the successful growth and development of the pathogen.

In the case of bacterial pathogens, similar interactions may be involved, although infection is usually dependent upon successful entry into host tissues via wounds and the feeding activities of insect vectors. Insect vectors are also essential for the transmission of many plant viruses.

The actions of ozone on the host or pathogen may involve the participation of the superoxide anion radical and hydrogen peroxide, as discussed in Section 6.2.2. Rapid increases in the levels of $O_2{}^-$ were detected in the hypersensitive reaction induced by infection of Burley 21 tobacco with *Pseudomonas syringae* pv *syringae* (Adam et al., 1989). In addition, Apostol et al. (1989) reported a rapid burst of H_2O_2 released by cultured plant cells upon treatment with elicitors of defence responses. Hence $O_2{}^-$ and H_2O_2 may well be intimately involved in the interactive responses of ozone, host, and pathogen, with the diversity of responses reflecting the degree to which the effects of the pathogen are limited to the sites of infection or spread to adjacent cells and modify their scavenging systems.

Following infection, ozone may influence the severity of the disease through its effects on the host plant. Conversely, the development of the pathogen may affect the susceptibility of the host plant. Since the early report of "protection" against "smog injury" afforded bean or sunflower leaves by infection with the rust fungi, *Uromyces phaseola* and *Puccinia helianthi*, respectively (Yarwood and Middleton, 1954), there have been numerous reports of reduced susceptibility to ozone being conferred by infection with viruses: tobacco mosaic virus on tobacco (Brennan and Leone, 1969), several viruses on bean (Phaseolus) (Davis and Smith, 1976); with bacteria: *Pseudomonas* on soybean (Pell et al., 1977), *Xanthomonas phaseoli* on bean (Temple and Bisessar, 1979); and with fungi: *Puccinia graminis* on wheat (Heagle and Key, 1973). The converse enhancement of the impact of ozone on host-plant growth has been observed with nematode infection (Bisessar and Palmer, 1984).

There is considerable evidence indicating that exposure to ozone can reduce infection, invasion, and sporulation of fungal pathogens, including "obligate" pathogens such as the rust fungi (Heagle, 1973). However, examples also exist of increased infection of ozone-injured plants (Manning et al., 1969). This diversity of response is not surprising in view of the complex interrelationships involved in which the genetics of the host and pathogen play important roles, as evidenced by the studies of Damicone et al. (1987) on soybean genotypes infected with *Fusarium oxysporum*, and of Weber et al. (1979) on parasitic nematode infestation of soybean and begonia.

We are far from possessing a clear understanding of the mechanisms involved in ozone-host-pathogen-environment interactions, but one generalization that can be made about the impact of ozone on disease is, to paraphrase Dowding (1988), pathogens that can benefit from injured host cells and disordered transport mechanisms will be enhanced by earlier exposure of the host to ozone, while those that depend on "healthy" host tissue will be disadvantaged.

6.5.2.2 Insects and Related Pests

Herbivorous insects and spider mites are major causes of crop loss, both through direct consumption of plant matter and because of their roles as disease vectors. Just as the interrelationships between plants and pathogens exhibit ecological complexity, so too it is with plant-insect relationships.

The effects of gaseous air pollutants in general, on plant-insect relationships have been the subject of several recent reviews (Alstad et al., 1982; Hughes and Laurence, 1984; Hughes, 1988; and Manning and Keane, 1988. However, most of our knowledge on the subject concerns pollutants other than ozone.

The topic can be subdivided into the influence of ozone on insect attack and population dynamics (whether direct or mediated by changes induced in the plant), and the converse effects of insect attack on plant response to ozone.

Host-plant resistance to attack may be modified through metabolic changes that affect feeding preference and insect behavior, development, and fecundity. Ozone-induced changes in both major and secondary metabolites (discussed in Section 6.2) may be qualitative and quantitative, and while there is abundant evidence that such changes can influence insect growth and development, there have been few experimental investigations of the specific effects of ozone. Trumble et al. (1987) reported that the tomato pinworm (*Keiferia lycopersicella*) developed faster on ozone-injured tomato plants, although fecundity and female longevity were unaffected. The Mexican bean beetle (*Epilachna varivestis*) preferentially selected soybean foliage that had been exposed to ozone (Endress and Post, 1985). Preference increased with increased exposure to ozone. The consequence of such preference is shown by the work of Chappelka et al. (1988), who found that, although no feeding preference existed between two soybean cultivars that differed in ozone-induced injury, the rates of larval growth were greatest on the foliage of ozone-treated plants of either cultivar.

Whittaker et al. (1989) investigated the behavior of the pea aphid, *Acyrthosiphon pisum*, on pea plants, and of the aphid, *Aphis rumecis*, and the beetle, *Gastrophysa viridula*, on *Rumex obtusifolius*. Increased mean relative growth rates (MRGR) of the pea aphid were only observed when feeding on plants previously exposed to the highest level of ozone used: 0.194 ppmv (7 h - 1d; 4 d). No significant effects on MRGR occurred as a result of exposure of infested plants. On *Rumex*, no significant effects of ozone exposure on *Aphis* MRGR were observed, but *Gastrophysa* laid more eggs, the eggs hatched faster, and the larvae developed more rapidly on ozone-treated plants (0.07 ppmv), leading to a four-fold increase in population density. The total productivity was 44.1 mg on controls vs. 62.2 mg on treated plants. Ozone-induced changes in the *Rumex* host plants resulted in the beetles consuming only 33 mm^2 of leaf tissue for every milligram of weight gain, compared with 68 mm^2 mg^{-1} of the controls.

A single report appears to exist with regard to insect attack modifying the effects of ozone on a herbaceous host. Rosen and Runeckles (1976) reported that the combination of extremely low-level ozone (0.02 ppmv) and infestation with the greenhouse whitefly (*Trialeurodes vaporariorum*) acted synergistically in inducing accelerated chlorosis and senescence of bean leaves. They speculated that the effect might be the result of the reaction of ozone with enhanced ethylene production resulting from whitefly injury.

Although progress continues to be made in observing ozone-plant-insect interactions, the information available is fragmentary and precludes a clear unravelling of the complexities of the relationships, a situation that will only be remedied by further systematic investigation.

6.5.2.3 Weed and Plant Competition

While pathogens and insect pests constitute obvious biotic factors that can interact with the ozone-plant relationship, an almost completely neglected biotic factor is that due to inter- and intraspecific competition for available resources. In the former case, the competition is usually between a crop and a weed species, although interspecific competition is also a feature of mixed plantings such as grass-legume forage plantings.

Although in many crop situations, competition from weeds contributes more to yield losses than any other factor, there appear to have been no studies of the ways in which ozone may influence such competition. However, there have been few studies of its effects in crop mixtures.

Bennett and Runeckles (1977b) investigated the effects of ozone on mixed plantings of annual ryegrass (*Lolium multiflorum*) and crimson clover (*Trifolium incarnatum*), using the replacement series approach of DeWit (1960). The relative crowding coefficient for ryegrass on clover (a measure of its competitive ability) rose from 0.72 in filtered air to 1.46 when the mixtures were exposed daily for 8 h to 0.09 ppmv ozone. The relative yield totals did not differ significantly from unity, indicating that with increasing exposure, depressed growth of clover was essentially replaced by increased growth of ryegrass, and that the two species were essentially competing for the same resources.

Other grass-legume studies have confirmed the fact that ozone exposure results in a shift in mixture biomass in favor of the grass species (Ashmore, 1984; Blum et al., 1983; Kohut et al., 1988). In keeping with these findings, Rebbeck et al. (1988) observed that a 40% reduction of the ambient ozone level (which averaged 0.048 ppmv over 12 daytime hours) resulted in a shift towards dominance by clover. In all of these studies the focus was on effects of ozone on yield and its components, rather than on the analysis of the nature of the competitive processes involved.

Intraspecific competition is a neglected issue, but is probably a contributor to the differences frequently observed between greenhouse or growth chamber studies and those undertaken in field situations. In most "laboratory" situations, plants are grown singly or spaced in pots, as a result of which root competition is abnormal, while the spacing of the pots rarely results in the closed or partially closed canopies typical of the field.

Although effects of ozone on competition may be relatively unimportant in many crop situations, the shifts in competitive ability, demonstrated in grass-legume mixtures, will certainly occur in natural communities and lead to shifts in composition. The nature of such shifts is not predictable from knowledge of the effects of ozone on the individual species growing in isolation or in pure stands. The more recent approaches to the analysis of plant competition, developed by Spitters (1983) and Jolliffe (1988), would lend themselves readily to investigation of the effects of ozone on such competition.

6.6 GENETIC VARIABILITY

A recurring theme throughout this chapter has been the diverse responses to ozone exhibited by different species and cultivars. Whether these differences are attributable to specific mechanisms, such as rapid stomatal closure, or high apoplastic levels of ascorbate, both of which influence the uptake and fate of ozone, they are phenotypic expressions of genotypes that have evolved naturally or have been selected by plant breeders.

Numerous classifications of species according to susceptibility to ozone have appeared (e.g., Guderian et al., 1985). However, such classifications rarely include differences among cultivars of a single species and rarely show consistencies in terms of responses to other pollutants. Such classifications are highly dependent upon the criteria used to define sensitivity, but they have some use in aiding the selection of species or cultivars that are likely to survive and grow well in areas subjected to ozone pollution.

The evidence for the genetic basis of susceptibility or resistance to air pollutants, in general, has been summarized by Roose et al. (1982). Resistance to ozone has been consciously incorporated into various plant breeding programs and may have been inadvertently incorporated into field selection trials for other characteristics. However, although a clear case can be made for the evolution of resistance to SO_2, there is only limited evidence of natural selection for ozone tolerance (Dunn, 1959).

6.7 PROTECTION AGAINST OZONE

The variation in susceptibility to ozone that is under genetic control permits the selection of resistant types as a means of minimizing the adverse effects of tropospheric ozone. Furthermore, since sensitivity is also influenced by numerous environmental factors, many of which can be managed (e.g., fertility, water availability, diseases, and pests), the judicious selection of appropriate cultural practices may also minimize the adverse impact of ozone on crops.

Normal crop production practices continue to make frequent use of various agricultural chemicals in weed, disease, and pest management programs, and early studies with tobacco showed that several fungicides and insecticides had antioxidant properties that could provide some degree of protection against ozone in the field (Guderian et al., 1985). The most widely applicable appears to be benomyl (methyl-1-butylcarbamoyl-2-benzimidazole carbamate) (Manning et al., 1974).

Many other natural and synthetic compounds have been found to be capable of conferring some degree of protection against ozone, at least on an experimental basis. These have ranged from the plant growth substances, kinetin and abscisic acid, to commercial antioxidants such as ascorbic acid derivatives and n-propyl gallate, the experimental ethylene diurea (EDU), whose mode of action is discussed in Section 6.2.2. While the weight of evidence points to the efficacy of EDU in reducing visible injury and preventing yield losses due to ozone, it has yet to be developed for commercial use (Manning, 1988).

6.8 THE FUTURE

This review of our knowledge of the effects of ozone on crops has deliberately placed the greatest emphasis on events at the molecular and cellular levels. This is not because effects on the whole plant, its parts, or plant populations are unimportant, but because they are consequences of events taking place within the cell. Observations of such effects are merely observations of the results of the collective integration of cellular events, and while they may lead to an empirical knowledge of the effects of ozone, they do not, of themselves, provide much more than a descriptive catalogue. This is not to say that such empirical knowledge is not useful in applications such as the estimation of the magnitudes of crop losses caused by ozone. But much remains to be learned about the ways in which such losses are modified by other environmental factors.

However, the knowledge that we have of many of the direct effects of ozone on crop growth, or of its interactive effects with other factors, will only become less fragmented (and less contradictory) with improved understanding of the mechanisms involved. Such mechanistic underpinnings will only be acquired through greater research efforts at the molecular and cellular levels.

The descriptive nature of much of the research on the effects of ozone on crops and other plant species is reflected in the published literature. A survey of over 500 publications on plant effects from the last decade reveals that almost three quarters of them contain phrases such as "effects of," "impact of," or "influence of" in their titles. While this may reflect a certain lack of imagination on the part of the authors, it may more disconcertingly reflect the fact that much research is still in the descriptive phase. Many of these publications include extensive discussions of possible mechanisms, but these are usually extrapolated from work with other species growing under different conditions and may or may not be relevant to the case at hand. As an example, a crucial issue in the uptake of ozone and its penetration into the cell hinges on the level of apoplastic ascorbate encountered en route; yet almost all of the detailed information on the topic relates to a single species, *Sedum album*. Although more information is now appearing on other species, one may still ask, "How representative is this of the flowering plants as a whole, and crops in particular?"

As has already been noted, the very nature of ozone as a highly reactive gas will require a greater focus on detail than has frequently been true in the past, in order to further our understanding of its mode of action. For example, gross tissue analyses provide descriptions of ozone-induced change, but little mechanistic information.

The lack of emphasis on mechanism is understandable in light of the mission orientation of the funding through which much ozone effects research has been and continues to be conducted in order to meet specific objectives. Even though the extensive National Crop Loss Assessment Network program had as one of its objectives the development of understanding of the mechanisms involved in crop response to ozone, the demands of obtaining data on which to base estimated yield-loss functions resulted in little investment in fundamental research aimed at elucidating mechanisms.

There are other reasons for advocating a shift in the focus of research. One is the need for such information in the development of improved mechanistic or process models of response. Several small models have been presented in this chapter, dealing with various pieces of the total picture of plant response, but as Luxmoore (1988) observed, there is a surprising scarcity of mechanistic crop-pollutant response models in general, to which we would add that this is particularly true of models of ozone response, notwithstanding the work of Schut (1985), Lieth and Reynolds (1986), King (1987) and Kobayashi et al. (1990).

Another reason is the growth of molecular biology and biotechnology, with their potential for genetic manipulation. The application of such methodologies to the development of plants with greater resistance to ozone is dependent upon the identification of the genes that regulate the specific biochemical and biophysical processes involved, which first must be identified. The work of Tepperman and Dunsmuir (1990) on transformed tobacco and tomato plants with enhanced levels of chloroplast superoxide dismutase, referred to in Section 6.2.2, illustrates the potential for genetic engineering to assist in improving our understanding of the mechanisms of ozone action, as well as in the development of ozone-tolerant plants.

In advocating the need for more research at the molecular and cellular levels, we are not overlooking the fact that much more information also needs to be collected on the effects at the plant, population, and community levels. Although essentially descriptive, it will nevertheless be invaluable for improving our understanding of the effects and will play a role in the development of better models of response.

Ongoing research is needed because the current levels of tropospheric ozone are largely anthropogenic in origin and will continue to impose an unwanted stress on crops and other vegetation for the foreseeable future. To this ozone stress need be added other consequences of human activity related to the planet as a whole: the continued enrichment of the earth's atmosphere with CO_2, with its implications for climatic change, and the depletion of the stratospheric ozone layer, with its implications for the incidence of solar ultraviolet radiation on the earth's surface (Krupa and Kickert, 1989). These present further sets of interactions that need to be investigated.

REFERENCES

Adam, A., Farkas, T., Somlyai ,G., Hevesi, M., and Kiraly, Z. (1989) Consequence of O_2^- generation during a bacterially induced hypersensitive reaction in tobacco: deterioration of membrane lipids. *Physiol. Mol. Plant Pathol.* **34**, 13–26.

Adedipe, N. O., Hofstra, G., and Ormrod, D. P. (1972) Effects of sulfur nutrition on phytotoxicity and growth responses of bean plants to ozone. *Can. J. Bot.* **50**, 1789–1793.

Adedipe, N. O., Khatamian, H., and Ormrod, D. P. (1973) Stomatal regulation of ozone phytotoxicity in tomato. *Z. Pflanzenphysiol.* **68**, 323–328.

Agrawal, M., Nandi, P. K., and Rao, D. N. (1983) Ozone and sulfur dioxide effects on *Panicum miliaceum* plants. *Bull. Torrey Bot. Club* **110**, 435–441.

Alscher, R. G. and Amthor, J. S. (1988) The physiology of free-radical scavenging: maintenance and repair processes. In *Air Pollution and Plant Metabolism* (eds., Schulte-Hostede, S., Darrall, N. M., Blank, L. W., and Wellburn, A. R.). Elsevier, London, 94–115.

Alstad, D. N., Edmunds, G. F., and Weinstein, L. H. (1982) Effects of air pollutants on insect populations. *Ann. Rev. Entomol.* **27**, 369–384.

Amthor, J. S. (1988) Growth and maintenance respiration in leaves of bean (*Phaseolus vulgaris* L.) exposed to ozone in open-top chambers in the field. *New Phytol.* **110**, 319–325.

Amthor, J. S. and Cumming, J. R. (1988) Low levels of ozone increase bean leaf maintenance respiration. *Can. J. Bot.* **66**, 724–726.

Amundson, R. G., Raba, R. M., Schoettle, A. W., and Reich, P. B. (1986) Response of soybeans to low concentrations of ozone. II. Effects on growth, biomass allocation, and flowering. *J. Environ. Qual.* **15**, 161–167.

Amundson, R. G., Kohut, R. J., Schoettle, A. W., Raba, R. M., and Reich, P. B. (1987) Correlative reductions in whole-plant photosynthesis and yield of winter wheat caused by ozone. *Phytopathology* **77**, 75–79.

Anderson, W. C. and Taylor, O. C. (1973) Ozone induced carbon dioxide evolution in tobacco callus cultures. *Physiol. Plant* **28**, 419–423.

Andreasson, L. E. and Vanngard, T. (1988) Electron transport in photosystems I and II. *Ann. Rev. Plant Physiol. Plant Mol. Biol.* **39**, 379–411.

Apostol, I., Heinstein, P. F., and Low, P. S. (1989) Rapid stimulation of an oxidative burst during elicitation of cultured plant cells. *Plant Physiol.* **90**, 109–116.

Aryan, A. P. and Wallace, W. (1985) Reversible inactivation of wheat leaf nitrate reductase by NADH involving superoxide ions generated by the oxidation of thiols and FAD. *Biochim. Biophys. Acta* **827**, 215–220.

Asada, K. and Badger, M. (1984) Photoreduction of $^{18}O_2$ and H_2O_2 with concomitant evolution of $^{16}O_2$ in intact spinach chloroplasts: evidence for scavenging of hydrogen peroxide by peroxidase. *Plant Cell Physiol.* **25**, 1169–1179.

Asada, K. and Takahashi, M. (1987) Production and scavenging of active oxygen in photosynthesis. In *Photoinhibition* (eds., Kyle, D. J., Osmond, C. B., and Arntzen, C. J.). Elsevier, Amsterdam, 227–287.

Ashmore, M. R. (1984) Effects of ozone on vegetation in the United Kingdom. In *Proc. International Workshop on the Evaluation and Assessment of the Effects of Photochemical Oxidants on Human Health, Agricultural Crops, Forestry, Materials, Visibility* (ed., Grennfelt, P.). Swedish Environmental Research Institute, Goteborg, 92–104.

Ashmore, M. R. and Onal, M. (1984) Modification by sulphur dioxide of the responses of *Hordeum vulgare* to ozone. *Environ. Pollut.* (Ser. A) **36**, 31–43.

Atkinson, C. J., Robe, S. V., and Winner, W. E. (1988) The relationship between changes in photosynthesis and growth for radish plants fumigated with SO_2 and O_3. *New Phytol.* **110**, 173–184.

Baker, E. A. (1982) Chemistry and morphology of plant epicuticular waxes. In *The Plant Cuticle* (eds., Cutler, D. F., Alvin, K. L., and Price, C. E.). Academic Press, London, 139–166.

Banwart, W. L., Porter, P. M., Ziegler, E. L., and Hassett, J. J. (1988) Growth parameter and yield component response of field corn to simulated acid rain. *Environ. Exp. Bot.* **28**, 43–51.

Barnes, J. D., Reiling, K., Davison, A. W., and Renner, C. J. (1988) Interactions between ozone and winter stress. *Environ. Pollut.* **53**, 235–254.

Barnes, J. D., Eamus, D., Davison, A. W., Ro-Poulsen, H., and Mortensen, L. (1990) Persistent effects of ozone on needle water loss and wettability in Norway spruce. *Environ. Pollut.* **63**, 345–363.

Barnes, R. (1972) Effects of chronic exposure to ozone on soluble sugar and ascorbic acid contents of pine seedlings. *Can. J. Bot.* **50**, 215–219.

Bazzaz, F. A. and Harper, J. L. (1977) Demographic analysis of the growth of *Linum usitatissimum*. *New Phytol.* **78**, 193–208.

Becker, K., Saurer, M., Egger, A., and Fuhrer, J. (1989) Sensitivity of white clover to ambient ozone in Switzerland. *New Phytol.* **112**, 235–243

Beckerson, D. W. and Hofstra, G. (1980) Effects of sulphur dioxide and ozone, singly or in combination, on membrane permeability. *Can. J. Bot.* **58**, 451–457.

Bennett, J. H., Lee, E. H., and Heggestad, H. E. (1984) Biochemical aspects of plant tolerance to ozone and oxyradicals: superoxide dismutase. In *Gaseous Air Pollutants and Plant Metabolism* (eds., Koziol, M. J. and Whatley, F. R.). Butterworths, London, 413–424.

Bennett, J. P. and Runeckles, V. C. (1977a) Effects of low levels of ozone on growth of crimson clover and annual ryegrass. *Crop Sci.* **17**, 443–445.

Bennett, J. P. and Runeckles, V. C. (1977b) Effects of low levels of ozone on plant competition. *J. Appl. Ecol.* **14**, 877–880.

Bennett, J. P., Oshima, R. J., and Lippert, L. F. (1979) Effects of ozone on injury and dry matter partitioning in pepper plants. *Environ. Exp. Bot.* **19**, 33–39.

Bennett, J. P., Resh, H. M., and Runeckles, V. C. (1974) Apparent stimulations of plant growth by air pollutants. *Can. J. Bot.* **52**, 35–41.

Bicak, C. J. (1978) Plant response to variable ozone regimes of constant dosage. M.Sc. Thesis, University of British Columbia, Vancouver B.C.

Bisessar, S. and Palmer, K. T. (1984) Ozone, antioxidant spray and *Meloidogyne hapla* effects on tobacco. *Atmos. Environ.* **18**, 1025–1027.

Black, V. J. (1984) The effect of air pollutants on apparent respiration. In *Gaseous Air Pollutants and Plant Metabolism* (eds., Koziol, M. J. and Whatley, F. R.). Butterworths, London, 231–248.

Black, V. J., Ormrod, D. P., and Unsworth, M. H. (1982) Effects of low concentration of ozone, singly and in combination with sulphur dioxide on net photosynthesis rates of *Vicia faba* L. *J. Exp. Bot.* **33**, 1302–1311.

Blum, U. and Heck, W. W. (1980) Effects of acute ozone exposures on snap bean at various stages of its life cycle. *Environ. Exp. Bot.* **20**, 73–85.

Blum, U., Heagle, A. S., Burns, J. C., and Linthurst, R. A. (1983) The effects of ozone on fescue-clover forage regrowth, yield and quality. *Environ. Exp. Bot.* **23**, 121–132.

Bolhar-Nordenkampf, H. R., Long, S. P., Baker, N. R., Oquist, G., Schreiber, U., and Lechner, E. G. (1989) Chlorophyll fluorescence as a probe of the photosynthetic competence of leaves in the field: a review of current instrumentation. *Functional Ecol.* **3**, 497–514.

Brennan, E. and Leone, I. A. (1969) Suppression of ozone toxicity symptoms in virus-infected tobacco. *Phytopathology* **59**, 263–264.

Brennan, T. and Anderson, L. E. (1980) Inhibition by catalase of dark-mediated glucose-6-phosphate dehydrogenase activation in pea chloroplasts. *Plant Physiol.* **66**, 815–817.

Brewer, P. F. and Heagle, A. S. (1983) Interactions between *Glomus geosporum* and exposure of soybeans to ozone or simulated acid rain in the field. *Phytopathology* **73**, 1035–1040.

Brewer, R. F., Guillemet, F. B., and Creveling, R. K. (1961) Influence of N-P-K fertilization on incidence and severity of oxidant injury to mangels and spinach. *Soil Sci.* **92**, 298–301.

Butler, L. K. and Tibbetts, T. W. (1979) Stomatal mechanisms determining genetic resistance to ozone in *Phaseolus vulgaris* L. *J. Am. Soc. Hort. Sci.* **104**, 213–216.

Bytnerowicz, A. and Taylor, O. C. (1983) Influence of ozone, sulfur dioxide and salinity on leaf injury, stomatal resistance, growth and chemical composition of bean plants. *J. Environ. Qual.* **12**, 397–405.

Cape, J. N., Paterson, I. S., Wellburn, A. R., Wolfenden, J., Melhorn, H., Freer-Smith, P. H., and Fink, S. (1988) *Early Diagnosis of Forest Decline.* Institute of Terrestrial Ecology, Merlewood Research Laboratory, Grange-over-Sands, England, 68.

Carnahan, J. E., Jenner, E. L., and Wat, E. K. W. (1978) Prevention of ozone injury to plants by a new protectant chemical. *Phytopathology* **68**, 1225–1229.

Cassab, G. I. and Varner, J. E. (1988) Cell wall proteins. *Ann. Rev. Plant Physiol. Plant Mol. Biol.* **39**, 321–353.

Castillo, F. J. and Greppin, H. (1986) Balance between anionic and cationic extracellular peroxidase activities in *Sedum album* leaves after ozone exposure. Analysis by high-performance liquid chromatography. *Physiol. Plant* **68**, 201–208.

Castillo, F. J. and Greppin, H. (1988) Extracellular ascorbic acid and enzyme activities related to ascorbic acid metabolism in *Sedum album* L. leaves after ozone exposure. *Environ. Exp. Bot.* **28**, 231–238.

Castillo, F. J., Miller, P. R., and Greppin, H. (1987) Extracellular biochemical markers of oxidant air pollution damage to Norway spruce. *Experientia* **43**, 111–115.

Castillo, F. J., Penel, C. L., and Greppin, H. (1984) Peroxidase release induced by ozone in *Sedum album* leaves; involvement of Ca^{2+}. *Plant Physiol.* **74**, 846–851.

Causton, D. R. and Venus, J. C. (1981) *The Biometry of Plant Growth.* Edward Arnold, London, 307.

Chameides, W. L. (1989) The chemistry of ozone deposition to plant leaves: the role of ascorbic acid. *Environ. Sci. Technol.* **23**, 595–600.

Chanway, C. P. and Runeckles, V. C. (1984a) The role of superoxide dismutase in the susceptibility of bean leaves to ozone injury. *Can. J. Bot.* **62**, 236–240.

Chanway, C. P. and Runeckles, V. C. (1984b) Effect of ethylene diurea (EDU) on ozone tolerance and superoxide dismutase activity in bush bean. *Environ. Pollut.* (Ser. A) **35**, 49–56.

Chappelka, A. H. and Chevone, B. I. (1989) Two methods to determine plant responses to pollutant mixtures. *Environ. Pollut.* **61**, 31–45.

Chappelka, A. H., Kraemer, M. E., Mebrahtu, T., Rangappa, M., and Benepal, P. S. (1988) Effects of ozone on soybean resistance to the Mexican bean beetle (*Epilachna varivestis* Mulsant). *Environ. Exper. Bot.* **28**, 53–60.

Chevone, B. I., Lee, W. S., Henderson, J. V., and Hess, J. L. (1989) Gas exchange rates and needle ascorbate content of eastern white pine exposed to ambient air pollution. In *Proc. 82nd Annual Meeting of the Air and Waste Management Association*, Anaheim, CA. Air & Waste Management Association, Pittsburgh.

Chevone, B. I. and Yang Y. S. (1985) CO_2 exchange rates and stomatal diffusive resistance in soybean exposed to O_3 and SO_2. *Can. J. Plant Sci.* **65**, 267–274.

Chimiklis, P. E. and Heath, R. L. (1975) Ozone-induced loss of intracellular potassium ion from *Chlorella sorokiniana. Plant Physiol.* **56**, 723–727.

Cooley, D. R. and Manning, W. J. (1987) The impact of ozone on assimilate partitioning in plants: a review. *Environ. Pollut.* **47**, 95–113.

Cooley, D. R. and Manning, W. J. (1988) Ozone effects on growth and assimilate partitioning in alfalfa, *Medicago sativa* L. *Environ. Pollut.* **49**, 19–36.

Coulson, C. and Heath, R. L. (1974) Inhibition of the photosynthetic capacity of isolated chloroplasts by ozone. *Plant Physiol.* **53**, 32–38.

Cowling, D. W. and Koziol, M. J. (1982) Mineral nutrition and plant response to air pollutants. In *Effects of Gaseous Air Pollution on Agriculture and Horticulture* (eds., Unsworth, M. H. and Ormrod, D. P.). Butterworth Scientific, London, 349–375.

Coyne, P. I. and Bingham, G. E. (1982) Variation in photosynthesis and stomata conductance in an ozone-stressed ponderosa pine stand: light response. *For. Sci.* **28**, 257-273.

Criegee, R. (1975) Mechanisms of ozonolysis. *Angew. Chem.* **14**, 745–760.

Czuba, M. and Ormrod, D. P. (1974) Effects of cadmium and zinc on ozone-induced phytotoxicity in cress and lettuce. *Can. J. Bot.* **52**, 645–649.

Damicone, J. P., Manning, W. J. and Herbert, S. J. (1987) Growth and disease response of soybeans from early maturity groups to ozone and *Fusarium oxysporum. Environ. Pollut.* **48**, 117–130.

Darrall, N. M. (1989) The effect of air pollutants on physiological processes in plants. *Plant Cell Environ.* **12**, 1–30.

Davis, D. D. and Smith, S. H. (1976) Reduction of ozone sensitivity of pinto bean by virus-induced local lesions. *Plant Dis. Reptr.* **60**, 31–34.

Davison, A. W., Barnes, J. D., and Renner, C. J. (1988) Interactions between air pollutants and cold stress. In *Air Pollution and Plant Metabolism* (eds., Schulte-Hostede, S., Darrall, N. M., Blank, L. W., and Wellburn, A. R.). Elsevier, London., 307–328.

Debus, R. J., Barry, B. A., Babcock, G. T., and McIntosh, L. (1988) Site-directed mutagenesis identifies a tyrosine radical involved in the photosynthetic oxygen-evolving system. *Proc. Natl. Acad. Sci.* **85**, 427–430.

Decleire, M., DeCat, W., deTemmerman, L., and Baeten, H. (1984) Changes of peroxidase, catalase, and superoxide dismutase activities in ozone-fumigated spinach leaves. *J. Plant Physiol.* **116**, 147–152.

DeWit, C. R. (1960) *On Competition.* Versl. Landbouwk. Onderzoek, 66.8, p. 82.

Dijak, M. and Ormrod, D. P. (1982) Some physiological and anatomical characteristics associated with differential ozone sensitivity among pea cultivars. *Environ. Exp. Bot.* **22**, 395–402.

Dizengremel, P. and Citerne, A. (1988) Air pollutant effects on mitochondria and respiration. In *Air Pollution and Plant Metabolism* (eds., Schulte-Hostede, S., Darrall, N. M., Blank, L. W., and Wellburn ,A. R.). Elsevier, London, 169–188.

Dominy, P. J. and Heath, R. L. (1985) Inhibition of the K^+-stimulated ATPase of the plasmalemma of pinto bean leaves by ozone. *Plant Physiol.* **77**, 43–45.

Dowding, P. (1988) Air pollutant effects on plant pathogens. In *Air Pollution and Plant Metabolism* (eds., Schulte-Hostede S., Darrall N. M., Blank L. W. and Wellburn A. R.). Elsevier, London, 329–355.

Downton, W. J. S., Loveys, B. R., and Grant, W. J. (1988a) Stomatal closure fully accounts for the inhibition of photosynthesis by abscisic acid. *New Phytol.* **108**, 263–266.

Downton, W. J. S., Loveys, B. R., and Grant, W. J. (1988b) Non-uniform stomatal closure induced by water stress causes putative non-stomatal inhibition of photosynthesis. *New Phytol.* **110**, 503–509.

Drolet, G., Dumbroff, E. B., Legge, R. L., and Thompson, J. E. (1986) Radical scavenging properties of polyamines. *Phytochemistry* **25**, 367–371.

Dugger, W. M., Jr. and Ting, I. P. (1970) Physiological and biochemical effects of air pollution oxidants on plants. *Rec. Adv. Phytochem.* **3**, 31–58.

Dunn, D. B. (1959) Some effects of air pollution on *Lupinus* in the Los Angeles area. *Ecology* **40**, 621–625.

Dunning, J. A., Heck, W. W., and Tingey, D. T. (1974) Foliar sensitivity of pinto bean and soybean to ozone as affected by temperature, potassium nutrition and ozone dose. *Water Air Soil Pollut.* **3**, 305–313.

Eaton, G. W. and Kyte, T. R. (1978) Yield component analysis in the cranberry. *J. Am. Soc. Hort. Sci.* **103**, 578–583.

Egneus, H., Mattiesen, U., and Kirk, M. (1975) Reduction of oxygen by the electron transport chain of chloroplasts during assimilation of carbon dioxide. *Biochim. Biophys. Acta* **408**, 252–268.

Elkiey, T. and Ormrod, D. P. (1979) Leaf diffusion resistance responses of three Petunia cultivars to ozone and/or sulfur dioxide. *JAPCA* **29**, 622–625.

Elkiey, T., Ormrod, D. P., and Pelletier, R. L. (1979) Stomatal and leaf surface features as related to ozone sensitivity of Petunia cultivars. *J. Am. Soc. Hort. Sci.* **104**, 510–514.

Elstner, E. F. (1982) Oxygen activation and oxygen toxicity. *Ann. Rev. Physiol.* **33**, 73–96.

Endress, A. G. and Grunwald, C. (1985) Impact of chronic ozone on soybean growth and biomass partitioning. *Agric. Ecosyst. Environ.* **13**, 9–23.

Endress, A. G. and Post, S. L. (1985) Altered feeding preference of Mexican bean beetle *Epilachna varivestis* for ozonated soybean foliage. *Environ. Pollut.* (Ser. A) **39**, 9–16.

Engle, R. L. and Gabelman, W. H. (1966) The effects of low levels of ozone on pinto beans, *Phaseolus vulgaris* L. *J. Am. Soc. Hort. Sci.* **91**, 304–309.

Evans, L. S. and Ting, I. P. (1974) Ozone sensitivity of leaves; relationship to leaf water content, gas transfer resistance and anatomical characteristics. *Amer. J. Bot.* **61**, 592–597

Feder, W. A. (1968) Reduction in tobacco pollen germination and tube elongation, induced by low levels of ozone. *Science* **160**, 1122.

Feder, W. A. (1970) Plant response to chronic exposure of low levels of oxidant type air pollution. *Environ. Pollut.* **1**, 73–79.

Finchk, B. F. and Kunert, K. J. (1985) Vitamins C and E: an antioxidative system against herbicide-induced lipid peroxidation in higher plants. *J. Agric. Food Chem.* **33**, 574–577.

Flagler, R. B. and Youngner, V. B. (1985) Ozone and sulfur dioxide effects on tall fescue. II. Alteration of quality constituents. *J. Environ. Qual.* **14**, 463–466.

Flagler, R. B., Patterson, R. P., Heagle, A. S., and Heck, W. W. (1987) Ozone and soil moisture deficit effect on nitrogen metabolism of soybean. *Crop Sci.* **27**, 1177–1184.

Floyd, R. A., West, M. S., Hogsett, W. E., and Tingey, D. T. (1989) Increased 8-hydroxyguanine content of chloroplast DNA from ozone-treated plants. *Plant Physiol.* **91**, 644–647.

Fong, F. (1985) Mechanisms of acute and chronic effects of ozone injury. In *Evaluation of the Scientific Basis for Ozone/Oxidant Standards* (ed., Lee, S. D.). Air Pollution Control Association, Pittsburgh, 107–114.

Forberg, E., Arnes, H., Nilsen S., and Semb, A. (1987) Effect of ozone on net photosynthesis in oat (*Avena sativa*) and duckweed (*Lemna gibba*). *Environ. Pollut.* **47**, 285–291.

Furukawa, A., Katase, M., and Ushijima, T. (1984) Inhibition of photosynthesis of poplar species and sunflower by O_3. *Res. Rept. Natl. Inst. Environ. Stud., Japan* **65** (1), 77–86.

Gaines, W. M. (1958) Methodology of feedback control. In *Handbook of Automation, Computations, and Control* Vol. 1, (eds., Gabbe, E. M., Ramo, S. and Woolridge, D. E.). John Wiley & Sons, New York, 19/01–19/21.

Gaspar, T. H., Penel, C. L., Castillo, F. J., and Greppin, H. (1985) A two-step control of basic and acidic peroxidases and its significance for growth and development. *Physiol. Plant* **64**, 418–423.

Ghanotakis, D. T. and Yocum, C. F. (1990) Photosystem II and the oxygen-evolving complex. *Ann. Rev. Plant Physiol. Plant Mol. Biol.* **41**, 255–276.

Giamalva, P., Church, D. F., and Pryor, W. A. (1985) A comparison of the rates of ozonation of biological antioxidants and oleate and linoleate esters. *Biochem. Biophys. Res. Comm.* **133**, 773–779.

Giaquinta, R. T. (1983) Phloem loading of sucrose. *Ann. Rev. Plant Physiol.* **34**, 347–387.

Goodyear, S. N. and Ormrod, D. P. (1988) Tomato response to concurrent and sequential NO_2 and O_3 exposures. *Environ. Pollut.* **51**, 315–326.

Govindjee, Kambara ,T. and Coleman, W. (1985) The electron donor side of photosystem II: the oxygen evolving complex. *Photochem. Photobiol.* **42**, 187–210.

Greitner, C. S. and Winner, W. E. (1988) Increases in $\sigma^{13}C$ values of radish and soybean plants caused by ozone. *New Phytol.* **108**, 489–494.

Grimes, H. D., Perkins, K. K., and Boss, W. F. (1983) Ozone degrades into hydroxyl radical under physiological conditions. A spin trapping study. *Plant Physiol.* **72**, 1016–1020.

Grunwald, C. and Endress, A. G. (1985) Foliar sterols in soybeans exposed to chronic levels of ozone. *Plant Physiol.* **77**, 245–247.

Grunwald, C. and Endress, A. G. (1988) Oil, fatty acid and protein content of seeds harvested from soybeans exposed to O_3 and/or SO_2. *Bot. Gaz.* **149**, 283–288.

Guderian, R. (ed.) (1985) *Air Pollution by Photochemical Oxidants*. Springer-Verlag, Berlin., 346.

Guderian, R., Tingey, D. T., and Rabe, R. (1985) Effects of photochemical oxidants on plants. In *Air Pollution by Photochemical Oxidants* (ed., Guderian R.). Springer-Verlag, Berlin, 127–333.

Guri, A. (1983) Variation in glutathione and ascorbic acid content among selected cultivars of *Phaseolus vulgaris* prior to and after exposure to ozone. *Can. J. Plant Sci.* **63**, 733–737.

Haas, J. H. (1970) Relation of crop maturity and physiology to air pollution incited bronzing of *Phaseolus vulgaris*. *Phytopathology* **60**, 407–410.

Haber, F. and Weiss, J. (1934) The catalytic decomposition of hydrogen peroxide by iron salts. *Proc. R. Soc. Lond.* A **147**, 332–351.

Haehnel, W. (1984) Photosynthetic electron transport in plants. *Ann. Rev. Plant Physiol.* **35**, 659–693.

Hallgren, J. E. (1984) Photosynthetic gas exchange in leaves affected by air pollutants. In *Gaseous Air Pollutants and Plant Metabolism* (eds., Koziol, M. J. and Whatley, F. R.). Butterworths, London, 147–159.

Halliwell, B. (1974) Superoxide dismutase, catalase and glutathione peroxidase: solutions to the problems of living with oxygen. *New Phytol.* **73**, 1075–1086.

Halliwell, B. (1982) Ascorbic acid and the illuminated chloroplast. *Am. Chem. Soc. Adv. Chem. Ser.* **200**, 263–274.

Harrison, B. H. and Feder, W. A. (1974) Ultrastructural changes in pollen exposed to ozone. *Phytopathology* **64**, 257–258.

Heagle, A. S. (1973) Interactions between air pollutants and plant parasites. *Ann. Rev. Phytopathol.* **11**, 365–388.

Heagle, A. S. (1979) Effects of growth media, fertilizer rate and hour and season of exposure on sensitivity of four soybean cultivars to ozone. *Environ. Pollut.* **18**, 313–322.

Heagle, A. S. (1982) Interactions between air pollutants and parasitic plant diseases. In *Effects of Gaseous Air Pollution on Agriculture and Horticulture* (eds., Unsworth, M. H. and Ormrod, D. P.). Butterworth Scientific, London, 333–348.

Heagle, A. S. and Key, L. W. (1973) Effect of *Puccinia graminis* f.sp. *tritici* on ozone injury in wheat. *Phytopathology* **63**, 609–613.

Heagle, A. S. and Letchworth, M. B. (1982) Relationships among injury, growth and yield responses of soybean cultivars exposed to ozone at different light intensities. *J. Environ. Qual.* **11**, 690–694.

Heagle, A. S., Body, D. F., and Pounds, E. K. (1972) Effect of ozone on yield of sweet corn. *Phytopathology* **62**, 683–687.

Heagle, A. S., Letchworth, M. B., and Mitchell, C. A. (1983) Injury and yield responses of peanuts to chronic doses of ozone and sulfur dioxide in open-top field chambers. *Phytopathology* **73**, 551–555.

Heagle, A. S., Philbeck, R. B., and Knott, W. M. (1979a) Thresholds for injury, growth and yield loss caused by ozone on field corn hybrids. *Phytopathology* **69**, 21–26.

Heagle, A. S., Spencer, S., and Letchworth, M. B. (1979b) Yield response of winter wheat to chronic doses of ozone. *Can. J. Bot.* **57**, 1999–2005.

Heagle, A. S., Miller, J. E., Heck, W. W., and Patterson, R. P. (1988b) Injury and yield responses of cotton to chronic doses of ozone and soil moisture deficit. *J. Environ. Qual.* **17**, 627–635.

Heagle A. S., Rebbeck J., Shafer S. R., Blum U. and Heck W. W. (1989) Effects of long-term ozone exposure and soil moisture deficit on growth of a Ladino clover-tall fescue pasture. *Phytopathology* **79**, 128–136.

Heagle, A. S., Flagler, R. B., Patterson, R. P., Lesser, V. M., Shafer, S. R., and Heck, W. W. (1987) Injury and yield response of soybean to chronic doses of ozone and soil moisture deficit. *Crop Sci.* **27**, 1016-1024.

Heagle, A. S., Kress, L. W., Temple, P. J., Kohut, R. J., Miller, J. E., and Heggestad, H. E. (1988a) Factors influencing ozone dose-yield response relationships in open-top field chamber studies. In *Assessment of Crop Loss from Air Pollutants* (eds., Heck, W. W., Taylor, O. C., and Tingey, D. T.). Elsevier, London, 141–179.

Heath, R. L. (1980) Initial events in injury to plants by air pollutants. *Ann. Rev. Plant Physiol.* **31**, 395–431.

Heath, R. L. (1984) Air pollutant effects on biochemicals derived from metabolism: organic, fatty and amino acids. In *Gaseous Air Pollutants and Plant Metabolism* (eds., Koziol, M. J. and Whatley, F. R.). Butterworths, London, 275–290.

Heath, R. L. (1987a) The biochemistry of ozone attack on the plasma membrane of plant cells. *Rec. Adv. Phytochem.* **21**, 29–54.

Heath, R. L. (1987b) Oxidant air pollutants and plant injury. In *Models in Plant Physiology and Biochemistry*, Vol. 3. (eds., Newman, D. W. and Wilson, K. G.). CRC Press, Boca Raton, FL, 63–66.

Heath, R. L. (1988) Biochemical mechanisms of pollutant stress. In *Assessment of Crop Loss from Air Pollutants* (eds., Heck, W. W., Taylor, O. C., and Tingey, D. T.). Elsevier, London, 259–286.

Heath, R. L. and Castillo, F. J. (1988) Membrane disturbances in response to air pollutants. In *Air Pollution and Plant Metabolism* (eds., Schulte-Hostede, S., Darrall, N. M., Blank, L. W., and Wellburn, A. R.). Elsevier, London, 55–75.

Heath, R. L. and Frederick, P. E. (1979) Ozone alteration of membrane permeability in *Chlorella*. I: permeability of potassium ion as measured by [86]rubidium tracer. *Plant Physiol.* **64**, 455–459.

Heck, W. W., Mudd, J. B., and Miller, P. R. (1977) Plants and microorganisms. In *Ozone and Other Photochemical Oxidants*, Vol. 2. National Academy of Sciences, Washington, D.C., 437–585.

Heck, W. W., Taylor, O. C., and Tingey, D. T. eds. (1988) *Assessment of Crop Loss from Air Pollutants*. Elsevier, London. 552

Heck, W. W., Cure, W. W., Rawlings, J. O., Zaragoza, L. J., Heagle, A. S., Heggestad, H. E., Kohut, R. J., Kress, L. W., and Temple, P. J. (1984a) Assessing impacts of ozone on agricultural crops. I. Overview. *JAPCA* **34**, 729–735.

Heck, W. W., Cure, W. W., Rawlings, J. O., Zaragoza, L. J., Heagle, A. S., Heggestad, H. E., Kohut, R. J., Kress, L. W., and Temple, P. J. (1984b) Assessing impacts of ozone on agricultural crops. II. Crop yield functions and alternative exposure statistics. *JAPCA* **34**, 810–817.

Heggestad, H. E., Anderson, E. L., Gosh, T. J., and Lee, E. H. (1988) Effects of ozone and soil water deficit on roots and shoots of field-grown soybeans. *Environ. Pollut.* **50**, 259–278.

Heggestad, H. E., Gosh, T. J., Lee, E. H., Bennett, J. H., and Douglas, L. W. (1985) Interaction of soil moisture stress and ambient ozone on growth and yields of soybeans. *Phytopathology* **75**, 472–477.

Hill, A. C. and Littlefield ,N. (1969) Ozone. Effect on apparent photosynthesis, rate of transpiration and stomatal closure in plants. *Environ. Sci. Technol.* **3**, 52–56.

Hofstra, G., Ali,A., Wukasch, R. T., and Fletcher, R. A. (1981) The rapid inhibition of root respiration after exposure of bean (*Phaseolus vulgaris* L.) plants to ozone. *Atmos. Environ.* **15**, 483–487.

Hogsett, W. E., Tingey, D. T., and Lee, E. H. (1988) Ozone exposure indices: concepts for development and evaluation of their use. In *Assessment of Crop Loss from Air Pollutants* (eds., Heck, W. W., Taylor, O. C. and Tingey, D. T.). Elsevier, London, 107–138.

Hoigne, J. and Bader, H. (1975) Ozonation of water: role of hydroxyl radicals as oxidizing intermediates. *Science* **190**, 782–784.

Holloway, P. J. (1982) Structure and histochemistry of plant cuticular membranes: an overview. In *The Plant Cuticle* (eds., Cutler, D. F., Alvin, K. L., and Price, C. E.). Academic Press, London, 1–32.

Hossain, M., Nakano, Y., and Asada, K. (1984) Monodehydroascorbate reductase in spinach chloroplasts and participation in regeneration of ascorbate for scavenging hydrogen peroxide. *Plant Cell Physiol.* **25**, 385–395.

Howell, R. K. (1970) Influence of air pollution on quantities of caffeic acid isolated from leaves of *Phaseolus vulgaris*. *Phytopathology* **60**, 1626–1629.

Howell, R. K. and Kremer, D. F. (1973) The chemistry and physiology of pigmentation in leaves injured by air pollution. *J. Environ. Qual.* **2**, 434–438.

Hughes, P. R. (1988) Insect populations on host plants subjected to air pollution. In *Plant-Stress-Insect Interactions* (ed., Heinricks, E. A.). John Wiley & Sons, New York, 249–319.

Hughes, P. R. and Laurence, J. A. (1984) Relationship of biochemical effects of air pollutants on plants to environmental problems: insect and microbial interactions. In *Gaseous Air Pollutants and Plant Metabolism* (eds., Koziol, M. J. and Whatley ,F. R.). Butterworths, London, 361–377.

Hunt, R. (1982) *Plant Growth Curves.* Edward Arnold, London, 248.

Hurwitz, B., Pell, E. J., and Sherwood, R. T. (1979) Status of coumestrol and 4′,7-dihydroxyflavone in alfalfa foliage exposed to ozone. *Phytopathology* **69**, 810–813.

Imlay, J. A. and Linn, S. (1988) DNA damage and oxygen radical toxicity. *Science* **240**, 1302–1309.

Irving, P. M., Kress, L. W., Prepejchal, W., and Smith, H. J. (1988) Studies on the interaction of ozone with drought stress or with sulfur dioxide on soybeans and corn. Tech. Prog. Rept. ANL-88-31 (January 1986 – October 1987), Argonne National Laboratory, Argonne, IL.

Ito, O., Mitsumori, F., and Totsuka, T. (1985a) Effects of NO_2 and O_3 alone or in combination on kidney bean plants (*Phaseolus vulgaris* L.): products of $^{13}CO_2$ assimilation detected by ^{13}C nuclear magnetic resonance. *J. Exp. Bot.* **36**, 281–289.

Ito, O., Okano, K., Kuroiwa M., and Totsuka, T. (1985b) Effects of NO_2 and O_3 alone or in combination on kidney bean plants (*Phaseolus vulgaris* L.): growth, partitioning of assimilates and root activities. *J. Exp. Bot.* **36**, 652–662.

Jablonski, P. P. and Anderson, J. W. (1982) Light-dependent reduction of hydrogen peroxide by ruptured pea chloroplasts. *Plant Physiol.* **69**, 1407–1413.

Jensen, K. F. (1981) Ozone fumigation decreased the root carbohydrate content and dry weight of green ash seedlings. *Environ. Pollut.* (Ser. A) **26**, 147–152.

Johnston, J. W., Jr. and Shriner, D. S. (1986) Yield response of Davis soybeans to simulated acid rain and gaseous pollutants in the field. *New Phytol.* **103**, 695–707.

Johnston, J. W., Jr., Shriner, D. S., and Kinerley, C. K. (1986) The combined effects of simulated acid rain and ozone on injury, chlorophyll, and growth of radish. *Environ. Exp. Bot.* **26**, 107–113.

Jolliffe, P. A. (1988) Evaluating the effects of competitive interference on plant performance. *J. Theor. Biol.* **130**, 447–459.

Jolliffe, P. A. and Courtney, W. H. (1984) Plant growth analysis: additive and multiplicative components of growth. *Ann. Bot.* **54**, 243–254.

Jolliffe, P. A., Eaton, G. W., and Lovett, Doust J. (1982) Sequential analysis of plant growth. *New Phytol.* **92**, 287–296.

Jones, A. W., Mulchi, C. L., and Kenworthy, W. J. (1985) Nodule activity in soybean cultivars exposed to ozone and sulfur dioxide. *J. Environ. Qual.* **14**, 60–65.

Jones, H. G. (1985) Partitioning stomatal and non-stomatal limitations to photosynthesis. *Plant Cell Environ.* **8**, 95–104.

Kaiser, W. M. (1976) The effect of hydrogen peroxide on CO_2 fixation of isolated intact chloroplasts. *Biochim. Biophys. Acta* **440**, 474–478.

Keen, N. T. and Taylor, O. C. (1975) Ozone injury in soybeans. *Plant Physiol.* **55**, 731–733.

Kerstiens, G. and Lendzian, K. J. (1989) Interactions between ozone and plant cuticles. II. Water permeability. *New Phytol.* **112**, 21–27.

King, D. A. (1987) A model for predicting the influence of moisture stress on crop losses caused by ozone. *Ecol. Model.* **35**, 29–44.

Kobayashi, K., Miller, J. E., Flagler, R. B., and Heck, W. W. (1990) Modeling the effects of ozone on soybean growth. *Environ. Pollut.* **65**, 33–64.

Kohut, R. J., Lawrence, J. A., and Amundson, R. G. (1988) Effects of ozone and sulfur dioxide on yield of red clover and timothy. *J. Environ. Qual.* **17**, 580–585.

Koiwai, A. and Kisaki, T. (1976) Effect of ozone on photosystem II of tobacco chloroplasts in the presence of piperonyl butoxide. *Plant Cell Physiol.* **17**, 1199–1207.

Kondo, N. and Sugahara, K. (1984) Effects of air pollutants on transpiration rate in relation to abscisic acid content. *Res. Rept. Natl. Inst. Environ. Stud., (Japan)* **65**(1), 1–7.

Koukol, J. and Dugger, W. M., Jr. (1967) Anthocyanin formation as a response to ozone and smog treatment of *Rumex crispus* L. *Plant Physiol.* **42**, 1023–1024.

Koziol, M. J. and Whatley, F. R. (eds.) (1984) *Gaseous Air Pollutants and Plant Metabolism.* Butterworths, London, 466.

Koziol, M. J., Whatley, F. R., and Shelvey, J. D. (1988) An integrated view of the effects of gaseous air pollutants on plant carbohydrate metabolism. In *Air Pollution and Plant Metabolism* (eds., Schulte-Hostede, S., Darrall, N. M., Blank, L. W., and Wellburn, A. R.). Elsevier, London, 148–168.

Krause, C. R. and Weidensaul, T. C. (1978) Effects of ozone on the sporulation, germination and pathogenicity of *Botrytis cinerea*. *Phytopathology* **68**, 196–198.

Kress, L. W. and Miller, J. E. (1983) Impact of ozone on soybean yield. *J. Environ. Qual.* **12**, 276–281.

Kropff, M. J. (1987) Physiological effects of sulphur dioxide. I. The effect of SO_2 on photosynthesis and stomatal regulation of *Vicia faba* L. *Plant Cell Environ.* **10**, 753–760.

Krupa, S. and Kickert, R. N. (1987) An analysis of numerical models of air pollutant exposure and vegetation response. *Environ. Pollut.* **44**, 127–158.

Krupa, S. V. and Kickert, R. N. (1989) The greenhouse effect: impacts of ultraviolet-B (UV-B), carbon dioxide (CO_2), and ozone (O_3) on vegetation. *Environ. Pollut.* **61**, 263–393.

Krupa, S. V. and Nosal, M. (1989) Application of spectral coherence analysis to describe the relationships between ambient ozone exposure and crop growth. *Environ. Pollut.* **60**, 319–330.

Kuchino, Y., Mori, F., Kasai, H., Iwai, S., Miura, K., Ohtsuka, E., and Nishimura, S. (1987) Misreading of DNA templates containing 8-hydroxyguanosine at the modified base and at adjacent residues. *Nature* **327**, 77–79.

Kunert, K. J. and Ederer, M. (1985) Leaf aging and lipid peroxidation: the role of the antioxidants vitamins C and E. *Physiol. Plant.* **65**, 85–88.

Kyle, D. J., Baker, N. R., and Arntzen, C. J. (1983) Spectral characterization of photosystem I fluorescence at room temperature using thylakoid protein phosphorylation. *Photobiochem. Photobiophys.* **5**, 79–86.

Laisk, A. (1983) Calculations of leaf photosynthetic parameters considering the statistical distribution of stomatal apertures. *J. Exp. Bot.* **34**, 1627–1635.

Laisk, A., Kull, O. ,and Moldau, H. (1989) Ozone concentration in leaf intercellular air spaces is close to zero. *Plant Physiol.* **90**, 1163–1167.

Larsen, R. I. and Heck, W. W. (1976) An air quality data analysis system for interrelating effects, standards and needed source reductions: Part III. Vegetation injury. *JAPCA* **26**, 325–333.

Le Sueur-Brymer, N. M. and Ormrod, D. P. (1984) Carbon dioxide exchange rates of fruiting soybean plants exposed to ozone and sulphur dioxide singly or in combination. *Can. J. Plant Sci.* **64**, 69–75.

Lee, E. H. and Bennett, J. H. (1982) Superoxide dismutase. A possible protective enzyme against ozone injury in snap beans (*Phaseolus vulgaris* L.) *Plant Physiol.* **69**, 1444–1449.

Lee, E. H., Tingey ,D. T., and Hogsett, W. E. (1988) Evaluation of ozone exposure indices in exposure-response modeling. *Environ. Pollut.* **53**, 43–62.

Lee, E. H., Tingey, D. T., and Hogsett, W. E. (1989) Interrelation of experimental exposure and ambient air quality data for comparison of ozone exposure indices and estimating agricultural losses. Report for U.S. Environmental Protection Agency, Office of Air Quality Planning and Standards, Research Triangle Park, NC.

Lee, E. H., Jersey, J. A., Gifford, C., and Bennett, J. (1984) Differential ozone tolerance in soybean and snapbeans: analysis of ascorbic acid in O_3-susceptible and O_3-resistant cultivars by high-performance liquid chromatography. *Environ. Exp. Bot.* **24**, 331–341.

Lee, S. D. (ed.) (1985) *Evaluation of the Scientific Basis for Ozone/Oxidant Standards*. Air Pollution Control Association, Pittsburgh, 394.

Lee, T. T. (1967) Inhibition of oxidative phosphorylation and respiration by ozone in tobacco mitochondria. *Plant Physiol.* **42**, 691–696.

Leffler, H. R. and Cherry, J. H. (1974) Destruction of enzymatic activities of corn and soybean leaves exposed to ozone. *Can. J. Bot.* **52**, 1233–1238.

Lefohn, A. S. and Ormrod, D. P. (1984) A review and assessment of the effects of pollutant mixtures on vegetation_ Research recommendations. Report EPA-600/3-84-037, Corvallis Environmental Research Laboratory, U.S. Environmental Protection Agency, Corvallis, OR, 104.

Lefohn, A. S. and Runeckles, V. C. (1987) Establishing standards to protect vegetation — ozone exposure/dose considerations. *Atmos. Environ.* **21**, 561–568.

Lefohn, A. S. and Tingey, D. T. (1984) The co-occurrence of potentially phytotoxic concentrations of various gaseous air pollutants. *Atmos. Environ.* **18**, 2521–2526.

Lefohn, A. S., Runeckles, V. C., Krupa, S. V., and Shadwick, D. S. (1989) Important considerations for establishing a secondary ozone standard to protect vegetation. *JAPCA* **39**, 1039–1045.

Lefohn, A. S., Davis, C. E., Jones, C. K., Tingey, D. T., and Hogsett, W. E. (1987) Co-occurrence patterns of gaseous air pollutant pairs at different minimum concentrations in the United States. *Atmos. Environ.* **21**, 2435–2444.

Legge, A. H. and Krupa, S. V. (eds.) (1986) *Air Pollutants and Their Effects on the Terrestrial Ecosystem.* John Wiley & Sons, New York, 662.

Lehnherr, B., Grandjean, A., Machler, F., and Fuhrer, J. (1987) The effect of ozone in ambient air on ribulosebisphosphate carboxylase/oxygenase activity decreases photosynthesis and grain yield in wheat. *J. Plant Physiol.* **130**, 189–200.

Lehnherr, B., Machler, F., Grandjean, A., and Fuhrer J. (1988) The regulation of photosynthesis in leaves of field-grown spring wheat (*Triticum aestivum* L., cv Albis) at different levels of ozone in ambient air. *Plant Physiol.* **88**, 1115–1119.

Leone, I. A. (1976) Response of potassium-deficient tomato plants to atmospheric ozone. *Phytopathology* **66**, 734–736.

Leone, I. A. and Brennan, E. (1970) Ozone toxicity in tomato as modified by phosphorus nutrition. *Phytopathology* **60**, 1521–1524.

Letchworth, M. B. and Blum, U. (1977) Effects of acute ozone exposure on growth, nodulation and nitrogen content of ladino clover. *Environ. Pollut.* **14**, 303–312.

Levitt, J. (1972) *Responses of Plants to Environmental Stresses.* Academic Press, New York, 697.

Lieth, J. H. and Reynolds, J. F. (1986) Plant growth analysis of discontinuous growth data: a modified Richards function. *Sci. Hort.* **28**, 301–314.

Lorimer, G. H. (1981) The carboxylation and oxygenation of ribulose-1,5-bisphosphate: the primary events in photosynthesis and photorespiration. *Ann. Rev. Plant Physiol.* **32**, 349–383.

Luckner, M. (1980) Expression and control of secondary metabolism. In *Secondary Plant Products,* Encycl. Plant Physiol. New Ser., Vol. 8. (eds., Bell, E. A. and Charlwood, B. V.). Springer-Verlag, New York, 23–63.

Luxmoore, R. J. (1988) Assessing the mechanisms of crop loss from air pollutants with process models. In *Assessment of Crop Loss from Air Pollutants* (edited by Heck W. W., Taylor, O. C., and Tingey, D. T.). Elsevier, London, 417–444.

Maas, E. V., Hoffman, G. J., Rawlins, S. L., and Ogata, G. (1973) Salinity-ozone interactions on pinto bean: integrated response to ozone concentration and duration. *J. Environ. Qual.* **2**, 400–404.

Macdowall, F. D. H. (1965) Stages of ozone damage to respiration of tobacco leaves. *Can. J. Bot.* **43**, 419–427.

Macdowall, F. D. H., Mukammal, E. I., and Cole, A. F. W. (1964) Direct correlation of air-polluting ozone and tobacco weather-fleck. *Can. J. Plant Sci.* **44**, 410–417.

Mackay, C. E., Senaratna, T., McKersie, B. D., and Fletcher, R. A. (1987) Ozone induced injury to cellular membranes in *Triticum aestivum* L. and protection by the triazole S-3307. *Plant Cell Physiol.* **28**, 1271–1278.

Manning, W. J. (1988) EDU: a research tool for assessment of the effects of ozone on vegetation. In *Proc. 81st Annual Meeting of the Air Pollution Control Association, Dallas, TX.* Air Pollution Control Association, Pittsburgh.

Manning, W. J. and Keane, K. D. (1988) Effects of air pollutants on interactions between plants, insects, and pathogens. In *Assessment of Crop Loss from Air Pollutants* (eds., Heck, W. W., Taylor, O. C., and Tingey, D. T.). Elsevier, London, 365–386.

Manning, W. J., Feder, W. A., and Vardaro, P. M. (1974) Suppression of oxidant injury by benomyl: effects on yields of bean cultivars in the field. *J. Environ. Qual.* **3**, 1–3.

Manning, W. J., Feder, W. A., Perkins, I., and Glickman, M. (1969) Ozone injury and infection of potato leaves by *Botrytis cinerea*. *Plant Dis. Rep.* **53**, 691–3.

Mansfield, T. A. and Freer-Smith, P. H. (1984) The role of stomata in resistance mechanisms. In *Gaseous Air Pollutants and Plant Metabolism* (eds., Koziol, M. J. and Whatley, F. R.). Butterworths, London, 131–146.

Mansfield, T. A. and McCune, D. C. (1988) Problems of crop loss when there is exposure to two or more gaseous pollutants. In *Assessment of Crop Loss from Air Pollutants* (edited by Heck W. W., Taylor O. C. and Tingey D.T.). Elsevier, London, 317–344.

Martin, B., Bytnerowicz, A., and Thorstenson, Y. R. (1988) Effects of air pollutants on the composition of stable isotopes, $\delta^{13}C$, of leaves and wood, and on leaf injury. *Plant Physiol.* **88**, 218–223.

Matters, G. L. and Scandalios, J. G. (1987) Synthesis of isozymes of superoxide dismutase in maize leaves in response to O_3, SO_2 and elevated O_2. *J. Exp. Bot.* **38**, 842–852.

McCool, P. M. (1988) Effect of air pollutants on mycorrhizae. In *Air Pollution and Plant Metabolism* (eds., Schulte-Hostede, S., Darrall, N. M., Blank, L. W., and Wellburn, A. R.). Elsevier, London, 356–365.

McCool, P. M. and Menge, J. A. (1983) Influence of ozone on carbon partitioning in tomato: potential role of carbon flow in regulation of the mycorrhizal symbiosis under conditions of stress. *New Phytol.* **94**, 241–247.

McCool, P. M., Menge, J. A., and Taylor, O. C. (1979) Effects of ozone and HCl gas on the development of the mycorrhizal fungus *Glomus fasciculatus* and growth of "Troyer" citrange. *J. Am. Soc. Hort. Sci.* **104**, 151–154.

McCool, P. M., Menge, J. A., and Taylor, O. C. (1982) Effect of ozone injury and light stress on response of tomato to infection by the vesicular-arbuscular mycorrhizal fungus, *Glomus fasciculatus*. *J. Am. Soc. Hort. Sci.* **107**, 839–842.

McIlveen, W. D., Spotts, R. A., and Davis, D. D. (1975) The influence of soil zinc on nodulation, mycorrhizae, and ozone-sensitivity of Pinto bean. *Phytopathology* **65**, 645–647.

McKersie, B. D., Beversdorf, W. D., and Hucl, P. (1982a) The relationship between ozone insensitivity, lipid-soluble antioxidants, and superoxide dismutase in *Phaseolus vulgaris*. *Can. J. Bot.* **60**, 2686–2691.

McKersie, B. D., Hucl, P., and Beversdorf, W. D. (1982b) Solute leakage from susceptible and tolerant cultivars of *Phaseolus vulgaris* following ozone exposure. *Can. J. Bot.* **60**, 73–78.

McLaughlin, S. B. and McConathy, R. K. (1983) Effects of SO_2 and O_3 on allocation of ^{14}C-labeled photosynthate in *Phaseolus vulgaris*. *Plant Physiol.* **73**, 630–635.

McLaughlin, S. B. and Taylor, G. E. (1981) Relative humidity: important modifier of pollutant uptake by plants. *Science* **211**, 167–169.

Mehler, A. H. (1951) Studies on reactions of illuminated chloroplasts. I. Mechanisms of the reduction of oxygen and other Hill reagents. *Arch. Biochem. Biophys.* **33**, 65–77.

Melhorn, H. and Wellburn, A. R. (1987) Stress ethylene formation determines plant sensitivity to ozone. *Nature* **327**, 417–418.

Melhorn, H., Senfert, G., Schmidt, A., and Kunert, K. I. (1986) Effect of SO_2 and O_3 on production of antioxidants in conifers. *Plant Physiol.* **82**, 336–338.

Menser, H. A. and Heggestad, H. E. (1966) Ozone and sulfur dioxide synergism: injury to tobacco plants. *Science* **153**, 424–425.

Meredith, F. I., Thomas, C. A., and Heggestad, H. E. (1986) Effect of the pollutant ozone in ambient air on lima beans. *J. Agric. Food Chem.* **34**, 179–185.

Mikami, B. and Ida, S. (1986) Reversible inactivation of ferredoxin-nitrate reductase from the cyanobacterium *Plectonema boryanum*. The role of superoxide anion and cyanide. *Plant Cell Physiol.* **27**, 1013–1021.

Miller, J. E. (1987) Effects of ozone and sulfur dioxide stress on growth and carbon allocation in plants. *Rec. Adv. Phytochem.* **21**, 55–100.

Miller, J. E. (1988) Effects on photosynthesis, carbon allocation and plant growth associated with air pollutant stress. In *Assessment of Crop Loss from Air Pollutants* (eds., Heck, W. W., Taylor, O. C., and Tingey, D. T.). Elsevier, London, 287–314.

Miller, J. E., Heagle, A. S., Vozzo, S. F., Philbeck, R. B., and Heck, W. W. (1989) Effects of ozone and water stress, separately and in combination, on soybean yield. *J. Environ. Qual.* **18**, 330–336.

Miyake, H., Furukawa, A., Totsuka, T., and Maeda, E. (1984) Differential effects of ozone and sulphur dioxide on the fine structure of spinach leaf cells. *New Phytol.* **96**, 215–228.

Mondovi, B., Rotilo, G., Finazzi-Agro, A., and Costa, M. T. (1967) Diamine oxidase inactivation by hydrogen peroxide. *Biochim. Biophys. Acta* **132**, 521–523.

Mooney, H. A. and Winner, W. E. (1988) Carbon gain, allocation and growth as affected by atmospheric pollutants. In *Air Pollution and Plant Metabolism* (eds., Schulte-Hostede, S., Darrall, N. M., Blank, L. W., and Wellburn, A. R.). Elsevier, London, 272–287.

Moser, T. J., Tingey, D. T., Rodecap, K. D., Rossi, D. J., and Clark, C. S. (1988) Drought stress applied during the reproductive phase reduced ozone-induced effects in bush bean. In *Assessment of Crop Loss from Air Pollutants* (eds., Heck, W. W., Taylor, O. C. and Tingey, D. T.).Elsevier, London., 345–364.

Moyer, J. W. and Smith, S. H. (1975) Oxidant injury reduction on tobacco induced by tobacco etch virus infection. *Environ. Pollut.* **9**, 103–106.

Mudd, J. B. (1982) Effects of oxidants on metabolic function. In *Effects of Gaseous Air Pollution on Agriculture and Horticulture* (eds., Unsworth, M. H. and Ormrod, D. P.). Butterworth Scientific, London, 189–201.

Mudd, J. B., McManus, T. T., Ongun, A., and McCullogh, T. E. (1971) Inhibition of glycolipid biosynthesis in chloroplasts by ozone and sulfhydryl reagents. *Plant Physiol.* **48**, 335–339.

Mukammal, E. I. (1965) Ozone as a cause of tobacco injury. *Agric. Meteorol.* **2**, 145–165.

Myhre, A., Forberg, E., Aarnes, H., and Nilsen, S. (1988) Reduction of net photosynthesis in oats after treatment with low concentrations of ozone. *Environ. Pollut.* **53**, 265–271.

Nakano, Y. and Asada, K. (1980) Spinach chloroplasts scavenge hydrogen peroxide on illumination. *Plant Cell Physiol.* **21**, 1295–1307.

Nakano, Y. and Asada, K. (1981) Hydrogen peroxide is scavenged by ascorbate-specific peroxidase in spinach chloroplasts. *Plant Cell Physiol.* **22**, 867–880.

Newman, D. W. and Wilson, K. G. (eds.) (1987) *Models in Plant Physiology and Biochemistry*, Vol. I, II, and III. CRC Press, Boca Raton, FL.

Nobel, P. S. (1983) *Biophysical Plant Physiology and Ecology*. W. H. Freeman, San Francisco, CA, 608.

Norby, R. J., Richter, D. D., and Luxmoore, R. J. (1985) Physiological processes in soybean inhibited by gaseous pollutants but not by acid rain. *New Phytol.* **100**, 79–85.

Nosal, M. (1983) Atmosphere-biosphere interface: probability analysis and an experimental design for studies of air-pollutant-induced plant response. Report 83/25 to Research Management Division, Alberta Environment, Edmonton, Alberta, 98.

Nouchi, I. and Toyama, S. (1988) Effects of ozone and peroxyacetyl nitrate on polar lipids and fatty acids in leaves of morning glory and kidney bean. *Plant Physiol.* **87**, 638–646.

Okano, K., Ito, O., Takaba, G., Shimizu, A., and Totsuka, T. (1984) Alteration of ^{13}C-assimilate partitioning in plants of *Phaseolus vulgaris* exposed to ozone. *New Phytol.* **97**, 155–163.

Omelian, J. A. and Pell, E. J. (1988) The role of photosynthetic activity in the response of isolated *Glycine max* mesophyll cells to ozone. *Can. J. Bot.* **88**, 745–749.

Ormrod, D. P. (1977) Cadmium and nickel effects on growth and ozone sensitivity of pea. *Water Soil Air Pollut.* **8**, 263–270.

Ormrod, D. P. and Beckerson, D. W. (1986) Polyamines as antiozonates for tomato. *Hort. Sci.* **21**, 1070–1071.

Ormrod, D. P., Adedipe, N. O., and Hofstra, G. (1973) Ozone effects on growth of radish plants as influenced by nitrogen and phosphorus nutrition and by temperature. *Plant Soil* **39**, 437–439.

Ormrod, D. P., Tingey, D. T., Gumpertz, M. L., and Olszyk, D. M. (1984) Utilization of a response-surface technique in the study of plant responses to ozone and sulfur dioxide mixtures. *Plant Physiol.* **75**, 43–48.

Oshima, R. J., Braegelmann, P. K., Flagler, R. B., and Teso, R. R. (1979) The effects of ozone on the growth, yield and partitioning of dry matter in cotton. *J. Environ. Qual.* **8**, 474–479.

Oshima, R. J., Taylor, O. C., Braegelmann, P. K., and Baldwin, D. W. (1975) Effect of ozone on the yield and plant biomass of a commercial variety of tomato. *J. Environ. Qual.* **4**, 463–464.

Pauls, K. P. and Thompson, J. E. (1981) Effects of *in vitro* treatment with ozone on the physical and chemical properties of membranes. *Physiol. Plant.* **53**, 255–262.

Peiser, G. and Yang, S. F. (1985) Biochemical and physiological effects of SO_2 on nonphotosynthetic processes in plants. In *Sulfur Dioxide and Vegetation* (eds., Winner, W. A., Mooney, H. A., and Goldstein, R. A.). Stanford University Press, Stanford, CA, 148–161.

Pell, E. J. and Brennan, E. (1973) Changes in respiration, photosynthesis, adenosine 5′-triphosphate, and total adenylate content of ozonated pinto bean foliage as they relate to symptom expression. *Plant Physiol.* **51**, 378–381.

Pell, E. J. and Pearson, N. S. (1983) Ozone-induced reduction in quantity of ribulose-1,5-bisphosphate carboxylase in alfalfa foliage. *Plant Physiol.* **73**, 185–187.

Pell, E. J., Lukezic, F. L., Levine, R. G., and Weisberger, W. C. (1977) Response of soybean foliage to reciprocal challenges by ozone and a hypersensitive-response-inducing Pseudomonad. *Phytopathology* **67**, 1342–1345.

Peters, J. L., Castillo, F. J., and Heath, R. L. (1989) Alteration of extracellular enzymes in pinto bean leaves upon exposure to air pollutants, ozone and sulfur dioxide. *Plant Physiol.* **89**, 159–164.

Pryor, W. A., Giamalva, D. H., and Church, D. F. (1984) Kinetics of ozonation. II. Amino acids and model compounds in water and comparison to rates in nonpolar solvents. *J. Am. Chem. Soc.* **106**, 7094–7100.

Purvis, A. C. (1978) Differential effects of ozone on *in vivo* nitrate reduction in soybean cultivars. I. Response to exogenous sugars. *Can. J. Bot.* **56**, 1540–1544.

Queiroz, O. (1988) Air pollution, gene expression and post-translational enzyme modifications. In *Air Pollution and Plant Metabolism* (eds., Schulte-Hostede, S., Darrall, N. M., Blank, L. W., and Wellburn, A. R.). Elsevier, London, 238–254.

Radford, P. J. (1967) Growth analysis formulae — their use and abuse. *Crop Sci.* **1**, 171–175.

Raschke, K. (1979) Movements of stomata. In *Physiology of Movements*. Encycl. Plant Physiol., New Ser., Vol. 7. (eds., Haupt, W. and Feinleib, M. R.). Springer-Verlag, Berlin, 383–441.

Rawlings, J. O. and Cure, W. W. (1985) The Weibull function as a dose-response model to describe ozone effects on crop yields. *Crop Sci.* **25**, 807–814.

Rawlings, J. O., Lesser, V. M., and Dassel, K. A. (1988) Statistical approaches to assessing crop losses. In *Assessment of Crop Loss from Air Pollutants* (eds., Heck, W. W., Taylor, O. C. and Tingey, D. T.). Elsevier, London., 389–416.

Rebbeck, J. and Brennan, E. (1984) The effect of simulated acid rain and ozone on the yield and quality of glasshouse-grown alfalfa. *Environ. Pollut.* (Ser. A) **36**, 7–16.

Rebbeck, J., Blum, U., and Heagle, A. S. (1988) Effects of ozone on the regrowth and energy reserves of a ladino clover-tall fescue pasture. *J. Appl. Ecol.* **25**, 659–681.

Reich, P. B. (1983) Effects of low concentrations of O_3 on net photosynthesis, dark respiration and chlorophyll contents in aging hybrid poplar leaves. *Plant Physiol.* **73**, 291–296.

Reich, P. B. (1987) Quantifying plant response to ozone: a unifying theory. *Tree Physiol.* **3**, 63–91.

Reich, P. B. and Amundson, R. G. (1984) Low level O_3 and/or SO_2 exposure causes a linear decline in soybean yield. *Environ. Pollut.* (Ser. A) **34**, 345–355.

Reich, P. B. and Amundson, R. G. (1985) Ambient levels of ozone reduce net photosynthesis in tree and crop species. *Science* **230**, 566–570.

Reich, P. B. and Lassoie, J. P. (1984) Effects of low level O_3 exposure on leaf diffusive conductance and water use efficiency in hybrid poplar. *Plant Cell Environ.* **7**, 661–668.

Reich, P. B., Schoettle, A. W., and Amundson, R. G. (1985) Effects of low concentrations of O_3, leaf age and water stress on leaf diffusive conductance and water use efficiency in soybean. *Physiol. Plant* **63**, 58–64.

Reich, P. B., Schoettle, A. W., Raba, R. M., and Amundson, R. G. (1986) Response of soybean to low concentrations of ozone. I. Reductions in leaf and whole plant net photosynthesis and leaf chlorophyll content. *J. Environ. Qual.* **15**, 31–36.

Rich, S. and Turner, N. C. (1972) Importance of moisture on stomatal behaviour of plants subjected to ozone. *JAPCA* **22**, 718–721.

Richards, F. J. (1959) A flexible growth function for experimental use. *J. Exp. Bot.* **10**, 290–300.

Rivett, A. J. (1986) Regulation of intracellular protein turnover: covalent modification as a mechanism of marking proteins for degradation. *Curr. Top. Cell. Regul.* **28**, 291–337.

Roberts, T. M. (1984) Effects of air pollutants on agriculture and forestry. *Atmos. Environ.* **18**, 629–652.

Robinson, D. C. and Wellburn, A. R. (1983) Light-induced changes in the quenching of 9-aminoacridine fluorescence by photosynthetic membranes due to atmospheric pollutants and their products. *Environ. Pollut.* (Ser. A) **32**, 109–120.

Robinson, J. M. (1988) Does O_2 photoreduction occur within chloroplasts *in vivo*? *Physiol. Plant* **72**, 666–680.

Roose, M. L., Bradshaw, A. D., and Roberts, T. M. (1982) Evolution of resistance to gaseous air pollutants. In *Effects of Gaseous Air Pollution in Agriculture and Horticulture* (eds., Unsworth, M. H. and Ormrod, D. P.). Butterworth Scientific, London, 379–409.

Rosen, P. M. and Runeckles, V. C. (1976) Interaction of ozone and greenhouse whitefly in plant injury. *Environ. Conserv.* **3**, 70–71.

Rowland, A. J., Borland, A. M., and Lea, P. J. (1988) Changes in amino-acids, amines and proteins in response to air pollutants. In *Air Pollution and Plant Metabolism* (eds., Schulte-Hostede, S., Darrall ,N. M., Blank, L. W., and Wellburn, A. R.). Elsevier, London, 189–221.

Rowland-Bamford, A. J., Coghlan, S., and Lea, P. J. (1989) Ozone-induced changes in CO_2 assimilation, O_2 evolution and chlorophyll *a* fluorescence transients in barley. *Environ. Pollut.* **59**, 129–140.

Rowlands, J. R., Gause, E. M., Rodriguez, C. F., and McKee, H. C. (1970) Electron spin resonance studies of vegetation damage. Final Report, SWRI Project 05-2622-01, Southwest Research Institute, San Antonio, TX, 67.

Runeckles, V. C. (1982) Relative death rate: a dynamic parameter describing plant response to stress. *J. Appl. Ecol.* **19**: 295–303.

Runeckles, V. C. (1987) Exposure, dose, vegetation response and standards: will they ever be related? In *Proc. 80th Annual Meeting of the Air Pollution Control Association, New York*, Air Pollution Control Association, Pittsburgh.

Runeckles, V. C. and Palmer, K. (1987) Pretreatment with nitrogen dioxide modifies plant response to ozone. *Atmos. Environ.* **21**, 717–719.

Runeckles, V. C. and Rosen, P. M. (1977) Effects of ambient ozone pretreatment on transpiration and susceptibility to ozone injury. *Can. J. Bot.* **55**, 193–197.

Runeckles, V. C. and Wright, E. F. (1989) Exposure-yield response models for crops. In *Proc. 82nd Annual Meeting of the Air and Waste Management Association, Anaheim, CA.* Air and Waste Management Association, Pittsburgh.

Sakaki, T., Kondo, N. and Sugahara, K. (1983) Breakdown of photosynthetic pigments and lipids in spinach leaves with ozone fumigation: role of active oxygens. *Physiol. Plant* **59**, 28–34.

Sakaki, T., Ohnishi, J., Kondo, N., and Yamada, M. (1985) Polar and neutral lipid changes in spinach leaves with ozone fumigation: triacylglycerol synthesis from polar lipids. *Plant Cell Physiol.* **26**, 253–262.

Salin, M. L. (1988) Toxic oxygen species and protective systems of the chloroplast. *Physiol. Plant* **72**, 681–689.

Saran M., Michel C. and Bors W. (1988) Reactivities of free radicals. In *Air Pollution and Plant Metabolism* (eds., Schulte-Hostede, S., Darrall, N. M., Blank, L. W., and Wellburn, A. R.). Elsevier, London., 76–93.

Schaedle, M. and Bassham, J. A. (1977) Chloroplast glutathione reductase. *Plant Physiol.* **59**, 1011–1012.

Schreiber, U. and Bilger, W. (1987) Rapid assessment of stress effects on plant leaves by chlorophyll fluorescence measurements. In *Plant Response to Stress. Functional Analysis in Mediterranean Ecosystems* (edited by Tenhunen J. D., Caterino F. M., Lange O. L. and Oechel W.C.), NATO ASI Series G, Vol. 15. Springer-Verlag, Berlin, 27–53.

Schreiber, U., Vidaver, W., Runeckles, V. C., and Rosen, P. (1978) Chlorophyll fluorescence assay for ozone injury in intact plants. *Plant Physiol.* **61**, 80–84.

Schulte-Hostede, S., Darrall, N. M., Blank, L. W., and Wellburn, A. R., (eds.) (1988) *Air Pollution and Plant Metabolism*. Elsevier, London, 381.

Schut, H. E. (1985) Models for the physiological effects of short O_3 exposures on plants. *Ecol. Model.* **30**, 175–207.

Schwab, W. G. W. and Komor, E. (1978) A possible mechanistic role of the membrane potential in proton-sugar cotransport in *Chlorella. FEBS Lett.* **87**, 157–160.

Shafer, S. R. (1988) Influence of ozone and simulated acid rain on microorganisms in the rhizosphere of *Sorghum. Environ. Pollut.* **51**, 131–152.

Sheng, S. and Chevone, B. I. (1988) Gas exchange response of soybean cultivars to short term exposure of sulfur dioxide and ozone. *Phytopathology* **78**, 1513.

Shimazaki, K. (1988) Thylakoid membrane reactions to air pollutants. In *Air Pollution and Plant Metabolism* (edited by Schulte-Hostede, S., Darrall, N. M., Blank, L. W., and Wellburn, A. R.). Elsevier, London., 116–133.

Sigal, L. L. and Johnston, J. W., Jr. (1986) Effects of acidic rain and ozone on nitrogen fixation and photosynthesis in the lichen *Lobaria pulmonaria* (L.) Hoffm. *Environ. Exp. Bot.* **26**, 59–64.

Sinclair, T. R. (1986) Water and nitrogen limitations in soybean grain production. I. Model development. *Field Crops Res.* **15**, 125–141.

Skarby, L. and Pell, E. J. (1979) Concentrations of coumestrol and 4',7- dihydroxyflavone in four alfalfa cultivars after exposure to ozone. *J. Environ. Qual.* **8**, 285–286.

Smith, T. A. (1985) Polyamines. *Ann. Rev. Plant Physiol.* **36**, 117-143.

Spitters, C. J. T. (1983) An alternative approach to the analysis of mixed cropping experiments. I. Estimation of competition effects. *Neth. J. Agric. Sci.* **31**, 1–11.

Staehelin, J. and Hoigne, J. (1982) Decomposition of ozone in water. Rate of initiation by hydroxide ion and hydrogen peroxide. *Environ. Sci. Tech.* **16**, 676–681.

Staehelin, J. and Hoigne, J. (1985) Decomposition of ozone in water in the presence of organic solutes acting as promoters and inhibitors of radical chain reactions. *Environ. Sci. Tech.* **19**, 1206–1213.

Stevens, T. M., Boswell, G. A., Jr., Adler, R., Ackerman, N. R., and Kerr, J. S. (1988) Induction of antioxidant enzyme activities by a phenylurea derivative, EDU. *Toxicol. Appl. Pharmacol.* **96**, 33–42.

Sugahara, K., Ogura, K., Takimoto, M., and Kondo, N. (1984) Effects of air pollutant mixtures on photosynthetic electron transport systems. *Res. Rept. Natl. Inst. Environ. Stud., (Japan)* **65**(1), 155–164.

Sutton, R. and Ting, I. P. (1977) Evidence for repair of ozone induced membrane injury: alteration in sugar uptake. *Atmos. Environ.* **11**, 273–275.

Takemoto, B. K., Bytnerowicz, A., and Olszyk, D. M. (1988a) Depression of photosynthesis, growth, and yield in field-grown green pepper (*Capsicum annuum* L.) exposed to acidic fog and ambient ozone. *Plant Physiol.* **88**, 477–482.

Takemoto, B. K., Bytnerowicz, A., and Olszyk, D. M. (1989) Physiological responses of field-grown strawberry (*Fragaria X ananassa* Duch.) exposed to acidic fog and ambient ozone. *Environ. Exp. Bot.* **29**, 379–386.

Takemoto, B. K., Hutton, W. J., and Olszyk, D. M. (1988b) Responses of field-grown *Medicago sativa* L. to acidic fog and ambient ozone. *Environ. Pollut.* **54**, 97–107.

Takemoto, B. K., Shriner, D. S., and Johnston, J. W., Jr. (1987) Physiological responses of soybean (*Glycine max* L. Merr) to simulated acid rain and ambient ozone in the field. *Water Soil Air Pollut.* **33**, 373–384.

Takemoto, B. K., Olszyk, D. M., Johnson, A. G., and Parada, C. R. (1988c) Yield- responses of field-grown crops to acidic fog and ambient ozone. *J. Environ. Qual.* **17**, 192–197.

Tanaka K., Otsubo, T., and Kondo, N. (1982) Participation of hydrogen peroxide in the inactivation of Calvin-cycle SH enzymes in SO_2-fumigated spinach leaves. *Plant Cell Physiol.* **23**, 1009–1018.

Tanaka, K., Saji, H., and Kondo, N. (1988) Immunological properties of spinach glutathione reductase and inductive biosynthesis of the enzyme with ozone. *Plant Cell Physiol.* **29**, 637–642.

Tanaka, K., Suda, Y., Kondo, N., and Sugahara, K. (1985) O_3 tolerance and the ascorbate-dependent H_2O_2 decomposing system in chloroplasts. *Plant Cell Physiol.* **26**, 1425–1431.

Taylor, G. E., Tingey, D. T., and Ratsch, H. C. (1982) Ozone flux in *Glycine max* (L.) Merr.: sites of regulation and relationship to leaf injury. *Oecologia* (Berlin) **53**, 179–186.

Taylor, G. E., Jr., Ross-Todd, B. M., and Gunderson, C. A. (1988) Action of ozone on foliar gas exchange in *Glycine max* L. Merr: a potential role for endogenous stress ethylene. *New Phytol.* **110**, 301–307.

Temple, P. J. (1986) Stomatal conductance and transpirational responses of field-grown cotton to ozone. *Plant Cell Environ.* **9**, 315–321.

Temple, P. J. and Benoit, L. F. (1988) Effects of ozone and water stress on canopy temperature, water use, and water use efficiency of alfalfa. *Agron. J.* **80**, 439–447.

Temple, P. J. and Bissesar, S. (1979) Response of white bean to bacterial blight, ozone, and antioxidant protection in the field. *Phytopathology* **69**, 101–103.

Temple, P. J., Taylor, O. C., and Benoit, L. F. (1985a) Cotton yield responses to ozone as mediated by soil moisture and evapotranspiration. *J. Environ. Qual.* **14**, 55–60.

Temple, P. J., Taylor, O. C., and Benoit, L. F. (1985b) Effects of ozone on yield of two field-grown barley cultivars. *Environ. Pollut.* (Ser. A) **39**, 217–225.

Temple, P. J., Kupper, R. S., Lennox, R. L., and Rohr, K. (1988b) Physiological and growth responses of differentially irrigated cotton to ozone. *Environ. Pollut.* **53**, 255–263.

Temple, P. J., Lennox, R. W., Bynerowicz, A. and Taylor, O. C. (1987) Interactive effects of simulated acidic fog and ozone on field-grown alfalfa. *Environ. Exp. Bot.* **27**, 409–417.

Temple, P. J., Benoit, L. F., Lennox, R. W., Regan, C. A., and Taylor, O. C. (1988a) Combined effects of ozone and water stress on alfalfa growth and yield. *J. Environ. Qual.* **17**, 108–113.

Tepperman, J. M. and Dunsmuir, P. (1990) Transformed plants with elevated levels of chloroplastic SOD are not more resistant to superoxide toxicity. *Plant Mol. Biol.* **14**, 501–511.

Thomson, W. W., Nagahashi, J. and Platt, K. (1974) Further observations on the effects of ozone on the ultrastructure of leaf tissue. In *Air Pollution Effects on Plant Growth* (edited by Dugger M.), ACS Symposium Series 3. American Chemical Society, Washington, D.C., 83–93.

Tingey, D. T. and Hogsett, W. E. (1985) Water stress reduces ozone injury via a stomatal mechanism. *Plant Physiol.* **77**, 944–947.

Tingey, D. T. and Taylor, G. E., Jr. (1982) Variation in plant response to ozone: a conceptual model of physiological events. In *Effects of Gaseous Air Pollution on Agriculture and Horticulture* (eds., Unsworth, M. H. and Ormrod, D. P.). Butterworth Scientific, London., 113–138.

Tingey, D. T., Fites, R. C., and Wickliff, C. (1973a) Ozone alteration of nitrate reduction in soybean. *Physiol. Plant* **29**, 33–38.

Tingey, D. T., Fites, R. C., and Wickliff, C. (1975) Activity changes in selected enzymes from soybean leaves following ozone exposure. *Physiol. Plant* **33**, 316–320.

Tingey, D. T., Hogsett, W. E., and Lee, E. H. (1989) Analysis of crop loss for alternative ozone exposure indices. In *Atmospheric Ozone Research and its Policy Implications* (eds., Schneider, T., Lee, S. D., Wolters, G. J. R., and Grant, L. D.). Elsevier, Amsterdam, 219–227.

Tingey, D. T., Reinert, R. A., Wickliff, C., and Heck, W. W. (1973b) Chronic ozone or sulfur dioxide exposures, or both, affect the early vegetative growth of soybean. *Can. J. Plant Sci.* **53**, 875–879.

Tingey, D. T., Wilhour, R. G., and Standley, C. (1976) The effect of chronic ozone exposures on the metabolite content of ponderosa pine seedlings. *Forest Sci.* **22**, 234–241.

Tomlinson, H. and Rich, S. (1969) Relating lipid content and fatty acid synthesis to ozone injury of tobacco leaves. *Phytopathology* **59**, 1284–1286.

Treshow, M. (ed.) (1984) *Air Pollution and Plant Life*. Wiley, Chichester, England, 486.

Trimble, J. L., Skelly, J. M., Tolin, S. A., and Orcutt, D. M. (1982) Chemical and structural characteristics of the needle wax of two clones of *Pinus strobus* differing in sensitivity to ozone. *Phytopathology* **72**, 653–656.

Troiano, J., Colavito, L., Heller, L., McCune, D. C., and Jacobson, J.S. (1983) Effects of acidity of simulated acid rain and its joint action with ambient ozone on measures of biomass and yield in soybean. *Environ. Exp. Bot.* **23**, 113–119.

Troughton, J. H. (1979) $\delta^{13}C$ as an indicator of carboxylation reactions. In *Photosynthesis* Encycl. Plant Physiol. New Ser. Vol. 6 (eds., Gibbs, M. and Latzko, E.). Springer-Verlag, Berlin., 140–149.

Trumble, J. T., Hare, J. D., Musselman, R. C., and McCool, P. M. (1987) Ozone-induced changes in host-plant suitability: interactions of *Keiferia lycopersicella* and *Lycopersicon esculentum*. *J. Chem. Ecol.* **13**, 203–218.

U.S. EPA. (1986) Air Quality Criteria for Ozone and Other Photochemical Oxidants. EPA/600/8-84/020cF, Vol. III. Environmental Criteria and Assessment Office, U.S. Environmental Protection Agency, Research Triangle Park, NC.

Unsworth, M. H. and Ormrod D,. P. (eds.) (1982) *Effects of Gaseous Air Pollution in Agriculture and Horticulture*. Butterworth Scientific, London, 532.

Unsworth, M. H., Lesser, V. M., and Heagle, A. S. (1984) Radiation interception and the growth of soybeans exposed to ozone in open-top field chambers. *J. Appl. Ecol.* **21**, 1058–1079.

Vaartnou, M. (1988) EPR Investigation of free radicals in excised and attached leaves subjected to ozone and sulphur dioxide air pollution. Ph.D. Thesis, University of British Columbia, Vancouver, B.C.

Vermaas, W. F. J., Pakrasi, H. B., and Arntzen, C. J. (1987) Photosystem II and inhibition by herbicides. In *Models in Plant Physiology and Biochemistry,* Vol. 1 (eds., Newman, D. W. and Wilson, K. G.). CRC Press, Boca Raton, FL, 9–12.

Walmsley L., Ashmore, M. R., and Bell, J. N. B. (1980) Adaptation of radish *Raphanus sativus* L. in response to continuous exposure to ozone. *Environ. Pollut.* (Ser. A) **23**, 165–177.

Weber, D. E., Reinert, R. A., and Barker, K. R. (1979) Ozone and sulfur dioxide effects on reproduction and host-parasite relationships of selected plant-parasitic nematodes. *Phytopathology* **69**, 624–628.

Wellburn, A. R. (1984) The influence of atmospheric pollutants and their cellular products upon photophosphorylation and related events. In *Gaseous Air Pollutants and Plant Metabolism* (eds., Koziol, M. J. and Whatley, F. R.). Butterworth Scientific, London, 203–221.

Wellburn, A. (1988) *Air Pollution and Acid Rain: the Biological Impact.* Longmans Scientific and Technical, Harlow, England, 274.

Whittaker, J. B., Kristiansen, L. W., Mikkelsen, T. N., and Moore, R. (1989) Responses to ozone of insects feeding on a crop and a weed species. *Environ. Pollut.* **62**, 89–101.

Winner, W. E., Gillespie, C., Shen, W. S., and Mooney, H. A. (1988) Stomatal responses to SO_2 and O_3. In *Air Pollution and Plant Metabolism* (eds., Schulte-Hostede, S., Darrall, N. M., Blank, L. W.,and Wellburn, A. R.). Elsevier, London, 255–271.

Wyse, R. E. (1985) Sinks as determinants of assimilate partitioning: possible sites for regulation. In *Phloem Transport* (eds., Cronshaw, J., Lucas, W. J., and Giaquinta, R. J.). Alan R. Liss, New York, 197–209.

Yang, Y. S., Skelly, J. M., Chevone B. I. and Birch J. B. (1983) Effects of short-term ozone exposure on net photosynthesis, dark respiration and transpiration of three eastern white pine clones. *Environ. Int.* **9**, 265–269.

Yarwood, C. E. and Middleton, J. T. (1954) Smog injury and rust infection. *Plant Physiol.* **29**, 393–395.

Yocum, C. F. (1987) Oxygen evolution by photosystem II: polypeptide structure and inorganic constituents. In *Models in Plant Physiology and Biochemistry* Vol. 1 (eds., Newman, D. W. and Wilson, K. G.). CRC Press, Boca Raton, FL, 3–5.

Younglove, T., McCool, P. M., Musselman, R. C., and Smith, E. D. (1988) The use of transfer function models to describe the relationship between pollutant exposure and leaf stomatal opening. *Environ. Exp. Bot.* **28**, 239–248.

Zwoch, I., Knorre, U., and Schaub, H. (1990) Influence of SO_2 and O_3 singly or in combination, on ethylene synthesis in sunflower. *Environ. Exp. Bot.* **30**, 193–205.

CHAPTER 7

Tree Responses to Ozone

Arthur H. Chappelka, School of Forestry, Auburn University, Auburn, AL

Boris I. Chevone, Department Plant Pathology, Physiology and Weed Science, Virginia Polytechnic and State University, Blacksburg, VA

7.1 INTRODUCTION

Trees are, and have been, used by man for many diverse purposes. In ancient times they provided food, fuel, and shelter necessary for survival. In recent history, trees have produced lumber for building materials, and wood for paper products and fuel. Within the last 20 to 30 years, air pollutants, including ozone (O_3), have become a potential threat to forest production in the U.S. and other countries. Recent observations indicate that unexplained declines have occurred in the growth rate of several tree species (McLaughlin, 1985). Ozone and other natural or man-made factors have been implicated as causal or contributing agents in these declines.

The purpose of this chapter is to discuss the effects of O_3 on tree health. Included are biochemical and physiological effects leading to alterations in growth and biomass production, interactions of O_3 exposure with other abiotic and biotic stresses, influence of physical factors, and genetic variability in tree sensitivity to O_3. This chapter will conclude with a section on future research directions.

Table 7.1. Response of Net Photosynthesis to Ozone Fumigation in Selected Coniferous and Hardwood Tree Species

Species	P_n response	Reference
Acer saccharum	inhibition	Reich et al.,1986
Populus hybrids	inhibition	Reich, 1983
Quercus rubra	no effect	Reich et al.,1986
Quercus alba	no effect	Foster et al., 1990
Lirodendron tulipifera	no effect	Roberts, 1990
Picea rubrens	no effect	Taylor et al., 1986
Pinus elliottii	inhibition	Barnes, 1972
Pinus ponderosa	inhibition	Miller et al., 1969
Pinus rigida	inhibition	Barnes, 1972
Pinus strobus	inhibition	Barnes 1972; Botkin et al., 1972; Yang et al., 1983a,b; Reich et al.; 1987; Barnes, 1972
Pinus taeda	inhibition	Barnes, 1972; Hanson et al., 1988; Sasek and Richardson, 1989
Picea abies	inhibition	Keller and Hasler, 1987
Pinus sylvestris	no effect	Barnes 1972; Skärby et al., 1987

7.2   EFFECTS OF OZONE

The effects of O_3 on trees are many and diverse. These effects originate with cellular injury that causes metabolic changes and, eventually, alterations in growth if O_3 dose is sufficient and plant protective or repair mechanisms are overcome. It is therefore important to understand O_3 effects on biochemical and physiological mechanisms in order to better understand subsequent responses in tree growth and productivity.

7.2.1   Metabolic and Biochemical Processes

Ozone can react with a diverse array of biological compounds and metabolites that are normally present in plant cells (Mudd, 1982; Heath, 1984; Pryor et al., 1984; Giamalva et al., 1985). Research in this area with regard to woody perennials is almost nonexistent. The data presented in Chapter 6 are of a more general nature, but applicable to trees, and the reader is referred to it for a detailed discussion of this subject.

7.2.2   Gas Exchange Functions

7.2.2.1   Net Photosynthesis Short-Term Responses

Ozone can inhibit the rate of net carbon assimilation in a variety of conifer and hardwood tree species (see Table 7.1). Early investigations of O_3 effects on photosynthesis (Miller et al., 1969; Barnes, 1972; Botkin et al., 1972) distinguished between short- and long-duration fumigations, concentration-dependent changes in CO_2 exchange rate, and variations in response among genotypes of a particular species. In later years O_3 inhibition of CO_2 uptake was observed to be modified by numerous environmental variables (light, humidity, ambient CO_2 concentration, other gaseous pollutants) and other biological factors (leaf age, plant water status, respiration rate), in addition to the genotype (Coyne and Bingham, 1981, 1982; Reich, 1983; Yang et al., 1983a; Reich and Lassoie, 1984; Lee et al., 1990). These studies have served to emphasize the complexity of interactions involved in O_3 impairment of net photosynthesis in trees.

Characteristics of the exposure regime are of considerable importance in determining the degree of the photosynthetic response to O_3. Exposure duration and O_3 concentration, the two most significant components of the fumigation profile, have been extensively investigated (Lefohn and Runeckles, 1987; Krupa and Nosal, 1989). However, interest at present also centers on other aspects of the exposure profile, which relate to temporal and dynamic fluctuations of the O_3 concentration (Lefohn et al., 1987a; Krupa and Nosal, 1989). Some of these features include the frequency and extent of peak concentrations, the rate of the concentration increase and decline, and the duration of respite (recovery) periods between peak exposures.

The length of the exposure period, from a biological perspective, has two primary functional phases: (1) short-term or diurnal, and (2) long-term or seasonal. A short-term fumigation typifies the peak segment of a summertime diurnal O_3 cycle, where concentrations increase for 2 to 4 h (or more) from lower nocturnal levels. The maximum concentrations are generally centered during the afternoon (1400 h). Plant response to this fumigation profile should favor metabolic processes that utilize endogenous metabolite or enzyme pools for O_3 detoxification or repair of injured chloroplasts. While *de novo* synthesis of proteins and enzyme substrates may occur during the later stages of short-term exposures, this response is probably more typical of longer-term, lower-concentration exposure regimes. Under growing season fumigation profiles, plant response may involve various metabolic processes that enhance the photosynthetic capacity of the plant to accommodate oxidative stress.

A time-course study characterizing the inhibition of net photosynthesis in eastern white pine during a short-term O_3 exposure (see Figure 7.1) is one of the few such investigations that has been conducted with a tree species (Yang et al., 1983a,b). The P_n response curve declined initially during the first hour, and then a stable CO_2 exchange rate was maintained for the remaining 3 h of the fumigation period. The occurrence of such a steady-state P_n identifies an equilibrium between the internal O_3 flux and the detoxification rate of O_3 or other reactive metabolic product(s) that limit photosynthesis.

The inhibition of P_n was concentration dependent, becoming more severe as O_3 concentration increased from 0.10 to 0.30 μl l^{-1}. A steady-state photosynthetic rate was attained within the first hour of the initiation of fumigation, regardless of O_3 concentration. The inhibition was not permanent, no visible

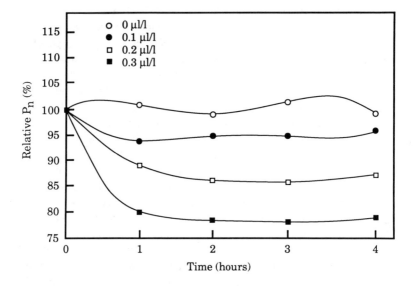

Figure 7.1. Net photosynthesis in eastern white pine (*Pinus strobus* L.) exposed to ozone for 4 h at concentrations of 0, 0.10, 0.20, and 0.30 $\mu l\ l^{-1}$ (after Yang et al., 1983a).

needle injury was evident, and P_n began to recover within 1 h after O_3 exposure.

The initial events occurring during the inhibition of photosynthesis are of interest for investigating the site and mode of action of O_3 in affecting both leaf gas exchange physiology and chloroplast function. This is especially important in assigning O_3 effects to stomatal and mesophyll components.

7.2.2.2 Stomatal Conductance

Stomata constitute the primary route of entry of O_3 into the leaf interior. Flux analysis has indicated that a major portion of diffusive resistance to O_3 resides with the stomata (and the leaf boundary layer in field conditions). The mesophyll comprises only 10 to 20% of the total leaf resistance at low to moderate (<0.20 $\mu l\ l^{-1}$) O_3 concentrations (Taylor et al., 1982). Reich (1987) suggested that at typical ambient O_3 levels (0.08 to 0.15 $\mu l\ l^{-1}$), leaf gas-phase resistance almost exclusively controls uptake. Thus, he proposed that sensitivity among tree species could be explained by differences in leaf conductance. However, Taylor et al. (1982) demonstrated that internal O_3 flux did not correlate well with visible injury of the leaf. Although leaf conductance is the primary controller of O_3 uptake, mesophyll metabolism is perhaps a more important factor in regulating flux and determining plant sensitivity.

Limited information is available concerning the effects of short duration (<1 d) O_3 exposures on stomatal conductance in tree species. Nevertheless, for both short-term and long-term fumigations, no evidence exists at present to support a direct effect of O_3 on guard-cell function (membranes) resulting in a stress avoidance behavior (Coyne and Bingham, 1981; Reich and Lassoie, 1984;

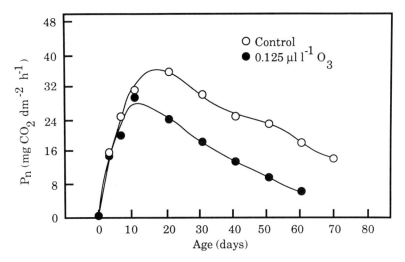

Figure 7.2. The effect of ozone exposure on light-saturated net photosynthesis in hybrid poplar leaves of various ages. Rooted cuttings were exposed to 0.25 (control) or 0.125 μl l⁻¹ ozone for approximately 5.5 h daily for 9 weeks (after Reich, 1983).

Jensen and Roberts, 1986; Keller and Hälser, 1987). Rather, O_3 effects on stomatal behavior may be mediated predominantly through alterations in the net photosynthetic rate (Coyne and Bingham, 1981; Reich, 1987). Recent studies (albeit with *Glycine max* L.) have shown that O_3 initially inhibits P_n, and subsequently, stomatal conductance is lowered (Sheng and Chevone, 1988). These results support the concept that O_3 flux into a leaf can be regulated by an interactive process involving P_n, C_i, and stomatal conductance. The development and maintenance of a reduced steady-state photosynthetic rate observed during short-term O_3 exposures (Yang et al., 1983a,b) suggest the attainment of an equilibrium condition between O_3 influx and P_n inhibition.

7.2.2.3 Net Photosynthesis Long-Term Responses

While short-duration O_3 experiments provide insights into the initial events disrupting the photosynthetic process, long-term studies are important for (1) determining the cellular metabolic capacity to withstand oxidative stress and (2) relating alterations in net CO_2 exchange to tree growth. The rate of P_n over time, during extended fumigation periods, has been investigated in only a few conifer and hardwood tree species (Coyne and Bingham, 1981, 1982; Reich, 1983; Yang et al., 1983b; Keller and Hasler, 1987). However, a depression in the maximum rate, and a slight increase in the normal decline of P_n that occurs with aging were observed with O_3 exposure (see Figure 7.2). This effect becomes more severe as O_3 concentration increases (Yang et al., 1983b) and is more evident in sensitive genotypes than in tolerant genotypes (Coyne and Bingham, 1981). Ozone exposure also increases the rate of foliar senescence (Reich, 1983), which is most pronounced in sensitive genotypes (Coyne and Bingham, 1981).

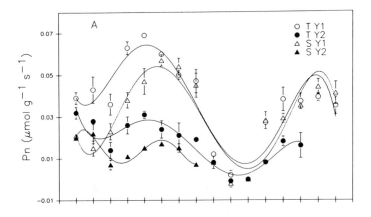

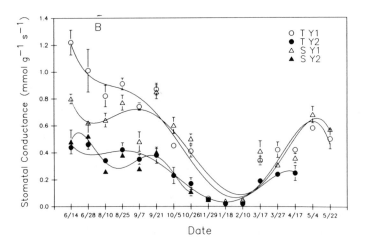

Figure 7.3.   Net photosynthesis and stomatal conductance (water vapor) in an oxidant-sensitive and tolerant genotype of eastern white pine under field conditions. T and S represent tolerant and sensitive genotypes, respectively. Y1 and Y2 represent current-year and 1-year-old needles, respectively.

Ozone apparently interferes with chloroplast function more than with stomatal function. Coyne and Bingham (1981, 1982) observed that losses in photosynthetic capacity of ponderosa pine were greater than reductions in stomatal conductance, suggesting more severe injury to mesophyll components of the CO_2 diffusion pathway than to the stomata. A recent study with eastern white pine (Chevone et al., 1989) supports the contention that oxidant resistance resides within the mesophyll in this species. Net photosynthesis and stomatal conductance were measured periodically in one oxidant-sensitive and oxidant-tolerant genotype classified by its needle appearance, under field conditions during the growing season and into the winter months (see Figure 7.3). The net

CO_2 exchange rate was consistently greater in the current-year needles of the tolerant genotype through late September and in one-year-old needles throughout the measurement period. Similarly, stomatal conductance to water vapor was also higher in both needle age classes in the oxidant-tolerant tree. This genotype had the capacity to maintain a high photosynthetic rate under conditions favorable for a greater internal flux of pollutant, indicating the involvement of a mesophyll mechanism(s) in oxidant resistance. Coyne and Bingham (1982) also observed that P_n and g_s were higher in O_3-tolerant than in O_3-sensitive ponderosa pine trees. However, these differences in gas exchange rates were evident only in needles 1 year of age and older.

7.2.2.4 Stomatal Conductance and Transpiration

The relationship between stomatal conductance and O_3 exposure over extended periods of time is complex, and no cause/effect association has yet emerged. Relative humidity, light intensity, O_3 concentration, leaf age, and genotype are some of the known factors that influence the effect of O_3 on stomatal conductance (Reich and Lassoie, 1984; Jensen and Roberts, 1986; Keller and Hasler, 1987). Long-term exposure does, however, appear to alter the response of stomata to certain stimuli. Reich and Lassoie (1984) observed that hybrid poplar leaves were less able to control water loss and withstand desiccation after fumigation than were nonexposed leaves. Keller and Hasler (1987) reported that stomata in one genotype of Norway spruce were less responsive to light intensity changes after O_3 exposure. These studies suggest that O_3 may directly affect stomatal function as well as alter conductance through a limitation of CO_2 fixation.

7.2.2.5 Photosynthetic Processes Affected by Ozone

Ozone appears to initially affect photosynthesis by interfering with mesophyll (chloroplast) processes. Stomatal closure, which occurs commonly during fumigation at higher concentrations, is probably a function of increased C_i resulting from reduced CO_2 fixation. The sequence of events leading to P_n inhibition is poorly understood.

Ozone exposure (long-term) reduced both the incident quantum yield and the light saturated P_{nmax} in hybrid poplar leaves (Reich, 1983). The effect on quantum yield remained nearly constant over time, whereas maximum photosynthesis decreased as O_3 exposure duration increased. Leaf chlorophyll content decreased with O_3 stress and was linearly correlated with decreases in P_n; however, inhibition of other photosynthetic processes could not be excluded as photosynthetic limitations. Nevertheless, it appears that light-dependent reactions involving electron generation and transfer through the photosystems are less sensitive to O_3 than either the metabolism associated with CO_2 reduction and the regeneration of ribulose bisphosphate or the loss of chlorophyll itself.

At present, there is insufficient information to define short- or long-term mechanism(s) responsible for O_3-induced inhibition of photosynthesis. The chloroplast is a complex energy capture, transfer, and storage system that is highly regulated. The oxidation/reduction state of many metabolic components is critical to the proper functioning of the chloroplast; the presence of an oxidizing agent, such as O_3, can perturb the redox potential. Ascorbate, glutathione, and sulfhydryl residues on Calvin-cycle enzymes, especially the light-activated enzymes that are regulated by the oxidation state of these SH groups, have the greatest potential for reaction with O_3. Oxidation of these compounds eventually results in the formation of water or oxygen and effectively detoxifies O_3. However, such a reduction of O_3 in the chloroplast requires the reducing potential of electrons generated through the photolysis of water. Failure to remove O_3 or other toxic metabolites could lead to photosynthetic collapse. Disruption of photosynthesis presents one of the most intriguing and challenging aspects concerning the toxicity of O_3 to plant cells.

7.2.3 Growth

Ozone effects originate with initial injury at the cellular level and manifest themselves as reductions in growth. Early literature pertaining to O_3 effects on plants reported response based on visible foliar injury (Davis and Wilhour, 1976). Foliar injury, however, is not directly related to growth reductions in trees exposed to long-term ambient or near-ambient concentrations of O_3, with the exception of certain areas of the western U.S. that are exposed to unusually high concentrations of O_3 (Miller, 1983; Reich and Amundson, 1985; Chappelka et al., 1988a,b). In a recent review article, Pye (1988) states the probable reason for this occurrence. Visible foliar injury indicates lost photosynthetic area following a failure of cellular homeostasis. This type of injury occurs most commonly during acute exposures. Relatively low-concentrations of O_3 can reduce photosynthetic efficiency and increase respiration without the loss of homeostasis. Because short-duration, low-concentration exposures cause minimal effects on photosynthesis, subtle injuries are observed as a cumulative result of longer-term exposures. Since trees are perennials and are exposed to some amount of O_3 continuously, primary interest has involved the study of alterations in growth and productivity under chronic O_3 exposure regimes, irrespective of visible symptom expression.

7.2.3.1 Greenhouse Studies

A major portion of research on tree response to O_3 has been conducted in greenhouse experiments with seedlings. The reasons for this are obvious; seedlings are easy to manage, are adapted for pot studies, and can be used in indoor fumigation facilities, such as continuously stirred tank reactors (CSTRs). Results from greenhouse studies need to be interpreted with caution, however. Plants differ morphologically and physiologically from those grown in the field, and pollutant response can be altered (Lewis and Brennan, 1977). Several factors that make it difficult to relate laboratory studies to field situations include: soil nutrient status, water availability, tree age, temperature, and duration and fluctuations in O_3 concentrations under ambient conditions.

A summary of the major findings regarding O_3 effects on biomass production and height growth, from controlled fumigation studies are shown in Tables 7.2 and 7.3, respectively. In general, O_3 exposure resulted in less total biomass (see Table 7.2) and height growth (see Table 7.3). Response was quite variable and was influenced by many factors, including environmental conditions before, during, and after fumigation, and seedling genotype.

7.2.3.2 Field Studies

Since O_3 is ubiquitous and tree response is altered by many other factors (light, nutrition, moisture), it has proven difficult to determine whether O_3 significantly affects tree growth and productivity in the field. The various methodologies used for studying the effects of O_3 on crops and trees are discussed in Chapter 4 and will only be mentioned briefly in this section.

Benoit et al. (1983) evaluated the effect of oxidant air pollution, using dendrochronological techniques, on the radial growth of eastern white pine differing in O_3 sensitivity in Virginia. Reduced radial growth was observed for trees in all sensitivity classes during the period 1955 to 1978. Mean annual growth was significantly less for the O_3-sensitive trees compared with trees exhibiting no visible symptoms.

Radial growth of ponderosa pine was compared for periods of relatively low air pollution (1910 to 1940) and high air pollution (1941 to 1971) in the San Bernardino National Forest in southern California (Miller, 1977). The average annual radial growth was 0.20-mm less for the period of high pollution than for the period of low pollution. This loss in growth was attributed primarily to O_3 that was transported from the Los Angeles urban area. Peterson et al. (1987) found that mean annual radial growth of Jeffery pines in the Sequoia and Kings Canyon National Parks in the western U.S. symptomatic of O_3 stress was 11% less than for trees not exhibiting symptoms. The majority of these differences were observed after 1965, coincident with increasing levels of oxidant pollution in that geographic area.

All of these studies indicate that radial growth of trees has declined at sites where symptoms of visible O_3 injury have been observed. Unfortunately, a sufficient database of ambient O_3 concentrations is not available in most forested areas in the U.S. and other countries. Therefore, it is difficult to establish a relationship between O_3 exposure and changes in mature tree growth on a regional basis. A rural air quality monitoring network, including O_3 measurements, is needed in order to acquire enough data to effectively use dendrochronological techniques for establishing cause-effect relationships.

Several studies have been conducted using open-top chambers to establish a cause-effect relationship between O_3 exposure and tree growth. Duchelle et al. (1982), Chevone et al. (1983), and Wang et al. (1986) all reported growth decreases over a 2 to 3 year period when comparing trees exposed to ambient air with those exposed to charcoal-filtered air (see Tables 7.2 and 7.3). Differences in growth were attributed to ambient O_3 concentrations. Such studies support the hypothesis that O_3 affects tree seedling growth in the field; however, it is not known what role the "chamber effect" has in altering response (Chapter 4).

Table 7.2. Effects of Controlled O_3 Fumigation on Biomass in Tree Seedlings[a]

Species	Exposure Duration (d)	Conc. ($\mu l\ l^{-1}$)	Dose ($\mu l\ l^{-1}$-h)	Biomass change (%)	References and notes
Acer saccharinum	60	0.05	36	8	Jensen, 1983
		0.10	72	0	CSTR light 15 h
		0.20	144	−64[b]	
Acer saccharum	28	0.05	8	−3	Kress and Skelly, 1982
		0.10	17	−7	CSTR consecutive days
		0.15	25	−41[b]	Light 16 h
	50	0.06	21	−2	Reich et al., 1986
		0.09	32	−2	CHAM control 0.03 µ/l
		0.12	42	−8	5 d/week; light 16 h/d
Fraxinus americana	28	0.05	8	22	Kress and Skelly, 1982a
		0.10	17	9[b]	CSTR consecutive days
		0.15	25	−17[b]	Light 16 h/d
	25	0.05	5	3	Chappelka and Chevone, 1986
		0.10	10	−7	CSTR 5 d/week
		0.15	15	−19	Exposed 4 h/d
	9	0.10	4	14	McClenahen, 1979
		0.20	7	9	CHAM 1 d/week
		0.30	1	−17	
		0.40	14	−6	
Fraxinus pennsylvania	28	0.05	8	−14	Kress and Skelly, 1982a
		0.10	17	−28	CSTR consecutive days
		0.15	25	−33	Light 16 h/d
Liquidambar styraciflua	28	0.05	8	−8	Kress and Skelly, 1982a
		0.10	17	−19	CSTR consecutive days
		0.15	25	−24[b]	Light 16 h/d
Liriodendron tulipifera	28	0.05	8	41	Kress and Skelly, 1982a
		0.10	17	5	CSTR consecutive days
		0.15	25	18	Light 16 h/d
	35	0.07	15	14	Mahoney et al., 1984 CSTR consecutive days
	30	0.10	12	−9	Chappelka et al., 1985 CSTR 5 d/week
Picea sitchensis	126	0.10	76	−14	Wilhour and Neely, 1977 OTC stem weight only
Pinus contorta var. contorta	126	0.10	76	−6	Wilhour and Neely, 1977
Pinus contorta var. murrayana		0.10	76	−8	OTC stem weight only

Table 7.2. Continued

Species	Exposure Duration (d)	Conc. ($\mu l\ l^{-1}$)	Dose ($\mu l\ l^{-1}$-h)	Biomass change (%)	References and notes
Pinus echinata	22	0.08	7	−14	Chevone et al., 1984
		0.08	7	−15[b]	CSTR 2 d/week; two soils
Pinus elliottii	112	0.08	122	−19[b]	Hogsett et al., 1985
var. densa		0.10	155	−48[b]	CHAM O$_3$ varied diurnally
Pinus elliottii		0.08	122	−21[b]	
var. elliottii		0.10	155	−50[b]	
Pinus jeffreyi	126	0.10	76	−2	Wilhour and Neely, 1977
Pinus lambertiana		0.10	76	0	OTC stem weight only
Pinus monticola		0.10	76	−9[b]	
Pinus ponderosa		0.10	76	−21[b]	
Pinus radiata		0.10	76	0	
Pinus rigida	28	0.05	8	−8	Kress and Skelly, 1982a
		0.10	17	−19	CSTR consecutive days
		0.15	25	−24[b]	Light 16 h/d
Pinus taeda	22	0.08	7	−10	Chevone et al., 1984
		0.08	7	−21[b]	CSTR 2 d/week; two soils
(wild type)	28	0.05	8	−14	Kress and Skelly, 1982a
		0.10	17	−22[b]	CSTR consecutive days
		0.15	25	−28[b]	Light 16 h/d
(6 –13 x 208)	28	0.05	8	7	Kress and Skelly, 1982a
		0.10	17	−6	CSTR consecutive days
		0.15	25	−14[b]	Light 16 h/d
Pinus virginiana	28	0.05	8	2	Kress and Skelly, 1982a
		0.10	17	−3	CSTR consecutive days
		0.15	25	−13	Light 16 h/d
Platanus occidentalis	28	0.05	8	−23	Kress and Skelly, 1982a
		0.10	17	−61[b]	CSTR consecutive days
		0.15	25	−69[b]	Light 16 h/d
Populus deltoides	17	0.06	6	−10[b]	Reich et al., 1984
x trichocarpa		0.08	8	−10[b]	CHAM control 0.048 $\mu l\ l^{-1}$ Exposed 17 of 27 d
	13	0.06	5	−9[b]	Reich et al., 1984
		0.08	6	−14[b]	CHAM control 0.044 $\mu l\ l^{-1}$ Exposed 13 or 20 d
	62	0.05	9	3[b]	Reich and Lassoie, 1985
		0.09	20	−10[b]	CHAM control 0.025 $\mu l\ l^{-1}$
		0.13	34	−13[b]	
	30	0.15	36	−60[b]	Jensen and Dochinger, 1974
	30	0.15	36	−54[b]	CHAM exposed 5 d/week

Table 7.2. Continued

Species	Duration (d)	Conc. (μl l^{-1})	Dose (μl l^{-1}-h)	Biomass change (%)	References and notes
Populus tremuloides					
clone 1	494	0.04	297	−17[b]	Wang et al., 1986
clone 2		0.04	297	−14[b]	OTC control 0.025 μl l^{-1}
					Tops harvested annually
Pop. x euroamer.					
var. *Dorskamp*	161	0.04	79	−6	Mooi, 1980
var. *Zeeland*	161	0.04	79	0	CHAM stem weight
					reduced 12 and 4%
Prunus serotina	9	0.10	4	11	McClenahen, 1979
		0.20	7	1	CHAM exposed a d/week
		0.30	11	6	
		0.40	14	−2	
Pseudotsuga	126	0.10	76	−15	Wilhour and Neely, 1977
menziesii					OTC stem weight only
Quercus phellos	28	0.05	8	−2	Kress and Skelly, 1982a
		0.10	17	−11	CSTR consecutive days
		0.15	25	−13	Light 16 h/d
Quercus rubra	63	0.04	19	3	Reich et al., 1986
		0.08	35	1	OTC control 0.024 μl l^{-1}
	50	0.07	25	5	Reich et al., 1986
		0.12	42	1	CHAM control 0.025 μl l^{-1}
					5 d/week

[a] Dose is the product of average concentration × days of fumigation × hours fumigated per day. Duration refers to actual fumigation days. Response is expressed as percentage change over control. CHAM, CSTR, and OTC refer to chamber, continuously stirred tank reactor, and open-top chamber, respectively (after Pye, 1988).
[b] Significant difference from the control at the 0.05 level.

Several exposure-response studies have been reported for loblolly pine seedlings exposed to ambient and above-ambient concentrations of O_3 in open-top chambers (Shafer et al., 1987; Adams et al., 1988; Kress et al., 1988; Shafer and Heagle, 1989; Edwards et al., 1990). In all studies, seedlings exposed to the highest O_3 concentrations exhibited less growth. Results were highly variable, however, with some families exhibiting more growth in nonfiltered air than in charcoal-filtered treatments, and other families showing the opposite response.

In summary, O_3 can alter tree growth in certain geographic areas. However, at this time, the magnitude of this impact on a regional or national basis remains unknown.

7.2.4   Effects on Biomass Partitioning

Ozone, in addition to causing reductions in the amount of dry matter produced, can alter partitioning of carbon in tree seedlings (Tingey et al., 1976; Jensen, 1981; Scherzer and McClenahen, 1989; Adams et al., 1990; Spence et al., 1990) and in mature trees (Wilkinson and Barnes, 1973; McLaughlin et al., 1982). An in-depth discussion on the impact of O_3 on assimilate partitioning in plants is provided by Cooley and Manning (1987).

Jensen (1981) studied the root carbohydrate content and above-ground biomass of one-year-old green ash exposed to 0.50 μl l^{-1} O_3, 8 h d^{-1}, 5 d week^{-1} for up to 6 weeks. Ozone-treated seedlings exhibited significantly less stem and leaf dry weight, less root starch, and reduced sugar content. Lower root reserves were hypothesized to contribute less above-ground biomass. Scherzer and McClenahen (1989) observed similar results with pitch-pine seedlings. Root starch content tended to decrease with increasing O_3 concentrations, with controls exhibiting significantly higher levels than 0.15, 0.20, or 0.30 μl l^{-1}-treated seedlings.

Tingey et al. (1976) exposed 1-week-old ponderosa pine to ^{1} O_3, 6 h d^{-1} for 20 weeks, to determine metabolite levels in shoots and roots. Ozone-treated plants contained high amounts of N and free amino acids in the roots. Levels of soluble sugars, starch, and phenolics increased in shoots and decreased in roots of seedlings to which O_3 was applied. The authors attributed their results to metabolite retention in the foliage and disruption in transport to the roots. Several researchers (Kress and Skelly, 1982b; Hogsett et al., 1985; Chappelka and Chevone, 1986) have reported that root growth of certain tree species can be depressed more than shoot growth, supporting the findings of Tingey et al. (1976).

Adams et al. (1990) using ^{14}C found that loblolly pine seedlings exposed to two times ambient O_3 over one growing season exhibited lower rates of CO_2 assimilation and increased respiration rates, and allocated less photosynthate to fine roots. Spence et al. (1990) exposed loblolly pine seedlings to 0.120 μ l^{-1} O_3, 7h d^{-1}, 5 d week^{-1} for 12 weeks. Seedlings were labeled with ^{11}C in order to characterize changes in allocation patterns. Ozone-treated seedlings exhibited a 16% reduction in CO_2 assimilation, 11% decrease in the speed of phloem transport, a 40% reduction in the amount of photosynthate in the phloem, and a 45% decrease in photosynthate allocation to the roots.

Ozone is also known to cause shifts in allocation patterns in mature trees. Wilkinson and Barnes (1973) exposed eastern white and loblolly pine branches to 0.00, 0.10, or 0.20 μl l^{-1} O_3 continuously for up to 21 d to study ^{14}C-fixation patterns in response to O_3. The primary changes in ^{14}C distribution due to O_3 exposure were reductions in soluble sugars, increases in sugar phosphates, and an increase in free amino acids. McLaughlin et al. (1982), using ^{14}C, demonstrated that the foliage and branches of oxidant-sensitive eastern white pines retained significantly more photosynthate than did intermediate or tolerant trees,

Table 7.3. Effects of Controlled O_3 Fumigation on Height Growth in Tree Seedlings[a]

Species	Exposure Duration (d)	Exposure Conc. ($\mu l\ l^{-1}$)	Exposure Dose ($\mu l\ l^{-1}$-h)	Biomass change (%)	References and notes
Acer rubrum	109	0.30	261	−25	Jensen, 1973 CSTR exposed 5 d/week
Maine	42	0.25	84	−17	Dochinger and
New Brunswick				−32	Townsend, 1979
Ohio				−36[b]	CHAM 3 seed sources
Acer saccharinum	109	0.30	261	−73[b]	Jensen, 1973 CHAM exposed 5 d/week
Acer saccharum	109	0.30	261	−64[b]	Jensen, 1973 CHAM exposed 5 d/week
Alnus glutinosa	109	0.30	261	17	Jensen, 1973 CHAM exposed 5 d/week
Betula alleghaniensis	74	0.25	149	−8	Jensen and Masters, 1975 CHAM exposed 5 d/week Control 0.02–0.03 $\mu l\ l^{-1}$
Betula papyrifera	74	0.25	149	−34	Jensen and Masters, 1975 CHAM exposed 5 d/week Control 0.02–0.03 $\mu l\ l^{-1}$
Fraxinus americana	9	0.10	4	13	McClenahen, 1979 CHAM exposed 1 d/week
		0.20	7	1	
		0.30	11	0	
		0.40	14	1	
	109	0.30	261	−8	Jensen, 1973 CHAM exposed 5 d/week
Fraxinus pennsylvania	297	0.05	153	−67[b]	Duchelle et al, 1982 OTC exposed 24 h/d Ambient O_3 in winter
	109	0.30	261	−35	Jensen, 1973 CHAM exposed 5 d/week
Juglans nigra	109	0.30	261	−22	Jensen, 1973 CHAM exposed 5 d/week
Larix leptolepis	74	0.25	149	4	Jensen and Masters, 1975 CHAM exposed 5 d/week Control 0.02–0.03 $\mu l\ l^{-1}$
Liriodendron tulipifera	109	0.30	261	0	Jensen, 1973 CHAM exposed 5 d/week
	297	0.05	153	−44	Duchelle et al., 1982 OTC exposed 24 h/d
	35	0.07	15	−2	Mahoney et al., 1984 CSTR consecutive days

Table 7.3.   Continued

Species	Exposure Duration (d)	Exposure Conc. $(\mu l \, l^{-1})$	Exposure Dose $(\mu l \, l^{-1}\text{-}h)$	Biomass change (%)	References and notes
Picea glauca	74	0.25	149	−7	Jensen and Masters, 1975 CHAM exposed 5 d/week
Populus deltoides × *trich.*	30	0.15	36	−35[b]	Jensen and Dochinger, 1974
				−37[b]	CHAM exposed 5 d/week Terminal vs. basal cuttings
	102	0.15	184	8	Patton, 1981
				−3	CSTR cuttings of different
				16	hybrids
				−23	
Populus max. × *berol.*	102	0.15	184	17[b]	Patton, 1981
Populus max. trich.				19	CSTR cuttings of different
Pop. nigra × *lauri.*				−5	hybrids
Prunus serotina	9	0.10	4	16	McClenahen, 1979
		0.20	7	5	CHAM exposed 1 d/week
		0.30	11	3	
		0.40	14	−28[b]	
Robinia pseudoacacia	297	0.05	153	−18	Duchelle et al., 1982 OTC exposed 24 h/d Ambient O_3 in winter
Tsuga canadensis	297	0.05	153	−23	Duchelle et al., 1982 OTC exposed 24 h/d
	500	0.07	264	−40	Chevone et al., 1983 OTC exposed 24 h/d Ambient O_3 in winter

[a] Dose is the product of average concentration × days of fumigation × hours fumigated per day. Duration refers to actual fumigation days. Response is expressed as percentage change over control. CHAM, CSTR, and OTC refer to chamber, continuously stirred tank reactor, and open-top chamber, respectively (after Pye, 1988).
[b] Significant difference from the control at the 0.05 level.

indicating that the export of photosynthate to the stems and roots of sensitive trees was inhibited.

These results indicate that in addition to having an effect on overall plant growth, O_3 can cause alterations in the way biomass is partitioned. Changes induced by ambient O_3 concentrations would therefore be expected to cause reductions in carbohydrate levels, especially in the roots. Depletions in these reserves could cause reduced vigor of the root systems and enhance the susceptibility of trees to other forms of stress, such as drought and root diseases.

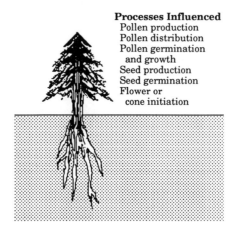

Processes Influenced
Pollen production
Pollen distribution
Pollen germination
   and growth
Seed production
Seed germination
Flower or
   cone initiation

Figure 7.4.   Ozone effects on reproduction in trees.

7.2.5 Effects on Reproduction

Air pollution influences the reproductive capacity of plants by causing reductions in photosynthesis and shifts in the partitioning of photosynthate (Cooley and Manning, 1987). Critical events in reproductive cycles may be altered by air pollution exposure, as shown in Figure 7.4.

Ambient O_3 exposure of ponderosa and Jeffrey pines growing in the San Bernardino Mountains of California reduced seed production (Luck, 1980). Severely injured trees produced significantly fewer cone crops over a 6-year period than did uninjured trees. Benoit et al. (1983) collected pollen from eastern white pine growing along the Blue Ridge Parkway in Virginia. Percent germination was significantly reduced in wet pollen exposed to O_3 (0.15 µl l^{-1}, 4 h). Dry pollen production and germination was unaffected by O_3 fumigation. Reasons for this discrepancy were not explained, but may be related to increased permeability of O_3 in the wet pollen.

The available literature indicates O_3 has the potential to influence reproduction in trees: directly by affecting reproductive structures, or indirectly by altering host metabolism. Much more research is needed, however, in order to determine the overall significance of these effects on forest ecosystems.

7.3 INTERACTIONS OF O_3 AND OTHER STRESS FACTORS

Trees grow in a dynamic environment and are continuously exposed to a wide variety of climatic and edaphic conditions. Plant response can be altered by various abiotic and biotic factors, and O_3 can predispose plants to various pests and pathogens.

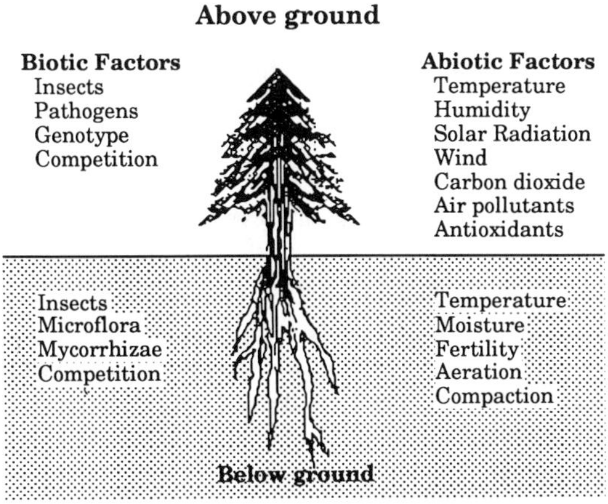

Above ground

Biotic Factors	Abiotic Factors
Insects	Temperature
Pathogens	Humidity
Genotype	Solar Radiation
Competition	Wind
	Carbon dioxide
	Air pollutants
	Antioxidants

Insects	Temperature
Microflora	Moisture
Mycorrhizae	Fertility
Competition	Aeration
	Compaction

Below ground

Figure 7.5.   Factors influencing tree response to ozone.

7.3.1 Abiotic Factors

Abiotic factors such as temperature, moisture, solar radiation, and nutrition can all have a significant influence on tree response to O_3 (see Figure 7.5). In addition, O_3, alone or in combination with other airborne contaminants in various physical forms (aerosols, gases, or precipitation), has been observed to alter tree response.

7.3.1.1 Exposure to Pollutant Mixtures

Ambient background air is composed of a variety of gases, aerosols, and suspended course particle matter. Polluted air also consists of these mixtures, but at higher concentrations and complexity. Concentrations of pollutant mixtures can be introduced into air masses via atmospheric mixing, emission into already contaminated air, or simultaneous mixing (Runeckles, 1984). The occurrence of components in these mixtures can be simultaneous, sequential, inverse, or any other random combination (Lefohn et al., 1987a,b), and may impact vegetation in a manner different from that of a single pollutant (Mahoney et al., 1984; Runeckles, 1984; Chappelka et al., 1985).

In relating the effects of air pollutants on vegetation, it is necessary to consider multiple, stochastic pollutant combinations using proper patterns of occurrence, in designing experiments relevant to what occurs in the ambient environment. This is especially important in studying the effects on trees, since they are perennials and are exposed to pollutant mixtures of various combinations on a frequent basis over many years.

7.3.1.1.1 Combined Effects of O_3 and Gaseous Pollutants Several research groups have exposed trees to various combinations of O_3 and other phytotoxic gases. Ozone in combination with SO_2 and/or NO_2 generally results in the greatest suppression of growth (see Table 7.4). These findings are highly variable and are dependent upon the pollutants and concentrations used, the genetic makeup of the material (Kress and Skelly, 1982b; Yang et al., 1983a,b), the timing of fumigation (Jensen, 1985), and the species exposed (Kress and Skelly, 1982a).

Some areas where research findings are restricted are as follows: (1) seedlings were used in all studies, the majority of which included less than 1-year-old seedlings, and all but one study (Reich et al., 1984) were conducted in the laboratory; (2) "square-wave" exposures were used in all studies, except in one case (Reich et al., 1984); (3) pollutants were applied concurrently; (4) the only pollutant mixtures used were O_3, SO_2, and/or NO_2; and (5) mechanistic research was lacking.

The general purpose of the experiments described in Table 7.4 was to establish the fact that O_3 in combination with other gaseous pollutants can be very detrimental to tree growth. The technology and resources are now available to design experiments that are more realistic in nature.

In testing tree response to pollutant mixtures, one of the most important issues is the determination of synergistic or antagonistic effects. Very few studies to date have adequately determined these effects (Chappelka et al., 1985, 1988a,b). By definition, synergism is the joint action of different substances in producing an effect greater than the sum of the individual effects of the substances. A proper test using statistical contrasts would be $(O_3 - \text{control}) + (SO_2 - \text{control})$ vs. $[(O_3 + SO_2) - \text{control}]$, and $(O_3 - \text{control}) + (SO_2 - \text{control})$ vs. $[(O_{3/2} - SO_{2/2}) - \text{control}]$. If both tests were equal, then the effects would be additive. If they were different, then inspection of the sign of this difference would indicate whether the effects were synergistic or antagonistic. If only the first contrast was significantly different, then the combined effect of $O_3 + SO_2$ would be either greater or less than additive and not synergistic or antagonistic. This subject is discussed in detail by Oshima and Bennett (1979), Wellburn (1988), and Chappelka and Chevone (1989).

Only limited research has been conducted on mechanisms of pollutant interactions (Carlson, 1979; Constantinidou and Kozlowski, 1979; Boyer et al., 1986). All the studies used O_3 alone and in combination with SO_2, and all the studies investigated the effects of these pollutants on net photosynthesis. Two studies used unrealistic concentrations of the pollutants (Carlson, 1979; Constantinidou and Kozlowski, 1979), and these results have to be interpreted with caution. Boyer et al. (1986) found the combination of O_3 and SO_2 to depress net photosynthesis in eastern white pine more than did O_3 alone. However, results were variable. No research to date has investigated the biochemical mechanisms of pollutant interactions in trees, at realistic concentrations and appropriate co-occurrence patterns. An excellent discussion of potential areas of research is provided in Wellburn (1988, pp. 200–204).

7.3.1.1.2 Combined Effects of O_3 and Acidic Deposition Results regarding combined effects of O_3 and acidic deposition on tree growth are not as definitive as those with O_3 in combination with gaseous pollutants (see Table 7.5) and are quite variable and dependent upon the environmental conditions imposed on them. Growth of yellow poplar (Chappelka et al., 1985, 1988b), white ash (Chappelka and Chevone, 1986), giant sequoia (Temple, 1988), and paper birch (Keane and Manning, 1989) exposed to O_3 and simulated acidic mist or rain exhibited significant treatment interactions, with growth decreasing as solution acidity increased, except in paper birch (Keane and Manning, 1989). In contrast, no interactive effects occurred on the growth of eastern white pine (Reich et al., 1987), Scots pine (Skeffington and Roberts, 1985), loblolly pine (Kress et al., 1988), red spruce (Taylor et al., 1986), sugar maple, northern red oak (Reich et al., 1986), Jeffrey pine (Temple, 1988), and green and white ash (Elliott et al., 1987), exposed to O_3 and acidic deposition.

Experimental differences that may have influenced the results are tree species, age, genotype, composition of rain or mist solution, and soil type. Chappelka et al. (1985, 1988b) and Chappelka and Chevone (1986) found significant interactions using a soilless potting mixture that had a high cation exchange capacity (CEC) and a low buffering capacity. In general, no significant O_3-rain interactions were observed when trees were planted in native soil in pots or in the ground (Taylor et al., 1986; Reich et al., 1987; Elliott et al., 1987; Kress et al., 1988). Although not mentioned in any of the studies, natural soils generally have lower CECs and higher buffering capacity than "soilless" mixtures, which may have an influence on uptake of H^+ ions and therefore affect plant response.

Reich et al. (1987) observed that soil type influenced the results; there was a strong acid-rain — soil-type interaction. Growth either increased or decreased with varying pH treatments, depending on the soil type in which the seedlings were grown. Kress et al. (1988) planted trees in native soils in the field; however, only first-year results are available at this time. Results may vary as trees are exposed to these multiple pollutants over longer durations (>1 year). Kress et al. (1988) observed less growth at the higher O_3 treatments. No O_3-rain interaction was found, and response varied with genotype.

One hypothesis defining the relationship between O_3 and acidic deposition is that O_3 accelerates the leaching of mobile cations from leaves by adversely affecting membrane integrity. The loss of these minerals may produce nutrient deficiencies in the trees (Prinz et al., 1982). In support of this theory, Krause et al. (1983) reported that foliar leaching of NH_4^+, Mg^{2+}, Ca^{2+}, K^+, and NO_3^- from Norway spruce saplings increased significantly when ozonated trees were treated with simulated acidic mists. Similar observations were reported by Skeffington and Roberts (1985) with Scots pine. These results are now being reevaluated, however, due to the findings of Brown and Roberts (1988). They report that results from previous experiments (Krause et al., 1983; Skeffington and Roberts, 1985) may be incorrect due the production of N_2O_5 as a contaminant in the generation of O_3, using air passed through an electrical discharge O_3 generator. When Brown and Roberts (1988) used pure O_2 in the production of O_3, no contamination occurred, and no significant amounts of foliar nutrient leaching in Norway spruce were observed.

Table 7.4. Effect of Ozone in Combination with Other Gaseous Pollutants on Tree Growth

Tree	Pollutant		Exposure[a]		Response[b]	References
	Type	Conc. (μl l^{-1})	Type	Duration		
Pinus taeda	O_3, SO_2	0.10, 0.10	CSTR	28 d/6 h-d	GA, ht; LA, biomass	Kress and Skelly, 1982b
Pinus ridgida					LA, ht, biomass	
Pinus virginiana					A, ht., LA, biomass	
Liquidambar styraciflua					LA, ht, biomass	
Fraxinus americana					LA, ht, biomass	
Fraxinus pennsylvanica					LA, ht, biomass	
Quercus phellos					LA, ht, biomass	
Fraxinus americana	O_3, SO_2	0.10, 0.08	CSTR	30 d/4 h-d/5 d-wk	LA, ht., biomass	Chappelka et al., 1988a
Fraxinus pennsylvanica					LA, ht, biomass	
Acer saccharinum	O_3	(0.05, 0.10, 0.20)	CSTR	60 d/12 h-d	Additivity nottested all combinations of $O_3 + SO_2$ caused sig. reduction in ht. and biomass from controls	Jensen, 1983
	SO_2	0.10				
Liriodendron tulipifera	O_3	0.07	CSTR	35 d/6 h-d	$O_3 + SO_2$ and $O_3 + SO_2 + NO_2$ GA, ht, biomass	Mahoney et al., 1984
	SO_2	0.06				
	NO_2	0.10				
	O_3, SO_2	0.10, 0.08	CSTR	30 d/4 h-d/5 d-wk	A, ht, biomass	Chappelka et al., 1985

Species	Pollutants	Concentration	System[a]	Duration	Response[b]	Reference
Populus deltiodes x trich.	O_3, SO_2	0.10, 0.08	field gradient	20-27 d/6 h-d, 13 and 17 d during study	LA, ht, biomass	Reich et al., 1984
Populus deltoides *Fraxinus americana* *Liriodendron tulipifera*	O_3, SO_2	0.10, 0.20	CSTR	50 d/12 h-d	LA, ht, biomass LA, ht, biomass LA, ht, biomass	Jensen, 1981
Pinus taeda	O_3, SO_2	0.07, 0.06	CSTR	40 d/5 h-d	Variable, some families sensitive to O_3 + SO_2, others resistant	Winner et al., 1987
Pinus taeda	O_3 NO_2 SO_2	0.05 0.10 0.14	CSTR	28 d/6 h-d	O_3 + SO_2 caused a reduction in growth more than single pollutant, adding NO_2 had a slight stimulatory effect	Kress et al., 1982

[a] CSTR = continuously stirred tank reactor.
[b] A = additive, GA = greater than additive, LA = less than additive.

Table 7.5. Effect of Ozone in Combination with Acid Rain or Mist on Tree Growth

Tree	Pollutant		Exposure[a]		Response[b]	References
	Ozone Conc. ($\mu l\ l^{-1}$)	Acid Rain (pH)	Type	Duration		
Fraxinus americana F. pennsylvanica	0.10	3.0, 4.3, 5.6	CSTR Rain table (needles)	30 d/4 h-d/5 d-wk 12 d/1 h-d/2 d-wk	No combined effects for ht. or biomass	Chappelka et al., 1988a
F. americana	0.05, 0.10, 0.15	3.0, 4.3, 5.6	CSTR Rain table (needles)	25 d/4 h-d/5 d-wk 10 d/1 h-d/2 d-wk	The combined effect of O_3 and acid rain resulted sig. decreases in root growth at 0.05 and 0.10 $\mu l\ l^{-1}$ O_3 as rain pH decreased	Chappelka and Chevone, 1986
Liriodendron tulipifera	0.10	3.0, 4.3, 5.6	CSTR Rain table (needles)	30 d/4 h-d/5 d-wk 12 d/1 h-d/2 s-wk	O_3 exposure resulted in a linear decrease in leaf area and root biomass as rain pH decreased	Chappelka et al., 1985
Acer saccharum	0.03, 0.06,	3.0, 4.3, 5.6	CSTR	50 d/7 h-d/5 d-wk	No interactive	Reich et al., 1986

Species	O_3 (ppm)	pH	Exposure system	Duration	Results	Reference
	0.09, 0.12		Rain table (needles)	20 d/1.25 h-d/2 d-wk	effects between O_3 and acid rain observed	
Quercus rubra	0.02, 0.07,	3.0, 4.0, 5.0	CSTR Rain table (nozzle)	50 d/7 h-d/5 d-wk; 20 d/1.25 h-d/2 d-wk	No interactive effects between O_3 and acid rain observed	Elliott et al., 1987
F. americana F. pennsylvanica	Ambient	Ambient	OTC, EDU natural	2–4 months	No effects observed	
Betula papyrifera	0.06–0.08	3.5, 5.6	Indoor Chambers	60 d/7 h-d/5 d-wk; 24 d/0.10 h-d/2 d-wk	Treatments with pH rain increased growth of ozone-treated seedlings	Keane and Manning, 1989
Pinus strobus	0.02, 0.06, 0.10, 0.14	3.0, 3.5, 4.0, 5.6	CSTR Rain table (nozzle)	47 d/7 h-d/3 d-wk; 30 d/? h-d/2 d-wk	No interactions observed between O_3 × rain pH	Reich et al., 1986
P. taeda	CF, NF × 1.5, NF × 2.25, NF × 3.0	3.5, 5.2	OTC (nozzle)	Continuing	No interactions observed 1st year of study between O_3 × rain pH. Study still underway.	Kress et al., 1988

Table 7.5. Continued

	Pollutant		Exposure[a]		Response[b]	References
Tree	Ozone Conc. (μl l^{-1})	Acid Rain (pH)	Type	Duration		
P. jefferyi	0.10, 2.0	2.0, 2.7, 3.4, 4.1	CSTR outdoor mist chamber	18-24 d/4 h-d/ 3 d-wk 18-24 d/0.5 h/ 3 d-wk	No combined effects of O_3 and acid rain observed with P.jefferyi. 0.20 μl l^{-1} O_3 resulted in decreased root growth as pH of mist decreased	Temple, 1988
Sequoidendron giganteum						
Picea rubens	0.12	4.1, 5:1 rain 3.6, 5:1 mist	CSTR Rain table (nozzle) Mist chamber	32 d/4 h-d/2 d-wk 64 d/12 h-d/4 d-wk	No interactions between O_3 and acid rain or mist observed for any variable measured pH. Study still underway	Taylor et al., 1986

[a] CSTR = continuously stirred tank reactor, OTC = open top chamber, EDU = ethylene diurea.

Chappelka et al. (1985, 1988b) and Chevone et al. (1986) hypothesized that decreases in growth, photosynthesis, and chlorophyll content for O_3-treated plants, as simulated rain acidity increased, may be related to the chemical activity of O_3 at different solution pH. In alkaline solutions, O_3 is decomposed rapidly and is more soluble than oxygen; to the contrary, O_3 is more stable in acidic solutions (Anlauf et al., 1985; Horvath et al., 1985). Wolfenden and Wellburn (1986) found that plastidic sulfate increased and vacuolar pH decreased in barley leaves sprayed with acidic mists of pH 3 or 4. This supports the hypothesis that cellular pH can change with acidic deposition. The effect of O_3, in combination with acidic deposition on cellular pH, needs to be examined to test this hypothesis.

As mentioned previously, it is important to mimic temporal and spacial patterns of gaseous pollutants when designing experiments. This also applies to wet deposition. In all experiments reported in the literature, simulated rain or mist concentrations (pH and chemical constituents) and the periodicity of events were kept constant. Future experiments need to closely mimic ambient O_3 and precipitation events in a geographic area of concern.

The periodicity (both temporal and diurnal) of precipitation events in relation to O_3 episodes also needs to be determined. Chappelka et al. (1985) reported that laboratory fumigations of wet leaves with O_3 resulted in greater growth losses as compared with dry leaves. The ion constituents in rain of comparable values, pH, i.e., SO_4^{-2}, NO_3^-, NH_4^+, etc., have been reported as having significantly different effects on plant growth (Olson et al., 1987). This may be an important factor in determining the plant response to acidic precipitation.

A major factor in testing the effects of O_3 and acidic precipitation on vegetation is the determination of their combined effects on growth. Since studies designed to test such responses are factorial in nature, they can become complicated very rapidly. In only a few reports in the literature (Chappelka et al., 1985, 1988 a,b; Chappelka and Chevone, 1986) was it mentioned that factors within the interaction were analyzed statistically.

Due to the complicated nature of factorial designs, certain combinations of interest may be significant, even though the overall interaction is nonsignificant (Chew, 1976; Little, 1981). An alternative approach is the use of preplanned single *df* comparisons (Chappelka and Chevone, 1989). This can be accomplished by constructing contrasts in which one level of a factor is tested against all levels of another factor. The application of this technique is discussed in detail by Chappelka and Chevone (1989). This procedure generally applies to evenly spaced treatments, so it may be only applicable to exposure-response studies using "square-wave" exposures.

Another alternative, that may be more useful in field situations, is the use of response surfaces to illustrate the nature of the pollutant interaction. Here, the primary objective is to estimate the shape of the response surface over a specified range of levels of factors under investigation. The applicability of response surfaces in air pollution studies is discussed by Ormrod et al. (1984) and Allen et al. (1987).

7.3.1.2 Physical Factors

Trees grow in a dynamic environment and respond to changes in biological and physical factors throughout their growing cycle (see Figure 7.5). Environmental conditions, such as light, temperature, soil moisture, and fertility, govern how plants grow. Variation in one or more of these factors may affect plant sensitivity to O_3. In addition, O_3 stress can alter tree response to any one of these and other environmental conditions. Although these issues are critical to our understanding of how O_3 affects tree growth, very little information is available.

7.3.1.2.1 Light, Temperature, and Relative Humidity The ambient environment exerts a profound influence on the physiological status of a plant and, therefore, its response to O_3 (Tingey and Taylor, 1982). Some early studies with agronomic crops indicated that changes in temperature, light, and relative humidity resulted in alterations in response to O_3 (Heck et al., 1965).

Only a few studies are available that examined the effects of variable temperature, light, and humidity on tree response to O_3 (Costonis and Sinclair, 1969; Wilhour, 1971; Davis and Wood, 1973). General conclusions from these experiments were that visible injury was more severe when plants were exposed to relatively high temperatures (27 to 32°C) and high humidity (>80%), compared with other treatments.

Davis and Wood (1973) reported that Virginia pine seedlings exposed to light for periods of 24 h or longer prior to O_3 exposure exhibited no visible injury. Those seedlings that were maintained in the dark and then exposed to the pollutant in the light exhibited very high amounts of visible injury. Extended periods of light after fumigation showed no apparent effect on symptom development.

These observed differences in visible symptom expression can be, for the most part, attributed to modifications in the effective O_3 dose (the amount that enters the plant). Changes in temperature, light, and humidity all influence stomatal opening and, thus, internal metabolic processes such as photosynthesis and respiration (Mansfield et al., 1981). These factors therefore influence O_3 uptake and toxicity (McLaughlin and Taylor, 1981; Tingey and Taylor, 1982).

No studies to date, however, have examined the effects of changing environmental conditions on tree growth in the field, using ambient or near-ambient O_3 concentrations. McLaughlin and Taylor (1981) hypothesized that greater uptake due to higher relative humidity is responsible, in part, for the greater sensitivity of vegetation growing in the eastern U.S. compared with the more arid western states. These issues may become critical factors, in the context of recent concern over potential changes in global climate, such as increases in CO_2 concentrations, UV radiation, and fluctuations in temperature and rainfall patterns, and their effects on plant growth and productivity (Woodman and Furiness, 1988).

Recently, evidence has accumulated in support of an O_3–low-temperature interaction, leading to visible foliar injury for several tree species, including Norway spruce (Barnes and Davison, 1987; Brown et al., 1987), red spruce (Fincher et al., 1989), and Sitka spruce (Lucas et al., 1988). This could be caused by an impairment of the winter hardening process due to O_3 and is, in part, generally controlled, as exhibited by a wide range in clonal variation in combined studies with these stress factors (Barnes and Davison, 1987; Brown et al., 1987). The mechanism(s) behind this response are unknown, but several hypotheses can be formulated. These are discussed briefly in this chapter, but in more detail by Alscher et al. (1989a) and Davison et al. (1987).

Winter hardening in plants involves a series of well-orchestrated events that prepare cells for exposure to cold temperatures. These events involve ultrastructural changes, alterations in starch-to-sugar ratios, accumulation of antioxidants, and reductions in photosynthetic rates. Ozone may disrupt one or more of these physiological functions, which then alters the cold hardiness of the plant (Alscher et al., 1989a). Alscher et al. (1989b) has recently found an O_3-induced decrease in raffinose in red spruce. Oligosaccharides such as raffinose and sucrose have been implicated as cryoprotectants (Quinn and Williams, 1985).

Ozone has been reported to impair plant metabolism through the formation of free radicals (Chapters 5 and 6). Increased photooxidation of chlorophyll and free radical formation resulting from intense radiation and low temperatures (Oquist, 1986) may be accentuated by O_3 (Alscher et al., 1989a). Trees exposed to high concentrations of O_3 may be damaged due to depletion of antioxidant reserves by O_3.

Chappelka et al. (1990) recently observed injury mediated by a late-season frost-O_3 interaction on loblolly pine foliage growing in modified open-top chambers (Chappelka et al., 1989) in Alabama. First-year foliage (approximately 3 cm in length) exposed to above-ambient O_3 concentrations and a late-season (April) frost exhibited photobleaching of needle tips, turning necrotic with age; and incidence of injury was more severe in the O_3-sensitive genotype.

These findings may be related to declines in growth observed in forests in North America and Europe, where trees are periodically subjected to low temperatures in the winter and elevated O_3 concentrations in the summer months (McLaughlin, 1985; Blank, 1985). Due to recent concerns over changes in global climate (Schneider, 1988; Woodman and Furiness, 1988), these results have even further implications.

Increased concentrations of CO_2 may result in increases in photosynthesis and growth (Rogers et al., 1983). Elevated CO_2 concentrations and projected warmer temperatures (Schneider, 1988) may extend the growing season, especially in eastern North America (Woodman and Furiness, 1988). These factors may also result in changes in normal winter-hardening processes in trees. Autumn or late-spring frosts, combined with elevated O_3 concentrations and increased UV radiation due to stratospheric depletion of O_3, could magnify the injury.

7.3.1.2.2 Soil Moisture and Fertility Soil moisture and fertility have a profound effect on plant growth. These factors also consistently alter the O_3 response of herbaceous plants, as illustrated in Chapter 6. Tseng et al. (1988) exposed Fraser fir seedlings to O_3 and moisture stress in the greenhouse. Water stress caused significant decreases in biomass, transpiration rates, and needle conductance. Exposure to O_3 resulted in no significant changes in biomass, and no O_3-water-stress interactions were observed for this species.

Harkov and Brennan (1980) observed a significant increase in leaf diffusive resistance (lower conductance) with increasing water stress, and less visible O_3 injury in potted hybrid poplar seedlings. Their results indicated that stomata were closed in the water-stressed seedlings, allowing little O_3 uptake. Lee et al. (1990) reported that root hydraulic conductivity of red spruce seedlings was significantly reduced by aging and drought stress, regardless of O_3 treatment. Ozone, in combination with acidic rain (pH 3.0), resulted in lower photosynthesis and root hydraulic conductivity, with decreasing water potential, after two drought cycles.

These results demonstrate that water stress can cause alterations in O_3 sensitivity in trees. All the studies, however, were conducted in pots within a greenhouse. Further research is needed with trees of various ages growing in the field over several years and exposed to different levels of O_3 and moisture stress in order to verify the greenhouse results.

The significance of soil fertility on plant response to O_3 can only be evaluated when suboptimal, optimal, and supraoptimal amounts of various nutrients are considered. Generally, optimal N, P, or K levels enhance visible O_3 injury in herbaceous species (Chapter 6). Harkov and Brennan (1980) found that increasing levels of foliar N resulted in increased O_3 injury in hybrid poplar. Mineral nutrition can have a profound effect on growth and gas exchange in trees (Keller, 1972). It is therefore important to gain an understanding of the relationships between soil fertility and O_3 toxicity, since O_3 uptake is governed by gas exchange.

In addition to direct effects, O_3 may also have an influence on internal cycling of certain nutrients. Switzer and Nelson (1972), working with loblolly pine, found that 39% and 60% of the annual requirements for N and P could be accounted for by internal nutrient cycling. This enabled these trees to meet annual requirements for these nutrients while minimizing demands on soil reserves.

Ozone causes premature senescence in many tree species (Jensen, 1973 ; Kress and Skelly, 1982a). It is unclear, at this time, how internal nutrient translocation may be modified by this factor. Wright et al. (1991), in a study with loblolly pine, found that an increase in N concentration in stem tissues occurred with increasing O_3 concentration, indicating that the internal translocation of N had been enhanced as a result of O_3-induced premature senescence of foliage. This may be considered as part of repair or compensation processes, superseeded by senescence.

7.3.2 Biotic Factors

Biotic factors can influence tree response to O_3 (see Figure 7.5). Trees can be weakened and predisposed to other stresses, such as insects and plant pathogens.

In addition, other factors, including symbiotic relationships (mycorrhizae), can be affected. Trees also exhibit a wide range of sensitivities to O_3, and tree response can be altered among genotypes.

7.3.2.1 Insects and Pathogens

It is widely recognized that O_3 can alter plant response to both insect herbivory and plant pathogens, such as fungi and bacteria (Hughes and Laurence, 1984). In order to understand these interactions, cause-and-effect relationships need to be established, and underlying biochemical and physiological mechanisms need to be identified. Critical features of the O_3-plant-pest interaction, and the variables within each feature characterized, are shown in Table 7.6. This framework applies to both herbaceous plants (Chapter 6) and woody perennials.

7.3.2.1.1 Insects Air pollutants can alter the natural resistance of many plant species to insect attack (Hughes, 1988). This response is reflected by alterations in plant metabolism, that affect insect feeding preference, host nutritional quality, or host defenses to insect infestation (see Table 7.7). The majority of reports in the area of plant-pollutant-insect interactions have been correlative in nature (Alstad et al., 1982). In the past 10 years, however, there has emerged a growing body of literature, mostly with herbaceous crop species, reporting on the cause-effect relationships among insects and their hosts, when exposed to a known concentration of a particular pollutant (Chappelka and Kraemer, 1988). Several reviews discuss this topic in more detail than is included in this section (Alstad et al., 1982; Hughes and Laurence, 1984; Chappelka and Kraemer, 1988; Hughes, 1988).

There is strong evidence to suggest that chronic exposure to O_3 can decrease vitality of trees and predispose them to insect herbivory. Stark et al. (1968) related O_3 damage of ponderosa pine (growing in natural stands in the San Bernardino Mountains of California) to increased susceptibility by pine bark beetles. Other research has confirmed these findings (Dahlsten and Rowney, 1980; Miller, 1983). Smith (1981) speculated that, if O_3 concentrations were high enough, predisposition of southern pines to the southern pine beetle (*Dendroctonus frontalis*) may occur.

Braun and Flückiger (1989) observed that aphid (*Phyllaphis fagi*) populations developed at a faster rate when grown on European beech exposed to ambient O_3, compared with beech exposed to purified air in Switzerland. Jeffords and Endress (1984) reported that Gypsy moth larvae (*Lymantria dispar*) preferred to feed on O_3-treated white oak leaf discs in dual-choice feeding preference assays in the laboratory.

In several experiments with eastern cottonwood, Coleman and Jones (1988a) and Jones and Coleman (1988) reported that adults and larvae of the imported willow leaf beetle (*Plagiodera versicolora*) preferred and consumed more plant tissue that had been exposed to one acute dose ($0.20 \, \mu l \, l^{-1}$, 5 h) of O_3, compared with controls. Females preferred to oviposit (Jones and Coleman, 1988) and were

Table 7.6. Four Classes of Critical Features to Be Considered in Ozone-Plant-Pest Interactions

Physical and chemical characteristics of the environment
    Climate (temperature, humidity, wind, insolation)
    Site (soil, exposure)
    Geomorphology (landforms, parent material)
    Geochemistry (buffering capacity, hydrology, watershed)
    Light penetration (in gaps and layers)
    Pollutants (distribution in time and space)

Biological and chemical features of the individual organisms
    Quality of host plant
        Attractiveness (everything about plant that makes it attractive to insect or
            pathogen actively seeking host)
        Surface characteristics (especially of different organs of plants)
        Structural characteristics (toughness, architecture, way in which different parts of
            plant are assembled)
        Biochemical status (C, N, defense compounds)
        Plant history (stress, age, size)
    Vulnerability of host plant
        Physiological status (carbon, nitrogen)
        Resource availability (water, nutrients, light)
        Phenology (age determinant or indeterminant, juvenile, mature)
        Vigor (crown rates, sapwood area, root starch, storage reserves)
        Defense compounds (toxins, lignin, digestibilities)
    Insects and pathogens
        Dispersal and host-finding mechanisms
        Initial establishment processes
        Growth/utilization/biomass conversion
        Reproductive output and mechanisms (sexual, asexual)
        Detoxification/toxin production capacity (part of overall competitive abilities)
        Parasite/disease/hyperparasite load

Population level features
    Host
        Age and size distribution
        Spatial distribution and geographic extent
        Genetic frequency of sensitive individuals, structure of sensitivity, diversity of
            gene pools
        Infection status (history)
        Infestation
    Insect/pathogen
        Sex ratio as it affects number of females
        Age and size distribution
        Spatial distribution
        Density (inoculum and survival characteristics)
        Proportion virulent
        Fecundity and survivorship

Community and ecosystem level features
    Tight interspecific interactions
        Symbionts (e.g., mycorrhizae, N-fixers)

Table 7.6. Continued

Vectors
Natural enemies of pest
Diffuse interspecific interactions
    Species composition and distribution (especially plants) and interacting species
    Community succession status and age of community
    Microbial community dynamics (e.g., free-living N-fixers, decomposers)
    Competitive interactions (e.g., pest-pest)

[a] After Bedford (1987), Hughes (1988).

more fecund (Coleman and Jones, 1988a) on tissue not exposed to O_3. Beetle development rates and survivorship were not affected by any treatment.

Coleman and Jones (1988b) reported the effects of acute O_3 exposure of eastern cottonwood on survivorship, reproduction, and development of the aphid *Chaitophorus populicola*. In this study, aphid performance was not significantly altered by treatment with O_3.

Some areas where further research to define O_3-insect interactions are necessary include (1) a determination of a biochemical or physiological basis for the alteration of insect success due to the O_3 treatment; (2) field studies using ambient or near-ambient conditions to examine the effect of O_3 on pests during different developmental stages of tree growth; (3) multiple pollutant interaction studies, and (4) studies that examine multiple stresses, e.g., drought-O_3-insect interactions, because trees are exposed to a multitude of different environmental stresses.

7.3.2.1.2 Biotic Plant Pathogens The effects of air pollutants on biotic plant pathogens can be stimulatory, neutral, or inhibitory (Heagle, 1973; Smith, 1981). Ozone may affect tree host-pathogen interactions by altering plant tissue susceptibility, plant resistance, pathogen virulence, and inoculum density (Bruck and Shafer, 1984). These and other critical factors to be considered in O_3-plant-pest interactions are shown in Table 7.6.

Similar to the situation with insects, many reports concerning tree host-pathogen interactions have been correlative in nature (Manning and Keane, 1988). Many findings in the literature indicate that air pollutants can directly, or indirectly, affect plant disease development (Heagle, 1973, 1982; Manning 1975; Laurence, 1981; Bruck and Shafer, 1984; Hughes and Laurence, 1984; Hüttunen, 1984; Shafer, 1985; Manning and Keane, 1988). The majority of these reports pertain to pathogen-herbaceous crop systems (Chapter 6).

Ozone has been shown to weaken trees in natural stands and increase their susceptibility to invasion by plant pathogens, such as *Heterobasidion annosum*, the causal agent of "annosum root rot" on ponderosa and Jeffrey pines in California (Miller, 1983), and *Verticicladiella procera*, the causal agent of "procera root disease" on eastern white pine in Virginia (Lackner and Alexander, 1983; Skelly et al., 1983).

James et al. (1980a) reported that *H. annosum* more readily invaded freshly

Table 7.7. Major Mechanisms by Which Pollutants Might Alter Plants as Hosts for Insects

Host vulnerability to discovery
   Host density
   Behavioral cues (chemical and physical)
      Color
      Surface morphology
      Surface chemicals
      Volatile chemicals
      Nonvolatile primary and secondary
       metabolites
   Plant species diversity/community structure

Host nutritional quality
   Plant nutrition
   Primary metabolite levels
   Secondary metabolite levels
   Plant water balance
   Metabolic activity (hormesis)
   Plant hormones

Host defenses
   Constitutive
      Surface morphology
      Toughness
      Secondary metabolites
   Induced
      Phytoalexins
      Translocated induced compounds

Alteration of gene expression

[a] After Hughes (1988).

cut stumps of ponderosa and Jeffrey pine trees exhibiting visible O_3 symptoms than stumps of asymptomatic trees. Increased susceptibility was associated with decreased oleoresin exudation and decreased colonization by other fungal competitors. Ozone injury increased susceptibility of these trees to root infection and colonization by *H. annosum* (James et al., 1980b). These trees became infected more often than asymptomatic trees, both in the field and greenhouse.

Ozone may enhance disease development by fungi that are normally saprophytic in nature. *Lophodermium pinastri* and *Pullalaria pullulans* were commonly associated with eastern white pine foliar injury when inoculated and exposed to O_3 (0.07 µl l^{-1}, 4.5 h) exposure (Costonis and Sinclair, 1972). Weidensaul and Darling (1979) reported that O_3, alone and in combination with SO_2, caused an increase in infection and colonization of Scots pine needles by *Scirrhia acicola*, compared with controls. The amount of diseased tissue resulting from the combination of gases was greater than additive.

7.3.2.2 Effects on Mycorrhizal Associations

Mycorrhizae are essential for tree growth in the natural environment (St. John and Coleman, 1983; Kottke and Oberwinker, 1986). The association of plant roots and beneficial fungi improve nutrient uptake and translocation, water uptake, root morphology, and disease resistance, for the plant host. Sensitivity of mycorrhizal development to O_3 could affect forest tree growth and productivity.

Beneficial effects of mycorrhizae in ameliorating O_3 stress have been reported for loblolly pine colonized by the ectomycorrhizal fungi *Pisilithus tinctorius* and *Thelephora terrestris* (Carney et al., 1978; Garrett et al., 1982; Mahoney et al., 1985). Mahoney et al. (1985) reported that mycorrhizae altered O_3 effects on shoot and root growth, promoting root growth. Fumigated seedlings had a higher root-shoot ratio compared with unfumigated seedlings.

Carney et al. (1978) and Garrett et al. (1982) measured O_2 uptake in detached mycorrhizal and nonmycorrhizal root segments of loblolly pine exposed to 0.05 or 0.5 $\mu l\ l^{-1}\ O_3$ for 3 h. Nonmycorrhizal root segments exhibited less O_2 uptake, suggesting that *P. tinctorius* and *T. terrestris* may, to some degree, protect trees from air pollution. Garrett et al. (1982) reported that *T. terrestris* provided seedlings with the best protection to O_3, while *P. tinctorius*-colonized root segments were more resistant to SO_2, based on O_2 uptake rates. These results, although interesting in nature, are difficult to extrapolate to a "real-world" stituation, since O_3 does not penetrate the soil (Turner et al., 1973). Therefore, the influence of O_3 on mycorrhizae must occur due to changes in plant metabolism.

Ozone has been shown to have a direct effect on mycorrhizal associations (McCool et al., 1979; Reich et al., 1985, 1986; Stroo et al., 1988; Keane and Manning, 1989; Simmons and Kelly, 1989; Meier et al., 1990). McCool et al. (1979) reported that chlamydospore production by *Glomus fasiculatus* was decreased by an acute (1.0 $\mu l\ l^{-1}$, 4 h) O_3 exposure. Endomycorrhizal formation was reduced after fumigation with O_3 at 0.9 $\mu l\ l^{-1}$ for 6 h, once a week for 19 weeks.

In a series of related studies (Reich et al., 1985, 1986; Stroo et al., 1988), potted northern red oak and eastern white pine were exposed to O_3 and other air pollutants. Ozone exposure, in general, resulted in more ectomycorrhizae and an increased percent infection of northern red oak. Percent of infection exhibited a quadratic response for eastern white pine;: at lower doses, infection was increased; at doses above 40 $\mu l\ l^{-1}$-h percent infection decreased. Numbers of infected short roots generally decreased with increasing O_3 dose. In both studies, levels of ectomycorrhizal infection varied with the soil type.

Keane and Manning (1989) exposed 4-week-old paper birch, inoculated or not inoculated with *P. tinctorius*, to O_3 and simulated acid rain. The percent of ectomycorrhizal infection was altered by an O_3-acid rain interaction, with the greatest percent of infection occurring in seedlings treated with charcoal-filtered air and pH 5.6 simulated rain.

Simmons and Kelly (1990) and Meier et al. (1990), in independent studies with loblolly pine, found that, in general, O_3-treated seedlings exhibited fewer numbers of ectomycorrhizae, and the percent colonization of feeder roots was less than the controls. As mentioned previously, O_3 generally does not penetrate the soil surface (Turner et al., 1973); therefore, the influence on mycorrhizae must occur due to changes in plant metabolism. Decreases in photosynthesis or carbon allocation to the root systems could reduce the levels of carbohydrate available for mycorrhizae (Adams et al., 1990; Spence et al., 1990). Ozone may cause a modification in root exudates (qualitatively and quantitatively), which may alter mycorrhizal infection. In addition, there may be a differential sensitivity in fungi (Meier et al., 1990), which may affect nutrient and water uptake in O_3-stressed trees.

7.3.2.3 Genetic Variation

Plants have adapted over time to a wide variety of environmental conditions. Terrestrial vegetation also has demonstrated evolutionary responses to the presence of anthropogenic pollutants (Pitelka, 1988). Populations can differ in their capacity to develop resistance to air pollutants, such as O_3. Resistance is defined as a quantitative trait that enables a plant to grow, survive, and reproduce in the presence of a particular pollutant (Pitelka, 1988) and involves either avoidance or tolerance mechanisms. Avoidance may be an effective adaptation to acute pollutant episodes, but tolerance is the mechanism of resistance with long-term chronic pollution.

Trees differ in their capacity to develop resistance and are subject to genetic variation within a population (see Figure 7.6). Populations that lack this variability will not survive. The speed at which this evolutionary process occurs is controlled by several systems, generation time, seed banks, and timing of selection during a tree's life cycle (Pitelka, 1988). This adaptation does have an effect on fitness (see Figure 7.7). Resistant trees often are less able to compete for resources available for growth in pristine environments. Resistant genotypes experience a metabolic cost of tolerating a pollutant (energy reduction) or a reduced rate of resource allocation (stomatal closure).

A considerable amount of research has been reported on plant resistance to both soil and air pollution. The purpose of this section is to briefly discuss genetic variation to O_3, among and within species. Evolutionary responses of plants to air pollution are discussed in more detail by Karnosky and Houston (1973) and Pitelka (1988).

Numerous reports in the literature list tree species according to a sensitivity ranking. Plant material (source, age, genotype), criteria for ranking (foliar injury, growth), and methodologies used (exposure chambers, ambient air) all play an important role in results obtained from various experiments (Davis and Gerhold, 1976). Relative O_3 sensitivity rankings for several major forest tree species are shown in Table 7.8.

Problems associated with comparing results between studies are apparent.

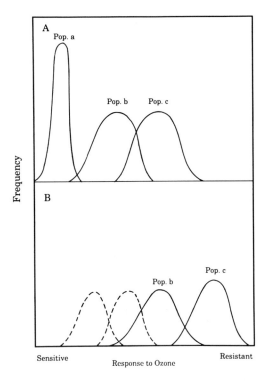

Figure 7.6.   A model illustrating the evolution of pollutant-resistant populations. Populations (b) and (c) possess latent variability in sensitivity and, over time, become more resistant to a pollutant. Population (a) lacks variability and does not survive (after Pitelka, 1988).

Several species have been classified into two or more different rankings (see Table 7.8). Plant sensitivity can be altered by many exogenous and endogenous factors, as discussed in this chapter. Caution against placing a great deal of faith in relative sensitivity rankings is warranted (Davis and Gerhold, 1976).

Reich (1987) quantified plant sensitivity to O_3, based on concentration, external dose (gas concentration times duration of exposure), or uptake. For an equivalent dose within a single growing season, hardwood species were found to be more sensitive to O_3 than were conifer species. These results are based on higher leaf conductance rates in hardwoods and, therefore, more O_3 uptake.

Harkov and Brennan (1979), using sensitivity rankings based on visible injury, proposed that slower-growing deciduous tree species in eastern North America (late successional and climax species) are less sensitive to ambient O_3 injury than rapidly growing tree species, typical of early successional stands. They hypothesized that, since early successional species have greater leaf conductance, they must have greater uptake of O_3. Generally, the faster growing species are more valuable for timber production, and altering their growth would

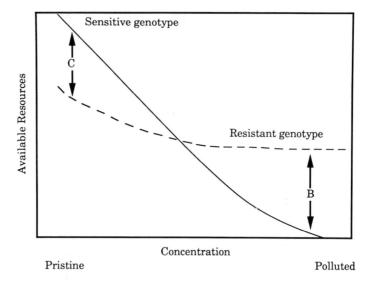

Figure 7.7.   Fitness costs of adaptation to ozone. The resistant genotype experiences a cost "c" in a pristine environment as a function of a metabolic cost or a reduced rate of resource acquisition, but obtains a benefit "b" in polluted environments due to tolerance to the particular pollutant in question. Available resources decline in the sensitive gentoype as pollutant concentrations increase (after Pitelka, 1988).

be of major economic importance if Harkov and Brennan's (1979) hypothesis is applicable regarding growth.

Within a species, tremendous variation in O_3 sensitivity can occur. Eastern white pine has been identified as one of the tree species most sensitive to O_3 (Berry, 1971), with some genotypes exhibiting "ultrasensitivity" (chlorotic dwarf). Other reports indicate that, in general, eastern white pine is relatively tolerant to O_3 (Davis and Wood, 1972; Anderson et al., 1988). This may be the result of an evolutionary adaptation to air pollution, with the most sensitive genotypes becoming extinct, as illustrated in Figure 7.6.

Quaking aspen is another species in which natural selection for O_3 tolerance has been observed (Berrang et al., 1986). The average O_3 injury was less for clones collected from national parks in the eastern U.S. which were subjected to the most pollution, based on ambient O_3 levels.

Green and white ash seedlings from different families and provenances in the eastern U.S. were fumigated with O_3 or sulfur dioxide (Karnosky and Steiner, 1981). Variation among provenances was greater than variation within a provenance. These results demonstrate both family and geographic differences in response to O_3 injury for green and white ash.

Genetic variation in loblolly pine to O_3 has been studied both in the greenhouse (Kress et al., 1982; Winner et al., 1987; Reinert et al., 1988) and in the field (Shafer et al., 1987; Kress et al., 1988; Adams et al., 1988). Differences in sensitivity were observed regarding visible injury and growth suppression;

Table 7.8. Sensitivity of Forest Species to Ozone[a]

Sensitive	Intermediate	Tolerant
Acer negundo	Abies concolor	Abies balsamea
Acer saccharinum	Acer negundo	Abies concolor
Acer saccharum	Almus sp.	Acer grandidentatum
Ailanthus altissima	Cercis canadensis	Acer platanoides
Amelanchier alnifolia	Juglans regia	Acer rubrum
Betula sp.	Larix decidua	Acer saccharum
Fraxinus americana	Larix leptolepis	Betula pendula
Fraxinus pennsylvanica	Libocedrus decurrens	Cornus florida
Gleditsia triacanthos	Liquidambar	Fagus sylvatica
Juglans regia	styraciflua	Ilex opaca
Larix decidua	Pinus attenuata	Juglans nigra
Liriodendron tulipifera	Pinus contorta	Juniperus
Pinus bankisiana	Pinus echinata	occidentalis
Pinus coulteri	Pinus elliottii	Nyssa sylvatica
Pinus jeffreyi	Pinus lambertiana	Persea americana
Pinus nigra	Pinus rigida	Picea abies
Pinus ponderosa	Pinus strobus	Picea glauca
Pinus radiata	Pinus sylvestris	Picea pungens
Pinus strobus	Pinus torreyana	Pinus resinosa
Pinus sylvestris	Pseudotsuga	Pinus sabiniana
Pinus taeda	macrocarpa	Pseduotsuga
Pinus virginiana	Pseudotsuga	menziesii
Platanus occidentalis	menziesii	var. menziesii
Populus maximowiczii	var. menziesii	Pyrus communis
× trichocarpa	Quercus coccinea	Quercus imbricaria
Populus tremuloides	Quercus palvestris	Quercus macrocarpa
Quercus alba	Quercus velutina	Quercus rubra
Quercus gambelii	Rhododendron spp.	Robinia
Rhus sp.	Salix babylonica	pseudoacacia
Salix babylonica	Syringa fulgaris	Rhododendron sp.
Sorbus aucuparia	Ulmus parrufolia	Sequoia
Tilia sp.		sempervirens
Tsuga canadensis		Sequoia giganteum
		Thuja occidentalis
		Tilia americana
		Tilia cordata
		Tsuga canadensis

[a] Classification of a species in more than one sensitivity group indicates conflicting reports in the literature (after Berry, 1971; Davis and Coppolino, 1974, 1976; Davis and Wood, 1968, 1972; Wood and Coppolino, 1972; Wood and Davis, 1969).

however, these indicators of genetic variation (visible injury and growth) were not correlated (Shafer et al., 1987; Winner et al., 1987; Adams et al., 1988).

Results included in this section indicate that trees can exhibit a wide range of sensitivity to O_3, both among and within species. Visible injury symptoms and growth reductions do not appear to be correlated. Although genetic variation is widely recognized, the mechanism(s) controlling this variation is unknown. In addition, more research is needed in order to determine the role other environmental factors play (genotype-environment interaction) in governing O_3 sensitivity in trees.

7.4 FUTURE DIRECTIONS

Ozone, as discussed in this chapter, can affect trees in many ways: disrupt metabolism, alter growth and productivity, and predispose plants to other abiotic and biotic stresses. Through approximately three decades of research, many aspects of O_3 effects on trees have been discovered. Only within the last few years, however, have tree responses to O_3 been intensively studied. We have discussed in this section some areas of study where we feel more O_3-effects research is necessary.

In order to better understand the effects of O_3 on tree response, it is important to determine O_3-induced effects on metabolic and biochemical processes as they relate to subsequent growth and productivity. An important area that is lacking is the determination of the mechanisms of O_3 toxicity and host tolerance. Future research is needed to establish the prevalence of apoplastic antioxidant systems in trees and to examine the relationship of these systems to O_3 tolerance.

Numerous researchers have reported that O_3 has a deleterious effect on photosynthesis and patterns of carbon allocation in plants. However, the relationship between photosynthetic response and translocation of photoassimilate has not been clearly demonstrated with trees. This association may be difficult to prove conclusively, due to the complexity of interactions involved in impairment of photosynthesis and translocation in trees. Ozone inhibition of plant growth can be modified by numerous environmental and biological factors. In addition, as trees age, the balance between photosynthetic and nonphotosynthetic tissues changes. For these and other reasons, mature tree response may be different than young-seedling response.

In order to accurately assess the response of terrestrial vegetation to O_3, experimental methodologies need to be developed that encompass various components of pollutant exposure and tree dynamics. A necessary step is a quantitative description of cause-effect relationships. Field assessment methods and exposure indices need to be developed that describe O_3-exposure–tree-response in a more realistic fashion, both in a temporal and spatial pattern. Both spatial and diurnal patterns of pollutants and daily physiological functions of the species examined need to be established. Also, whether pollutants co-occur sequentially or contemporarily in the area, and with what frequency, needs to be determined in order to obtain realistic data on growth. Studies need to be conducted for more than 1 year, since effects are cumulative and may not appear until the second growing season. Presently, exclusion methodologies are limited by tree size and research funding. Other methods, such as chamberless systems and modeling approaches, need to be developed in order to adequately study O_3 effects.

The ambient environment exerts a profound influence on the physiological status of a plant and, therefore, its response to O_3. No studies to date have examined the effects of changing environmental conditions on tree growth, using ambient or near-ambient O_3 concentrations. Results demonstrate that moisture stress and nutrient status can have a profound effect on tree response to O_3. The majority of these studies were conducted in pots within a greenhouse. Further research is needed with trees of various ages and genotypes, growing in the field over several years, in order to verify greenhouse results.

Biotic factors can influence tree response to O_3. Tree sensitivity to other stresses, such as insects and biotic plant pathogens, can be altered by exposure to O_3. However, knowledge is limited to only a few stress-tree systems under very controlled conditions. Research needs to be expanded to encompass a greater variety of interactions in more realistic field conditions. In addition, very little research has been conducted on mechanisms influencing these systems. The establishment of either a biochemical or physiological basis for the alteration of insect/pathogen success on trees is needed.

Results included in this chapter indicate that trees exhibit a wide range of sensitivity to O_3, both among and within species. Although genetic variation is widely recognized, the mechanism(s) controlling this variation is unknown. In addition, more research is needed in order to determine the role other environmental factors play (genotype-environment interaction) in governing O_3 sensitivity in trees.

Ozone effects in the natural environment are generally very subtle and complicated to interpret. The vast majority of research involves studies with individual trees. Models and exposure facilities need to be developed that encompass growth characteristics, nutrient and moisture fluxes, competitive effects, and realistic O_3 exposure indices in order to determine the response of forest stands and ecosystems to O_3.

APPENDIX

List of Common and Scientific Names of Major North American Trees

Common Name	Scientific Name
Alaska-cedar	*Chamaecyparis nootkatensis*
Alder	
red	*Alnus rubra*
Sitka	*Alnus sinuata*
Ash	
black	*Fraxinus nigra*
blue	*Fraxinus quadrangulata*
Carolina	*Fraxinus caroliniana*
green	*Fraxinus pennsylvanica*
Oregon	*Fraxinus latifolia*
pumpkin	*Fraxinus profunda*
white	*Fraxinus americana*
Aspen	
bigtooth	*Populus grandidentata*
quaking	*Populus tremuloides*
Baldcypress	*Taxodium distichum*
Basswood	
American	*Tilia americana*
white	*Tilia heterophylla*
Beech, American	*Fagus grandifolia*
Birch	
gray	*Betula populifolia*
paper	*Betual papyrifera*
river	*Betula nigra*
sweet	*Betula lenta*
yellow	*Betula alleghaniensis*
Blackgum (black tupelo)	*Nyssa sylvatica*
Boxelder	*Acer negundo*
Buckeye	
California	*Aesculus californica*
Ohio	*Aesculus glabra*
yellow	*Aesculus octandra*
Buckthorn, cascara	*Rhamnus purshiana*
Butternut	*Umbellularia californica*

California-laurel   *Juglans cinerea*
Cherry
  black   *Prunis serotina*
  pin   *Prunus pennsylvanica*
Chestnut, American   *Castanea dentata*
Chinkapin, golden   *Castanopsis chyrsophylla*
Cottonwood
  black   *Populus trichocarpa*
  eastern   *Populus deltoides* var.
                                              deltoides
  plains   *Populus deltoides* var.
                                              occidentialis
  swamp   *Populus heterophylla*
Cucumbertree   *Magnolia acuminata*
Cypress, Arizona   *Cupressus arizonica*
Dogwood
  flowering   *Cornus florida*
  Pacific   *Cornus nuttallii*
Douglas-fir   *Pseudotsuga menziesii* var.
                                              menziesii
  bigcone   *Pseudotsuga macrocarpa*
Rocky Mountain   *Pseudotsuya menziesii* var.
                                              glauca

Elm
  American   *Ulmus americana*
  cedar   *Ulmus crassifolia*
  rock   *Ulmus thomasii*
  September   *Ulmus serotina*
  slippery   *Ulmus rubra*
  winged   *Ulmus alata*
Fir
  Balsam   *Abies balsamea*
  bristlecone   *Abies bracteata*
  Fraser   *Abies fraseri*
  grand   *Abies grandis*
  noble   *Abies procera*
  Pacific silver   *Abies amabilis*
  red   *Abies magnifica* and *Abies*
                                              magnifica var. *shastensis*
  subalpine   *Abies lasiocarpa*
  white   *Abies concolor*
Hackberry   *Celtis occidentalis*
Hawthorn   *Crataegus* spp.
Hemlock
  eastern   *Tsuga canadensis*

mountain	*Tsuga mertensiana*
western	*Tsuga heterophylla*
Hickory	
bitternut	*Carya cordiformis*
mockernut	*Carya tomentosa*
nutmeg	*Carya myristicaeformis*
pignut	*Carya glabra*
shagbark	*Carya ovata*
shellbark	*Carya laciniosa*
swamp	*Carya leiodermis*
water	*Carya aquatica*
Honeylocust	*Gleditsia triacanthos*
Hophornbeam, eastern	*Ostrya virginiana*
Incense-cedar	*Libocedrus decurrens*
Juniper	
alligator	*Juniperus deppeana*
one-seed	*Juniperus monosperma*
Rocky Mountain	*Juniperus scopulorum*
Utah	*Juniperus osteosperma*
western	*Juniperus occidentalis*
Larch	
subalpine	*Larix lyallii*
western	*Larix occidentalis*
Locust, black	*Robinia pseudoacacia*
Madrone, Pacific	*Arbutus menziesii*
Magnolia, southern	*Magnolia grandiflora*
Maple,	
bigleaf	*Acer macrophyllum*
black	*Acer nigrum*
red	*Acer rubrum*
silver	*Acer saccharinum*
striped	*Acer pennsylvanicum*
sugar	*Acer saccharum*
vine	*Acer circinatum*
Oak	
Arizona white	*Quercus arizonica*
bear	*Quercus ilicifolia*
black	*Quercis velutina*
blackjack	*Quercus marilandica*
blue	*Quercus douglasii*
bluejack	*Quercus incana*
bur	*Quercus macrocarpa*
California black	*Quercus kelloggii*
California live	*Quercus agrifolia*
California white	*Quercus lobata*

canyon live	*Quercus chrysolepis*
cherrybark	*Quercus falcata* var. *pagodaefolia*
chestnut	*Quercus prinus*
chinkapin	*Quercus muehlenbergii*
interior live	*Quercus wislizenii*
laurel	*Quercus laurifolia*
live	*Quercus virginiana*
myrtle	*Quercus myrtifolia*
northern pin	*Quercus ellipsoidalis*
northern red	*Quercus rubra*
Nuttall	*Quercus nuttallii*
Oregon white	*Quercus garryana*
overcup	*Quercus lyrata*
pin	*Quercus palustris*
post	*Quercus stellata*
scarlet	*Quercus coccinea*
shingle	*Quercus imbricaria*
Shumard	*Quercus shumardii*
southern red	*Quercus falcata* var. *falcata*
swamp chestnut	*Quercus michauxii*
swamp white	*Quercus bicolor*
turkey	*Quercus laevis*
water	*Quercus nigra*
white	*Quercus alba*
willow	*Quercus phellos*
Pecan	*Carya illinoensis*
Persimmon, common	*Diospyros virginiana*
Pine	
Arizona	*Pinus ponderosa* var. *arizonica*
bishop	*Pinus muricata*
bristlecone	*Pinus aristata*
Coulter	*Pinus coulteri*
Digger	*Pinus sabiniana*
eastern white	*Pinus strobus*
foxtail	*Pinus balfouriana*
jack	*Pinus banksiana*
Jeffrey	*Pinus jeffreyi*
knobcone	*Pinus attenuata*
limber	*Pinus flexilis*
loblolly	*Pinus taeda*
lodgepole	*Pinus contorta*
longleaf	*Pinus palustris*
pitch	*Pinus rigida*

pond	*Pinus serotina*
ponderosa	*Pinus ponderosa*
red	*Pinus resinosa*
sand	*Pinus clausa*
shortleaf	*Pinus echinata*
slash	*Pinus elliottii*
spruce	*Pinus glabra*
sugar	*Pinus lambertiana*
Table-Mountain	*Pinus pungens*
Virginia	*Pinus virginiana*
western white	*Pinus monticola*
whitebark	*Pinus albicaulis*
Pinyon	*Pinus edulis*
Mexican	*Pinus cembroides*
Parry	*Pinus quadrifolia*
singleleaf	*Pinus monophylla*
Pondcypress	*Taxodium distichum* var. *nutans*
Poplar, balsam	*Populus balsamifera*
Port-Orford-cedar	*Chamaecyparis lawsoniana*
Redcedar	
eastern	*Juniferus virginiana*
western	*Thuja plicata*
Redwood	*Sequoia sempervirens*
Sassafras	*Sassafras albidum*
Sequoia, giant	*Sequoia gigantea*
Sourwood	*Oxydendrum arboreum*
Spruce	
black	*Picea mariana*
blue	*Picea pungens*
Engelmann	*Picea engelmannii*
red	*Picea rubens*
Sitka	*Picea sitchensis*
white	*Picea glauca*
Sugarberry	*Celtis laevigata*
Sweetbay	*Magnolia virginiana*
Sweetgum	*Liquidambar styraciflua*
Sycamore, American	*Platanus occidentalis*
Tupelo	
black	*Nyssa sylvatica*
swamp	*Nyssa sylvatica* var. *biflora*
water	*Nyssa aquatica*
Tamarack	*Larix laricina*
Tanoak	*Lithocarpus densiflorus*
Walnut, black	*Juglans nigra*

Waterlocust	*Gleditsia aquatica*
White-cedar	
Atlantic	*Chamaecyparis thyoides*
northern	*Thuja occidentalis*
Willow	*Salix* spp.
black	*Salix nigra*
Yellow-poplar	*Liriodendron tulipifera*
Yew, Pacific	*Taxus brevifolia*

REFERENCES

Adams, M. B., Kelly, J. M., and Edwards, N. T. (1988) Growth of *Pinus taeda* L. seedlings varies with family and ozone exposure levels. *Water Air Soil Pollut.* **38**, 137–150.

Adams, M. B., Edwards, N. T., Taylor, G. E., and Skaggs, B. L. (1990) Whole-plant ^{14}C-photosynthate allocation in *Pinus taeda:* seasonal patterns at ambient and elevated ozone levels. *Can. J. For. Res.* **20**, 152–158.

Allen, O. B., Marie, B. A., and Ormrod, D. P. (1987) Relative efficiency of factorial designs for estimating response surfaces with reference to gaseous pollutant mixtures. *J. Environ. Qual.* **16**, 316–320.

Alscher, R. G., Cumming, J. R., and Fincher, J. (1989a) Air pollutant-low temperature interactions in trees. In *Biological Markers of Air Pollution Stress and Damage in Forests.* Nation. Academic Press, New York, 341–345

Alscher, R. G., Amundson, R. G., Cumming, J. R., Fellows, S., Fincher, J., Rubin, G., VanLouken, P., and Weinstein, L. (1989b) Seasonal changes in the pigments, carbohydrates and growth of red spruce as affected by ozone. *New Phytol.* **113**, 211–224.

Alstad, D. N., Edmunds, G. F., Jr., and Weinstein, L. H. (1982) Effects of air pollutants on insect populations. *Annu. Rev. Entomol.* **27**, 369–384.

Anderson, R. L., Brown, H. D., Chevone, B. I., and McCartney, T. C. (1988) Occurrence of air pollution symptoms (needle tip necrosis and chlorotic motling) on eastern white pine in the southern Appalachian Mountains. *Plant Dis.* **72**, 130–132.

Anlauf, K. G., Bottenheim, J. W., Krice, K. A., Tellin, P., Wiebe, H. A., Scheff, H. I., Mackay, G. I., Branmar, R. S., and Gilbert, R. (1985) Measurements of atmospheric aerosols and photochemical products at a rural site in S.W. Ontario. *Atmos. Environ.* **19**, 1859–1870.

Arndt, U., Seufert, G., and Nobel, W. (1982) The contribution of ozone to the silver fir. (*Abies alba Mill*) disease complex-A hypothesis worth examining. *Staub-Reinhalt. Luft* **42**, 243–247.

Ashmore, M., Bell, N., and Rutter, J. (1985) The role of ozone in forest damage in West Germany. *Ambio* **14**, 81–87.

Barnes, R. (1972) Effects of chronic exposure to ozone on soluble sugar and ascorbic acid contents of pine seedlings. *Can. J. Bot.* **50**, 215–219.

Barnes, J. D. and Davison, A. W. (1987) The influence of ozone on the winter hardiness of Norway spruce. (*Picea abies.* [L.] Karst.) *New Phytol.* **108**, 159–166.

Bedford, B. L. (ed.) (1987) Modification of plant-pest interactions by air pollutants. Proc. International Workshop. ERC Report No. 117. Ecosystems Research Center, Ithaca, NY.

Benoit, L. F., Skelly, J. M., Moore, L. D., and Dochinger, L. S. (1982) Radial growth reductions of *Pinus strobus* L. correlated with foliar ozone sensitivity as an indicator of ozone induced losses in eastern forests. *Can. J. For. Res.* **12**, 673–678.

Benoit, L. F., Skelly, J. M., Moore, L. D., and Dochinger, L. S. (1983) The influence of ozone on *Pinus strobus* L. pollen germination. *Can. J. For. Res.* **13**, 184–187.

Berrang, P. D., Karnosky, D. F., Mickler, R. A., and Bennett, J. P. (1986) Natural selection for ozone tolerance in *Populus tremuloides. Can. J. For. Res.* **16**, 1214–1216.

Berry, C. R. (1964) Differences in concentrations of surface oxidant between valley and mountaintop conditions in the southern Appalachians. *J. Air Pollut. Control Assoc.* **14**, 238–239.

Berry, C. R. (1971) Relative sensitivity of red, jack and white pine seedlings to ozone and sulfur dioxide. *Phytopathology* **61**, 231–232.

Blank L. W. (1985) A new type of forest decline in Germany. *Nature* **314**, 311–314.

Botkin, D. B., Smith, W. H., Carlson, R. W., and Smith, T. L. (1972) Effects of ozone on white pine saplings: variation in inhibition and recovery of net photosynthesis. *Environ. Pollut.* **3**, 273–289.

Boyer, J. N., Houston, D. B., and Jensen, K. F. (1986) Impacts of Chronic SO_2, O_3 and $SO_2 + O_3$ exposures on photosynthesis of *Pinus strobus* clones. *Eur. J. For. Pathol.* **16**, 293–299.

Braun, S. and Flückiger, W. (1989) Effect of ambient ozone and acid mist on aphid development. *Environ. Pollut.* **56**, 177–187.

Brown, K. A. and Roberts, T. M. (1988) Effects of ozone on foliar leaching in Norway spruce (*Picea abies* L. Karst.): confounding factors due to NO_x production during ozone generation. *Environ. Pollut.* **55**, 55–73.

Brown, K. A., Roberts, T. M., and Blank, L. W. (1987) Interaction between ozone and cold sensitivity in Norway spruce: a factor contributing to the forest decline in central Europe. *New Phytol.* **105**, 149–155.

Bruck, R. I. and Shafer, S. R. (1984) Effects of acid precipitation on plant diseases, In *Direct and Indirect Effects of Acidic Precipitation on Vegetation* (ed., Linthurst R. A.). Butterworths, Stonehaven, MA, 19–32.

Carlson, R. W. (1979) Reduction in the photosynthetic rate of *Acer, Quercus and Fraxinus* species caused by sulphur dioxide and ozone. *Environ. Pollut.* **18**, 159–170.

Carney, J. L., Garrett, H. E., and Hedrick, H. G. (1978) Influence of air pollution gases on oxygen uptake of pine roots with selected ectomycorrhizae. *Phytopathology* **68**, 1160–1163.

Chappelka, A. H., III and Chevone, B. I. (1986) White ash seedling growth response to ozone and simulated acid rain. *Can. J. For. Res.* **16**, 786–790.

Chappelka, A. H., III and Chevone, B. I. (1989) Two methods used to determine tree responses to pollutants. *Environ. Pollut.* **61**, 31–45.

Chappelka, A. H., III and Kraemer, M. E. (1988) Effects of air pollutants on plant-insect interactions. *Proc. 81st Annuual Meeting of Air Pollutution Control Assoc.iation June 19–24, 1988, Dallas, TX,* 88-125.5. Air Pollution Control Association, Pittsburgh.

Chappelka, A. H., III, Chevone, B. I., and Burk, T. E. (1985) Growth response of yellow poplar (*Liriodendron tulipifera* L.) seedlings to ozone, sulfur dioxide and simulated acidic precipitation, alone and in combination. *Environ. Exp. Bot.* **25**, 233–244.

Chappelka, A. H., III, Chevone, B. I., and Burk, T. E. (1988a) Growth response of green and white ash to ozone, sulfur dioxide, and simulated acid rain. *For. Sci.* **34**, 1016–1029.

Chappelka, A. H., III, Chevone, B. I., and Seiler, J. R. (1988b) Growth and physiological responses of yellow-poplar seedlings exposed to ozone and simulated acidic rain. *Environ. Pollut.* **49**, 1–18.

Chappelka, A. H., III, Kush, J. S., Meldahl, R. S., and Lockaby, B. G. (1990) An ozone-low temperature interaction in loblolly pine (*Pinus taeda* L). *New Phytol.* **114**, 721-726.

Chappelka, A. H., III, Lockaby, B. G., Meldahl, R. S., and Kush, J. S. (1989) Atmospheric deposition effects on loblolly pine: development of an intensive field research site. *Proc. 5th Biennial So. Silvic. Research Conference, Nov. 1–2, 1988, Memphis, TN,* 57–60.

Chevone, B. I., Skelly, J. M., and Yang, Y. S. (1983) The influences of ambient pollutants on the growth of seedling forest trees and native vegetation. *Aquilo Ser. Bot.* **19**, 198–207.

Chevone, B. I., Yang, Y. S., and Reddick, G. S. (1984) Acidic precipitation and ozone effects on growth of loblolly and shortleaf pine seedlings. *Phytopathology* **74**, 756 (Abstr.).

Chevone, B. I., Herzfeld, D. E., Krupa, S. V., and Chappelka, A. H. (1986) Direct effects of atmospheric sulfate deposition on vegetation. *J. Air Pollut. Control Assoc.* **36**, 813–815.

Chevone B. I., Lee W. S., Anderson J. V. and Hess J. L. (1989) Gas exchange and needle acsorbate content of eastern white pine exposed to ambient air pollution. In *Proc. 82nd Annual Meeting Air and Waste Management Association, Anaheim, CA, June 25–30, 1989*, 89–89.2.

Chew, V. (1976) Comparing treatment means: a compendium. *Hort. Sci.,* **11**, 48–57.

Coleman, J. S. and Jones, C. G. (1988a) Acute ozone stress on eastern cottonwood (*Populus deltoides* Bont.) and the pest potential of the aphid, *Chaitophorus populicola* Thomas (Homoptera: Aphididae). *Environ. Entomol.* **17**, 207–212.

Coleman, J. S. and Jones, C. G. (1988b) Plant stress and insect performance: cottonwood, ozone and a leaf beetle. *Oecologia* **76**, 57–61.

Constantinidou, H. A. and Kozlowski, T. T. (1979) Effects of sulfur dioxide and ozone on *Ulmus americana* seedlings. II. Carbohydrates, proteins, and lipids. *Can. J. Bot.* **57**, 176–184.

Cooley, D. R. and Manning, W. J. (1987) The impact of ozone on assimilate partitioning in plants: a review. *Environ. Pollut.* **47**, 95–113.

Costonis, A. C. and Sinclair, W. A. (1969) Relationship of atmospheric ozone to needle blight of eastern white pine. *Phytopathology* **59**, 1566–1574.

Costonis, A. C. and Sinclair, W. A. (1972) Susceptibility of healthy and ozone-injured needles of *Pinus strobus* to invasion by *Lophordermium pinastri* and *Aureobasidium pullulans*. *Eur. J. For. Pathol.* **2**, 65–73.

Coyne, P. I. and Bingham, G. E. (1981) Comparative ozone dose response of gas exchange in a ponderosa pine stand exposed to long-term fumigation. *J. Air Pollut. Control Assoc.* **31**, 38–41.

Coyne, P. I. and Bingham, G. E. (1982) Variation in photosynthesis and stomatal conductance in an ozone-stressed ponderosa pine stand: light response. *For. Sci.* **28**, 257–273.

Dahlsten, D. L. and Rowney, D. L. (1980) Influence of air pollution on population dynamics of forest insects and on tree mortality. In *Effects of Air Pollutants on Mediterranean and Temperate Forest Ecosystems*, Proc. Symp. Gen. Tech. Rep. PSW-43, USDA Forest Service, Pacific Southwest Forest and Range Exp. Sta., Berkeley, CA, 125–130.

Davis, D. D. and Coppolino, J. B. (1974) Relative ozone susceptibility of selected woody ornamentals. *Hort. Sci.* **9**, 537–539.

Davis, D. D. and Coppolino, J. B. (1976) Ozone susceptibility of selected woody shrubs and vines. *Plant Dis. Rept.* **60**, 876–878.

Davis, D. D. and Gerhold, H. O. (1976) Selection of trees for tolerance of air pollutants. Better Trees for Metropolitan Landscaping Symposium Proceedings. USDA Forest Service Gen. Tech. Rep. NE-22, 61–66.

Davis, D. D. and Wilhour, R. (1976) Susceptibility of woody plants to sulfur dioxide and photochemical oxidants: a literature review. USEPA Ecol. Res. Ser. EPA-600/3-76-102.

Davis, D. D and Wood, F. A. (1968) Relative sensitivity of twenty-two tree species to ozone. *Phytopathology* **58**, 399 (Abstr.).

Davis, D. D. and Wood, F. A. (1972) The relative susceptibility of eighteen coniferous species to ozone. *Phytopathology* **62**, 14–19.

Davis, D. D. and Wood, F. A. (1973) The influence of environmental factors on the sensitivity of Virginia pine to ozone. *Phytopathology* **63**, 371–376.

Davis, D. D., Umback, D. M., and Coppolino, J. B. (1982) Susceptibility of tree and shrub species and responses of black cherry foliage to ozone. *Plant Disease* **65**, 904–907.

Davison, A. W., Barnes, J. D., and Renner, C. J. (1987) Interactions between air pollutants and cold stress. In *Air Pollution and Plant Metabolism* (eds., Schulte-Hostede, S., Darrall, N. M., Blank, L. W., and Wellburn, A. R.). Elsevier, London, 307–329.

Dochinger, L. S. and Townsend, A. M. (1979) Effects of roadside deicing salts and ozone on red maple progenies. *Environ. Pollut.* **19**, 229–237.

Duchelle, S. F., Skelly, J., M. and Chevone, B. I. (1982) Oxidant effects on forest tree seedling growth in the Appalachian Mountains. *Water Air Soil Pollut.* **18**, 363–373.

Edwards, N. T., Taylor, G. E., Adams, M. B., Simmons, G. L., and Kelly, J. M. (1990) Ozone, acidic rain and soil magnesium effects on growth and foliar pigments of *Pinus taeda* L. *Tree Physiol.* **6**, 95–104.

Elliott, C. L., Eherhandt, J. C., and Brennan, E. G. (1987) The effect of ambient ozone pollution and acidic rain on the growth and chlorophyll content of green and white ash. *Environ. Pollut.* **41**, 61–70.

Fincher, J., Cumming, J. R., Alscher, R. G., Rubin, G., and Weinstein, L. (1989) Long term ozone exposure affects winter hardiness of red spruce seedlings. *New Phytol.* **113**, 85–96.

Foster, J. R., Loats, K. V., and Jensen, K. F. (1990) Influence of two growing seasons of experimental ozone fumigation on photosynthetic characteristics of white oak seedlings. *Environ. Pollut.* **65**, 371–380.

Garrett, H. E., Carney, J. L., and Hedrick, H. G. (1982) The effects of ozone and sulfur dioxide on respiration of ectomycorrhizal fungi. *Can. J. For. Res.* **12**, 141–145.

Giamalva, P., Church, D. F., and Pryor, W. A. (1985) A comparison of the rates of ozonation of biological antioxidants and oleate and linoleate esters. *Biochem. Biophys. Res. Comm.* **133**, 773–779.

Hanson, P. J., McLaughlin, S. B., and Edwards, N. T. (1988) Net CO_2 exchange of *Pinus taeda* shoots exposed to variable ozone levels and rain chemistries in field and laboratory settings. *Physiol. Plant* **74**, 635–642.

Harkov, R. and Brennan, E. (1979) An ecophysiological analysis of the response of trees to oxidant pollution. *J. Air Pollut. Control Assoc.* **29**, 157–161.

Harkov, R. and Brennan, E. (1980) The influence of soil fertility and water stress on the ozone response of hybrid poplar trees. *Phytopathology* **70**, 991–994.

Heagle, A. S. (1973) Interactions between air pollutants and plant parasites. *Annu. Rev. Phytopathol.* **11**, 305–388.

Heagle, A. S. (1982) Interactions between air pollutants and parasite plant diseases. In *Effects of Gaseous Air Pollution in Agriculture and Horticulture* (eds., Unsworth, M. H. and Ormrod, D. P.). Butterworths, London, 333–348.

Heath, R. L. (1984) Air pollutant effects on biochemicals derived from metabolism: organic, fatty and amino acids. In *Gaseous Air Pollutants and Plant Metabolism* (eds., Koziol, M. J. and Whatley, F. R.). Butterworths, London, 275–290.

Heck, W. W., Dunning, J. A., and Hindawi, I. J. (1965) Interactions of environmental factors on the sensitivity of plants to air pollution. *J. Air Pollut. Control Assoc.* **15**, 511–515.

Hogsett, W. E., Plocher, M., Wildman, V., Tingey, D. T., and Bennett, J. P. (1985) Growth response of two varieties of slash pine seedlings to chronic ozone exposures. *Can. J. Bot.* **63**, 2369–2376.

Hogsett, W. E., Olszyk, D., Ormrod, D. P., Taylor, G. E., and Tingey, D. T. (1987) *Air Pollution Exposure Systems and Experimental Protocols,* Vols. 1 and 2. EPA 6001 3-87/037, Corvallis, OR.

Horvath, M., Bilitzky, L. and Huttner, J. (1985) *Ozone.* Elsevier, New York.

Hughes, P. R. (1988) Insect populations on host plants subjected to air pollution. In *Plant-Stress-Insect Interactions* (ed., Heinricks, E. A.). John Wiley & Sons, New York, 249–319.

Hughes, P. R. and Laurence, J. A. (1984) Relationship of biochemical effects of air pollutants on plants to environmental problems: insects and microbial interactions. In *Gaseous Air Pollutants and Plant Metabolism* (eds., Koziol, M. K. and Whatley, F. R.). Butterworths, London, 361–377.

Hüttunen, S. (1984) Interactions of disease and other stress factors with atmospheric pollution. In *Air Pollution and Plant Life* (ed., Treshow, M.). John Wiley & Sons, New York, 321–356. .

James, R. L., Cobb, F. W., Jr., Wilcox, W. W., and Rowney, D. L. (1980a) Effects of photochemical oxidant injury of ponderosa and Jeffrey pines on susceptibility of sapwood and freshly cut stumps to *Fomes annosus*. *Phytopathology* **70**, 704–708.

James, R. L., Cobb, F. W., Jr., Wilcox, W. W., and Rowney, D. L. (1980b) Effects of oxidant air pollution of pine roots to *Fomes annosus*. *Phytopathology* **70**, 560–563.

Jeffords, M. R. and Endress, A. G. (1984) Possible role of ozone in tree defoliation by the Gypsy moth (Lepidoptera: Lymantriidae). *Environ. Entomol.* **13**, 1249–1252.

Jensen, K. F. (1973) Response of nine forest tree species to chronic ozone fumigation. *Plant Dis. Rep.* **57**, 914–917.

Jensen, K. F. (1979) A comparison of height growth and leaf parameters of hybrid poplar cuttings growth in ozone-fumigated atmospheres. USDA Forest Service Research Paper NE-446, NE Forest Exp. Stn., Broomall, PA.

Jensen, K. F. (1981) Air pollutants affect the relative growth rate of hardwood seedlings. USDA Forest Service Research Paper NE-470, NE Forest Exp. Stn., Broomall, PA.

Jensen, K. F. (1983) Growth relationship in silver maple fumigated with O_3 and SO_2. *Can. J. For. Res.* **13**, 298–302.

Jensen, K. F. (1985) Response of yellow-poplar seedlings to intermittent fumigation. *Environ. Pollut.* (Ser. A) **38**, 138–191.

Jensen, K. F. and Dochinger, L. S. (1974) Responses of hybrid poplar cuttings to chronic and acute levels of ozone. *Environ. Pollut.* **6**, 289–295.

Jensen, K. F. and Masters, R. G. (1975) Growth of six woody species fumigated with ozone. *Plant Dis. Rep.* **59**, 760–762.

Jensen, K. F. and Noble, R. D. (1984) Impact of ozone and sulfur dioxide on net photosynthesis of hybrid poplar cuttings. *Can. J. For. Res.* **14**, 385–388.

Jensen, K. F. and Roberts, B. R. (1986) Changes in yellow-poplar stomatal resistance with SO_2 and O_3 fumigation. *Environ. Pollut.* **41**, 235–245.

Jones, C. G. and Coleman, J. S. (1988) Plant stress and insect behavior: cottonwood, ozone and the feeding and oviposition preference of a beetle. *Oecologia* **76**, 51–56.

Karnosky, D. F. and Houston, D. B. (1973) Genetics of air pollution tolerance of trees in the northeastern U.S. In *Proc. 26th Northeastern Forest Tree Improvement Conference, July 25–26, 1973*, Pennsylvania State University, State College, PA.

Karnosky, D. F. and Steiner, K. C. (1981) Provenance and family variation in response of *Fraxinus americana* and *F. pennsylvanica* to ozone and sulfur dioxide. *Phytopathology* **71**, 804–807.

Keane, K. D and Manning, W. J. (1989) Effects of ozone and simulated acid rain on birch seedling growth and formation of ectomycorrhizae. *Environ. Pollut.* **52**, 55–65.

Keller, T. (1972) Gaseous exchange of forest trees in relation to some edaphic factors. *Photosynthetica* **6**, 197–206.

Keller, T. and Häsler, R. (1987) Some effects of long-term ozone fumigations on Norway spruce. I. Gas exchange and stomatal response. *Trees* **1**, 129–133.

Kottke, I. and Oberwinkler, F. (1986) Mycorrhizae on forest tree-structure and functions. *Trees* **1**, 1–24.

Krause, G. H. M., Prinz, B., and Jung, K. D. (1983) Forest effects in West Germany. In *Air Pollution and the Productivity of the Forest: Proceedings of the Symposium* (eds., Davis, D. D., Miller, A. A., and Dochinger, L.). October 1983, Washington, D.C., Izaak Walton League of America, 297–332.

Krause, G. H. M., Arndt, U., Brandt, C. J., Bücker, J., Krenk, G., and Matzner, E. (1986) Forest decline in Europe: development and possible causes. *Water Air Soil Pollut.* **31**, 647–668.

Kress, L. W. and Skelly, J. M. (1982a) Response of several eastern forest tree species to chronic doses of ozone and nitrogen dioxide. *Plant Dis.* **66**, 1149–1152.

Kress, L. W. and Skelly, J. M. (1982b) Relative sensitivity of 18 full-sib families of *Pinus taeda* to O_3. *Can. J. For. Res.* **12**, 203–209.

Kress, L. W., Skelly, J. M., and Hinkelmann, K. H. (1982) Growth impact of ozone, NO_2 and/or SO_2 on *Pinus taeda*. *Environ. Monit. Assess.* **1**, 229–239.

Kress, L. W., Allen, H. L., Mudano, J. E., and Heck, W. W. (1988) Response of loblolly pine to acidic precipitation and ozone, In *Proceedings 81st Annual Meeting Air Pollution Control Assocation, 19–24 June, 1988, Dallas, TX,* Air Pollution Control Association, Pittsburgh, 88–70.5

Krupa, S. V. and Nosal, M. (1989) Effects of ozone on agricultural crops. In *Atmospheric Ozone Research and its Policy Implications* (eds., Schneider, T., Lee, S. D, Wolters, G. J. R., and Grant, L. D.). Elsevier, Amsterdam, 229–238.

Lackner, A. L. and Alexander, S. A. (1983) Root disease and insect infestations on air-pollution sensitive *Pinus taeda* and studies of pathogenicity of *Verticicladiella procera*. *Plant Dis.* **67**, 679–681.

Laurence, J. A. (1981) Effects of air pollutants on plant pathogen interactions. *Z. Pflanzenkr. Pflanzenschutz.* **87**, 156–172.

Lee, W. S., Chevone, B. I., and Seiler, J. R. (1990) Growth response and drought susceptibility of red spruce seedlings exposed to simulated acidic rain and ozone. *For. Sci.* **36**, 265–275.

Lefohn, A. S. (1984) A comparison of ambient ozone exposures for selected nonurban sites. In *Proc. 77th Annual Meeting of the Air Pollution Control Association in San Francisco, California, June 24–29, 1984.* Air Pollution Control Association, Pittsburgh, 84-104.1.

Lefohn, A. S. and Runeckles, V. C. (1987) Establishing standards to protect vegetation-ozone exposure/dose considerations. *Atmos. Environ.* **21**, 561–568.

Lefohn, A. S. and Tingey, D. T. (1984) The co-occurrence of potentially phytotoxic concentrations of various gaseous air pollutants. *Atmos. Environ.* **18**, 2521–2526.

Lefohn, A. S., Hogsett, W. E., and Tingey, D. T. (1987a) The development of sulfur dioxide and ozone exposure profiles that mimic ambient conditions in the rural souteastern U.S. *Atmos. Environ.* **21**, 659–669.

Lefohn, A. S., Davis, C. E., Jones, C. K., Tingey, D. T., and Hogsett, W. E. (1987b) Co-occurrence patterns of gaseous air pollutant pairs at different minimum concentrations in the U.S. . *Atmos. Environ.*, **21**, 2435–2444.

Lewis, E. and Brennan, E. (1977) A disparity in the ozone response of bean plants grown in a greenhouse, growth chamber or open-top chamber. *J. Air Pollut. Control Assoc.* **27**, 889–891.

Little, T. M. (1981) Interpretation and presentation of results. *Hort. Sci.,* **16**, 637–640.

Lucas, P. W., Cotham, D. A., Sheppard, L. J., and Francis, B. J. (1988) Growth responses and delayed winter hardening in Sitka spruce following summer exposure to ozone. *New Phytol.* **108**, 495–504.

Luck, R. F. (1980) Impact of oxidant air pollution on ponderosa and Jeffrey pine cone production. In *Proc. Symposium on Effects of Air Pollutants on Mediterranean and Temporate Forest Ecosystems* (ed., Miller, P. R.), June, 1979, Riverside, CA, Berkley, CA. U.S. Dept. of Agriculture Pacific Southwest Forest and Range Experiment Station; Gen. Tech. Rep. No. PSW-43. Available from: NTIS, Springfield, VA. PB81-133720, 240.

Mahoney, M. J., Skelly, J. M., Chevone, B. I., and Moore, L. D. (1984) Response of yellow-poplar (*Lirodendron tulipifera* L.) seedling shoot growth to low concentrations of O_3, SO_2, and NO_2. *Can. J. For. Res.* **14**, 150–153.

Mahoney, M. J., Skelly, J. M., Chevone, B. I., and Moore, L. D. (1985) Influence of mycorrhizae on the growth of loblolly pine seedlings exposed to ozone and sulfur dioxide. *Phytopathology.* **75**, 679–682.

Manning, W. J. (1975) Interactions between air pollutants and fungal, bacterial and viral plant pathogens. *Environ. Pollut.* **9**, 87–90.

Manning, W. J. and Keane, K. D. (1988) Effects of air pollutants on interactions between plants, insects, and pathogens. In *Assessment of Crop Loss from Air Pollutants* (eds., Heck, W. W., Taylor, O. C., and Tingey, D. T.). Elsevier, London, 365–386.

Mansfield, T. A., Travis, A. J., and Jarvis, R. G. (1981) Response to light and carbon dioxide. In *Stomatal Physiology* (edited by Jarvis P. G. and Mansfield T. A.). Cambridge University Press, Cambridge, 119–135.

McClenahen, J. R. (1979) Effects of ethylene diurea and ozone on the growth of tree seedlings. *Plant Dis. Rep.* **63**, 320–323.

McCool, P. M., Menge, J. A., and Taylor, O. C. (1979) Effects of ozone and HCl gas on the development of the mycorrhizal fungus *Glomus fasciculatus* and growth of Troyer citrange. *J. Am. Soc. Hort. Sci.* **104**, 151–154.

McLaughlin, S. B. (1985) Effects of air pollution on forests: a critical view. *J. Air Pollut. Control Assoc.* **35**, 512–534.

McLaughlin, S. B. and Taylor, G. E. (1981) Relative humidity: important modifier of pollutant uptake by plants. *Science* **211**, 167–169.

McLaughlin, S. B., McConathy, R. K., Duvick, D., and Mann, L. K. (1982) Effects of chronic air pollution stress on photosynthesis, carbon allocation, and growth of white pine trees. *For. Sci.* **28**, 60–70.

Meier S., Grand, L. F., Schoenberger, M. M., Reinert, R. A., and Bruck, R. I. (1990) Growth, ectomycorrhizae and nonstructural carbohydrates of loblolly pine seedlings exposed to ozone and soil water deficit. *Environ. Pollut.* **64**, 11–27.

Miller, P. R. (1977) Photochemical oxidant air pollutant effects on a mixed conifer forest ecosystem. Annual Progress Report, 1975–1976. EPA-600/3-77-104. NTIS No. PB 274 531/AS. U.S. Environmental Protection Agency, 338.

Miller, P. R. (1983) Ozone effects in the San Bernardino National Forest. In *Air Pollution and Productivity of the Forest* (eds., Davis, D. D., Miller, A. A., and Dochinger, L. S.). Izaak Walton League, Arlington, VA, 161–197.

Miller, P. R., Parmeter, J. R., Flick, B. H., and Martinez, C. W. (1969) Ozone dosage response of ponderosa pine seedlings. *J. Air Pollut. Contol Association.* **19**, 435–538.

Mooi, J. (1980) Influence of ozone on growth of two poplar cultivars. *Plant Dis.* **64**, 772–773.

Mudd, J. B. (1982) Effects of oxidants on metabolic function. In *Effects of Gaseous Pollutants in Agriculture and Horticulture* (eds., Unsworth, M. H. and Ormrod, D. P.). Butterworths, London, 182–203.

Olson, Jr., R. L., Winner, W. E., and Moore, L. D. (1987) Effects of pristine and industrial simulated acidic precipitation on greenhouse-grown radishes. *Environ. Exp. Bot.* **27**, 239–244.

Oquist, G. (1986) Effects of winter stress on chlorophyll organization and function in Scots pine. *J. Plant Physiol.* **122**, 169–179.

Ormrod, D. P., Tingey, D. T., Gumpertz, M. L., and Olszyk, D. M. (1984) Utilization of a response surface technique in the study of plant responses to ozone and sulphur dioxide mixtures. *Plant Physiol.* **75**, 43–48.

Oshima, R. J. and Bennett, J. P. (1979) Experimental design and analysis. In *Handbook of Methodology for the Assessment of Air Pollution Effects on Vegetation* (eds., Heck, W. W., Krupa, S. V., Linzon, S., and Fredrick, E. R.).Upper Midwest Section, Air Pollution Control Association., 4.1–4.22.

Patton, R. L. (1981) Effects of ozone and sulfur dioxide on height and stem specific gravity of *Populus* hybrids. USDA Forsest Service Research Paper. NE-471, NE Forest Experimental Station, Broomall, PA.

Peterson, D. L., Arbaugh, M. J., Wakefield, V. A., and Miller, P. R. (1987) Evidence of growth reduction in ozone-injured Jeffrey pine (*Pinus jeffreyi* Grev. and Balf.) in Sequoia and Kings Canyon National Parks. *J. Air Pollut. Control Assoc.* **37**, 906–912.

Phillips, S. O., Skelly, J. M., and Burkhart, H. E. (1977a) Growth fluctuations of loblolly pine (*Pinus taeda* L.) proximal to a periodic source of air pollution. *Phytopathology* **67**, 716–720.

Phillips, S. O., Skelly, J. M., and Burkhart, H. E. (1977b) Inhibition of growth in asymptomatic white pine exposed to fluctuating levels of air pollution. *Phytopathology* **67**, 721–725.

Pitelka, L. F. (1988) Evolutionary responses of plants to anthropogenic pollutants. *Trends. Ecol.* **3**, 233–236.

Prinz, B., Krause, G. H. M., and Stratmann, H. (1982) Vorlaufiger Bericht der Landesanstalt für Immissionsschutz uber Untensuchungenzur Aufklarung der Waldschöden in der Bundesrepublik Deutschland. LIS-Bericht No. 28; Landesanstalt fur Immissionsschutz des Landes NW, Wallneyer Stn. 6, 4300 Essex 1, p. 154.

Pryor, W. A., Giamalva, D. H., and Church, D. F. (1984) Kinetics of ozonation. II. Amino acids and model compounds in water and comparison to rates in nonpolar solvents. *J. Am. Chem. Soc.* **106**, 7094–7100.

Pye, J. M. (1988) Impact of ozone on the growth and yield of trees: a review. *J. Environ. Qual.* **17**, 347–360.

Reich, P. B. (1983) Effects of low concentrations of O_3 on net photosynthesis, dark respiration and chlorophyll contents in aging hybrid poplar leaves. *Plant Physiol.* **73**, 291–296.

Reich, P. B. (1987) Quantifying the response of plants to ozone: a unified explanation. *Tree Physiol.* **3**, 63–91.

Reich, P. B. and Lassoie, J. P. (1984) Effect of low level ozone exposure on leaf diffusive conductance and water-use efficiency in hybrid poplars. *Plant Cell Environ.* **7**, 661–668.

Reich, P. B. and Lassoie, J. P. (1985) Influence of low concentrations of ozone on growth, biomass partitioning and leaf senscence in young hybrid poplar plants. *Environ. Pollut.* **39**, 39–51.

Reich, P. B., Lassoie, J. P., and Amundson, R. G. (1984) Reduction in growth of hybrid poplar following field exposure to low levels of O_3 and/or SO_2. *Can. J. Bot.* **62**, 2835–2841.

Reich, P. B., Schoettle, A. W., and Amundson, R. G. (1986) Effects of O_3 and acidic rain on photosynthesis and growth in sugar maple and northern red oak seedlings. *Environ. Pollut.* **40**, 1–15.

Reich, P. B., Schoettle, A. W., Stroo, H. F., Troiano, J., and Amundson, R. G. (1985) Effects of O_3, SO_2, and acid rain on mycorrhizal infection in northern red oak seedlings. *Can. J. Bot.* **63**, 2049–2055.

Reich, P. B., Schoettle, A. W., Stroo, H. F., Troiano, J., and Amundson, R. G. (1987) Influence of O_3 and acid rain on white pine (*Pinus strobus*) seedlings grown in five soils. I. Net photosynthesis and growth. *Can. J. Bot.* **65**, 977–987.

Reinert, R. A., Schoenberger, M. M., Shafer, S. R., Eason, G., Horton, S. J., and Wells, C. (1988) Responses of loblolly pine half-sib families to ozone. In *Proc. 81st Annual Meeting of the Air Pollution Control Association, June 19–24, 1988, Dallas, TX.* Air Pollution Control Association, Pittsburgh, PA, 88-125.2.

Roberts, B. R. (1990) Physiological response of yellow-poplar seedlings to simulated acid rain, ozone fumigation, and drought. *For. Ecol. Manage.* **31**, 215–224.

Rogers, H. H., Bingham, G. E., Cure, J. D., Smith, J. M., and Surano, J. A. (1983) Responses of selected plant species to elevated carbon dioxide in the field. *J. Environ. Qual.* **12**, 569–574.

Runeckles, V. C. (1984) Impact of air pollutant combinations on plants. In *Air Pollution and Plant Life* (ed., Treshow, M.). John Wiley & Sons, New York, 239–258.

Sasek, T. W. and Richardson, C. J. (1989) Effects of chronic doses of ozone on loblolly pine: photosynthetic characteristics in the third growing season. *For. Sci.* **35**, 745–755.

Scherzer, A. J. and McClenahen ,J. R. (1989) Effects of ozone or sulfur dioxide on pitch pine seedlings. *J. Environ. Qual.* **18**, 57–61.

Schneider, S. H. (1988) Doing something about the weather. *World Monitor* **1**(3), 28–37.

Shafer, S. R. (1985) Effects of airborne chemicals on microorganisms. In *Air Pollutants Effects on Forest Ecosystems,* May 8–9, 1985, The Acid Rain Foundation, St. Paul, MN, 285–296.

Shafer ,S. R. and Heagle, A. S. (1989) Growth response of field-grown loblolly pine to chronic doses of ozone during multiple growing seasons. *Can. J. For. Res.* **19**, 821–231.

Shafer, S. R., Heagle, A. S., and Camberato, D. M. (1987) Effects of chronic doses of ozone on field-grown loblolly pine: seedlings responses in the first year. *J. Air Pollut. Control Assoc.* **37**, 1179–1184.

Sheng, S. and Chevone, B. I. (1988) Gas exchange response of soybean cultivars to short term exposure of sulfur dioxide and ozone. *Phytopathology* **78**, 1513 (Abstr.).

Simmons, G. L. and Kelly, J. M. (1989) Influence of O_3, rainfall activity, and soil Mg status on growth and ectomycorrhizal colonization of loblolly pine roots. *Water Air Soil Pollut.* **44**, 159–171.

Skärby, L., Troeng, E., and Botström, C. A. (1987) Ozone uptake and effects on transpiration, net photosynthesis, and dark respiration in Scots pine. *For. Sci.* **33**, 810–808.

Skeffinton, R. A. and Roberts, T. M. (1985) The affects of ozone and acid mists on Scots pine saplings. *Oecologica* **65**, 201–206.

Skelly, J. M., Yang, Y. S, Chevone, B. I., Long, S. J., Nellesson, J. E., and Winner, W. E. (1983) Ozone concentrations and their influence on forest species in the Blue Ridge Mountains of Virginia. In *Air Pollution and the Productivity of the Forest* (eds., Davis, D. D., Miller, A. A., and Dochinger, L. S.). October 4–5, 1983, Washington, D.C., 143–160.

Smith, W. H. (1981) *Air Pollution and Forests.* Springer-Verlag, New York, 379.

Spence, R. D., Rykiel, E. J., and Sharpe, P. J. H. (1990) Ozone alters carbon allocation in loblolly pine: assessment with carbon-11 labeling. *Environ. Pollut.* **64**, 93–106.

St. John, T. V. and Coleman, D. C. (1983) The role of mycorrhizae in plant ecology. *Can. J. Bot.* **61**, 1005–1014.

Stark, R. W., Miller, P. R., Cobb, F. W., Wood, D. L., and Parameter, J. R. (1968) Incidence of bark beetle infestation in injured trees. *Hilgardia* **39**, 121–126.

Stone, L. L. and Skelly, J. M. (1974) The growth of two forest tree species adjacent to a periodic source of air pollutants. *Phytopathology* **64**, 773–778.

Stroo, H. F., Reich, P. B., Schoettle, A. W., and Amundson, R. G. (1988) Effects of ozone and acid rain on white pine *Pinus strobus* seedlings grown in five soils. II. Mycorrhizal infection. *Can. J. Bot.* **66**, 1510–1516.

Switzer, G. L. and Nelson, L. E. (1972) Nutrient accumulation and cycling in loblolly pine (*Pinus taeda* L.) plantation ecosystems: The first twenty years. *Soil Sci. Soc. Am. Proc.* **36**, 143–147.

Taylor, G. E. and Norby, R. J. (1985) The significance of elevated levels of ozone on natural ecosystems of North American. In *Evaluation of the Scientific Basis for Ozone/Oxidants Standards* (ed., Lee, S. D.). Air Pollution Control Association, Pittsburgh, 152–184.

Taylor, G. E., Tingey, D. T., and Ratch, H. C. (1982) Ozone flux in *Glycine max* (L.) *Merr.*: sites of regulation and relationships to leaf injury. *Oecologia* **53**, 179–186.

Taylor, G. E., Norby, R. J., McLaughlin, S. B., Johnson, A. H., and Turner, R. S. (1986) Carbon dioxide assimilation and growth of red spruce (*Picea rubens Sarg.*) seedlings in response to ozone, precipitation chemistry, and soil type. *Oecologia* **70**, 163–171.

Temple, P. J. (1988) Injury and growth of Jeffrey pine and giant sequoia in response to ozone and acidic mist. *Environ. Exp. Bot.* **28**, 323–333.

Tingey, D. T. and Taylor, G. E. (1982) Variation in plant response to ozone: a conceptual model of physiological events. In *Effects of Gaseous Air Pollution in Agriculture and Horticulture* (eds., Unsworth, M. H. and Ormrod, D. P.). Buttersworths, London, 113–138.

Tingey, D. T., Wilhour, R. G., and Standley, C. (1976) The effect of chronic ozone exposures on the metabolic content of ponderosa pine seedlings. *For. Sci.* **22**, 234–241.

Tseng, E. C., Seiler, J. R., and Chevone, B. I. (1988) Effects of ozone and water stress on greenhouse-grown fraser fir seedling growth and physiology. *Environ. Exp. Bot.* **28**, 37–41.

Turner, N. C., Rich, S., and Waggoner, P. E. (1973) Removal of ozone by soil. *J. Environ. Qual.* **2**, 259.

Wang, D., Karnosky, D. F., and Bormann, H. F. (1986) Effects of ambient ozone on the productivity of *Populus tremuloides*. Mich. grown under field conditions. *Can. J. For. Res.* **16**, 47–55.

Weidensaul, T. C. and Darling, S. L. (1979) Effects of ozone and sulfur dioxide on the host-pathogen relationship of Scottish pine and *Scirrhia acicola*. *Phytopathology* **69**, 939–941.

Wellburn, A. (1988) *Air Pollution and Acid Rain: The Biological Impact*. Longman, London, 264.

Wilhour, R. G. (1971) The influence of ozone on white ash (*Fraxinus americana* L.). Penn. St. Univ. Cent. Environ. Stud. No. 188-71. Pennsylvania State University, State College.

Wilhour, R. G. and Neely, G. E. (1977) Growth response of conifer seedlings to low ozone concentrations. In Proc. Int. Conference Photo. Oxidant Pollution and its Control. USEPA Rep. No. 300/377-0106, 1169p., p. 635–645.

Wilkinson, T. G. and Barnes, R. L. (1973) Effects of ozone on $^{14}CO_2$ fixation patterns in pines. *Can. J. Bot.* **51**, 1573–1578.

Winner, W. E., Cotter, I. S., Powers, H. R., and Skelly, J. M. (1987) Screening loblolly pine seedling response to SO_2 and O_3: analysis of families differing in resistance to fusiform rust disease. *Environ. Pollut.* **47**, 205–220.

Wolfenden, J. and Wellburn, A. R. (1986) Cellular readjustment of barley seedlings to simulated acid rain. *New Phytol.* **104**, 97–109.

Wood, F. A. and Davis, D. D. (1969) Sensitivity to ozone determined for trees. *Sci. Agric.* **17**, 4–5.

Woodman, J. N. and Cowling, E. B. (1987) Airborne chemicals and forest health. *Environ. Sci. Technicol.* **21**, 120–126.

Woodman, J. N. and Furiness, C. S. (1988) Potential effects of climate change on U.S. forests: case studies of California and the Southeast. USEPA, Contract No. 68-03-3439. Office of Reserch and Development, Washington D.C.

Wright, L., Thornton, F., Meldahl, R., Lockaby, B. G., and Chappelka, A. H. (1991) Acid precipitation and ozone influences on nitrogen nutrition in young loblolly pine. *Soil Sci. Soc. Amer.* (in press).

Yang, Y. S., Skelly, J. M., Chevone, B. I., and Birch, J. B. (1983a) Effects of short-term ozone exposure on net photosynthesis, dark respiration and transpiration of three eastern white pine clones. *Environ. Int.* **9**, 265–269.

Yang, Y. S., Skelly, J. M., Chevone, B. I., and Birch, J. B. (1983b) Effects of long-term ozone exposure on photosynthesis and dark respiration of eastern white pine. *Environ. Sci. Tech.* **17**, 371–373.

CHAPTER 8

Ozone Standards and Their Relevance for Protecting Vegetation

Allen S. Lefohn, A.S.L. & Associates, Helena, MT

8.1. INTRODUCTION

Air quality standards are established to prevent or minimize the risk of adverse effects from air pollution to human health, vegetation, and materials. At present, ozone has been identified as the appropriate measure of photochemical oxidant pollution. In the U.S., the primary (health effects) and secondary (welfare effects) national ambient air quality standards (NAAQS) are both set at an hourly average concentration of 0.12 ppm, not to be exceeded on more than 1 d per year (Federal Register 44 FR 8202, 1979). There is no requirement that the primary and secondary standards be identical, nor is there any requirement that only a single expression of the standard be used (i.e., an average concentration for a single time period vs. multiple exceedances or integrated exposures).

In order to develop ozone standard(s) that provide an adequate measure of protection to vegetation, it is necessary to define, in as precise terms as possible, the relationship between ozone exposures and the potential for adverse effects on vegetation. Although the form of the standard should be made as simple as possible, it is essential that it be related, directly or indirectly, to identifiable adverse effects.

It is important to define clearly what the proposed standard is designed to protect. Tingey et al. (1989) have discussed the definition of "adverse' as the word relates to the standard-setting process. In some cases, only injury occurs to the plant; in other cases, growth reduction occurs. Tingey et al. (1989) point out that injury encompasses all measurable plant reactions, such as reversible changes in metabolism, reduced photosynthesis, leaf necrosis, leaf drop, altered quality, or reduced growth, that do not influence agronomic yield or reproduction. Damage, on the other hand, includes all effects that reduce the intended human use or the value of the plant or ecosystem. The U.S. EPA (1988) has made a distinction between the relative importance of foliar injury to vegetation and reduced crop yield. EPA (1988) decided to place greater emphasis on damage or yield loss than on injury.

For establishing a standard at a defensible form and level, several issues require resolution. Runeckles (1987) has suggested that the following need to be addressed:

- The distinction between exposure to ambient conditions and dose per se
- The dynamic nature of pollutant concentrations in ambient air (i.e., concentration, fluctuations, and episodicity)
- Selection of suitable measures for defining the "dose" term in dose-response relationships
- The establishment of toxicant-response relationships
- Temporal, environmental, genetic, and developmental influences on such response
- Interactive effects of concurrent and sequential exposure to other stresses, including those caused by other pollutants.

Many of the above concerns were addressed in previous chapters in this book. However, additional comment is necessary on the selection of suitable measures for defining the "dose" term in exposure/dose-response relationships. Any index that is selected as a surrogate for "dose" should (1) describe the most important exposure characteristics that elicit an adverse effect and (2) order itself properly when comparing the absolute value experienced in an experiment with the value calculated under actual ambient conditions.

Exposure indices are important because they form the linkage between air quality standards that are promulgated to protect specific targets and the actual dose that is responsible for eliciting an effect. Results have been reported in the literature, relating ozone exposure with vegetation effects. Although the perfect exposure index that can serve as a surrogate for dose does not exist, there are some ozone exposure indices that do relate fairly well with vegetation effects (see Chapter 3).

8.2 CONSIDERATIONS ASSOCIATED WITH DESCRIBING OZONE EXPOSURES

8.2.1 Capturing Important Components of Exposure

The appeal of a single number, such as the current form of the standard, to describe ambient ozone pollutant exposure is undeniable. However, there are problems associated with condensing data to such a point that the identification of the important components of exposure are eliminated. In Chapter 3 the problems associated with using long-term seasonal average concentrations were discussed.

Long-term seasonal average concentrations (e.g., 7- or 12-h average concentrations) do not correlate strongly, at most ozone monitoring sites, with the components of exposure regimes that are most important in affecting vegetation. In Chapter 3 it was mentioned that research results reported by the EPA and other investigators have illustrated that cumulative exposure indices appear to provide more promise than long-term average concentration exposure indices in relating exposures with vegetation effects.

Before an ozone standard can be established to protect vegetation, the actual pollutant levels below which plants will be protected must be identified. In Chapters 3 and 5 the importance of high hourly average ozone concentrations affecting vegetation growth was mentioned. Numerous studies exist, as referenced in Chapter 3, indicating that high concentrations are important in affecting vegetation injury and growth. Guderian et al. (1985) have pointed out that plant growth is influenced more by concentration than by exposure duration, when similar products of concentration and time are used. The authors concluded that during chronic exposures, injury increases with increasing concentration.

Although all plants are capable of being adversely affected by exposure to phytotoxic gases and particulates in polluted air, the nature of the response can be extremely variable. Runeckles and Wright (1988) have indicated the following features play important roles in determining target sensitivity:

- The species of plant
- The stage of development of the plant
- The nature of the pollutant or mix of pollutants
- The pattern of exposure to the pollutant(s), which involves consideration of the concentration and durations of exposure
- Environmental conditions in the soil, such as water availability and nutrition
- Environmental conditions in the ambient air, such as light intensity, temperature, humidity, and air movement
- Biological factors, such as the occurrence of pests and diseases, and competitive stresses exerted by individual plants on their neighbors.

Given the large variability in response, efforts have been made to estimate the ranges of ozone exposures that may result in injury and damage to vegetation. Guderian et al. (1985) have proposed maximum acceptable ozone concentrations for the protection of vegetation (Table 8.1). The authors' numerical values are based on the limiting values proposed by Jacobson (1977) and the dose-response values for definite injury levels developed by Heck and Brandt (1977). In general, the recommendations made by Guderian et al. (1985) appear to reinforce the belief that hourly average concentrations of 0.10 ppm and higher are required to elicit adverse effects on vegetation. The one exception is the recommendation that, for sensitive species, vegetation should not be exposed for more than 4 h to hourly average concentrations of 0.05 ppm. There is some inconsistency associated with these recommendations. Ozone hourly average concentrations of 0.05 ppm routinely occur at many "clean" site locations in the world (Lefohn et al., 1990a). The recommendation of 0.05 ppm for a 4-h exposure made by Guderian et al. (1985) suggests that the ozone exposure regimes that occur at many of these cleanest sites in the world may cause adverse effects on sensitive vegetation species. The occurrence of hourly average concentrations of 0.05 ppm are not necessarily associated with anthropogenic sources, and thus, using a threshold of 0.05 ppm may not be realistic for protecting sensitive species. Table 8.1 summarizes the recommendations made by the authors for hourly average concentrations for durations of exposures of 0.5, 1.0, 2.0, and 4.0 h. The information in the table provides an indication that long-term exposures consisting only of lower hourly average ozone concentrations will not necessarily produce adverse effects on vegetation.

Using NCLAN data, two retrospective analyses were undertaken to search for indices that perform adequately in describing the episodic occurrences (i.e., high hourly average concentrations) of ozone exposure. Lefohn et al. (1988) reported that, while none of the exposure indices consistently provided a best fit with the models tested, their analysis indicated that exposure indices that weight peak concentrations of ozone differently than lower concentrations of an exposure regime can be used in the development of exposure-response functions. In a more extensive analysis of NCLAN data, Lee et al. (1988) fitted more than 600 exposure indices to response data from seven crop studies. For most of the NCLAN experiments used in their analyses, they characterized the daily hourly mean ozone concentrations that were recorded over the 7-h period (0900 to 1559 h) by the original experimenters. Using mostly the 7-h windowed data provided by the NCLAN investigators, the authors concluded that the top-performing indices were those whose form (1) accumulated the hourly ozone concentrations over time, (2) used a sigmoid weighting scheme that emphasized concentrations of 0.06 ppm and higher, and (3) phenologically weighted the exposure. The authors suggested that lower concentrations should be included, but given lesser weight, in the calculation of the exposure index.

The results reported by Lee et al. (1988) agreed with the recommendations of Lefohn and Runeckles (1987) and the findings reported by Lefohn et al. (1988). In a subsequent analysis using NCLAN data, Lee et al. (1989) reported that the phenologically weighted cumulative impact indices, as well as the sigmoidally weighted integrated index centered at 0.062 ppm and the cumulative censored indices that integrated hourly average concentrations of 0.06 and 0.07 ppm or higher, performed at near optimal levels.

In a continuation of their investigation, Lee et al. (1991) summarized their findings using vegetation-effects data for 49 case studies obtained from 31 field experiments (involving 12 crops). Most of the data were generated from NCLAN investigators. Lee et al. (1991) evaluated the efficacy of the following four ozone exposure indices:

Table 8.1. Proposed Maximum Acceptable Ozone Concentrations for Protection of Vegetation[a]

Exposure duration (h)	Resistance Level (ppm)		
	Sensitive	Intermediate	Less sensitive
0.5	0.150	0.25	0.50
1.0	0.075	0.18	0.25
2.0	0.060	0.13	0.20
4.0	0.050	0.10	0.18

[a] Adapted from Guderian et al. (1985).

- The sum of all hourly average concentrations using a sigmoidally weighted function (SIGMOID)
- The sum of all hourly average concentrations greater than or equal to 0.06 ppm (SUM06)
- The 7-h average concentration calculated over the experimental period
- The second highest daily maximum concentration (the current form of the standard).

The authors concluded that although no single exposure index was best in describing the exposure-response relationship for the 49 case studies, the performance of the second highest daily maximum concentration exposure index (the current form of the U.S. Federal standard) was considerably worse than the other three indices. Lee et al. (1991) reported that the second highest daily maximum concentration index did not perform adequately because it (1) poorly related to plant growth, (2) ignored exposure duration, and (3) placed too much emphasis on a single peak 1-h concentration. The SIGMOID, SUM06, and 7-h average concentration indices were nearly equivalent in performance, with a slight preference for the cumulative indices (i.e., SIGMOID and SUM06).

The work by Lee et al. (1991) suggests that the correlation between the current form of the standard and the occurrences of elevated hourly average concentrations is weak. Lee et al. (1991) believed that the second-highest daily maximum concentration was an inappropriate index to use to protect vegetation from elevated ozone exposures because exposure regimes can experience similar second-highest daily maximum concentrations, but exhibit exposure patterns of widely diverse characteristics that contain from two to many peak concentrations. As a follow-up to the results reported by Lee et al. (1991), Tingey et al. (1991) evaluated several alternative ambient ozone standards that they believed would protect agricultural crops against adverse ozone exposures. The authors suggested that the SUM06 ozone exposure index be used as the form of a secondary standard to protect agricultural crops. The value of the SUM06 exposure parameter, as determined by Tingey et al. (1991), was calculated by summing hourly average concentrations across a fixed 3-month period (i.e.,

April to June, May to July, June to August, July to September, and August to October). Tingey et al. (1991) reported that a 3-month SUM06 value of 24.4 ppm-h was estimated to cause a 10% yield loss in half the cases they investigated.

Before one is willing to accept the use of a single-parameter index, such as the SUM06 index, it is important to investigate whether the index captures the important exposure components responsible for affecting vegetation. From results reported in the literature, we know that high hourly average concentrations are important. However, in the retrospective NCLAN studies cited above, there was no attempt to identify the specific ozone exposure regimes that elicited an adverse effect on vegetation. The retrospective studies focused on the adequacy of mathematical parameters to relate exposure with growth reduction. It was assumed that the mathematical parameters adequately correlated with those important exposure regimes that elicit an adverse effect. In applying a 24.4 ppm-h SUM06 value calculated over a 3-month period, Tingey et al. (1991) assumed that vegetation experiencing, under ambient conditions, a 24.4 ppm-h or higher cumulative exposure would behave in the same manner that vegetation in the NCLAN experimental chambers responded.

Ignoring for the moment the criticism that has been cited in the literature, concerning the limitations of applying NCLAN experimental results (see for example Lefohn et al., 1989), it is important to explore whether it might be possible to experience 3-month SUM06 values of 24.4 ppm-h without experiencing frequent occurrences of elevated hourly average concentrations. It is important to investigate (1) whether those sites that experience SUM06 values equal to or greater than 24.4 ppm-h also experience frequent high hourly average ozone concentrations and (2) whether those sites that exhibit SUM06 values less than 24.4 ppm-h over a 3-month period also experience exposure regimes that contain high hourly average concentrations. Simply put, the question is, "Is the occurrence of 3-month SUM06 values of 24.4 ppm-h or higher correlated with elevated hourly average concentrations?"

For investigating this question, ozone hourly average concentrations from each site in the EPA's AIRS database, as well as the EPA's National Dry Deposition Network (NDDN) and Mountain Cloud Chemistry Program (MCCP), were characterized by summing by month all hourly average concentrations greater than or equal to 0.06 ppm. For the period April to October, those sites found to experience a 3-month SUM06 cumulative value of 24.4 ppm-h or higher were organized into either (1) Metropolitan Statistical Areas (MSA), (2) Consolidated Metropolitan Statistical Areas (CMSA), or (3) non-MSA subdivisions.

To explore how those areas, which experienced SUM06 values of 24.4 ppm-h or higher, compared with areas designated as being in nonattainment by EPA, the results of the SUM06 analysis for 1987 and 1988 were compared with EPA's list of nonattainment areas for the period 1986 to 1988 and the results of the 1989 SUM06 analysis with the EPA's list of nonattainment areas for 1987 to 1989. Using the SUM06 index, those areas that experienced 24.4 ppm-h or higher over a continuous 3-month period were compared with those areas that previously had been identified as being in nonattainment. The 1987 and 1988 SUM06 results were compared with the 1986 to 1988 nonattainment list because the nonattainment areas for 1986 to 1988 represent a "worst case" scenario (in comparison to the 1987 to 1989 period). In 1989, the ozone exposures in many locations in the U.S. were mucher lower than in the 2 previous years. The 1989 SUM06 results were compared with the 1987 to 1989 nonattainment list because this was the only period in which the 1989 data were included.

As an example of the results of the analysis, Table 8.2 summarizes the ozone exposures that occurred in 1989 at several sites in the U.S. The highest SUM06 value calculated over a running 3-month period was used to rank the exposures that occurred at these locations. The following sites were designated by the U.S. EPA as being a subset of the locations that were in nonattainment areas for the period 1987 to 1989:

> Baton Rouge, LA
> Kountze, TX
> New Roads, LA
> Westlake, LA
> Santa Barbara, CA
> New Haven, CT
> Salt Lake City, UT
> Long Beach, CA
> Houston, TX
> E. Baton Rouge, LA
> El Paso, TX
> Iberville Parish, LA
> Bridgeport, CT
> Cape Elizabeth, ME
> Livermore,CA
> Hawthorne, CA
> Lynwood, CA

Although each of the monitoring sites experienced maximum hourly average concentrations exceeding the current National Air Quality Standard for ozone, none of these sites had a SUM06 exposure value over the 3-month period that was greater than some of the sites that did not violate the ozone standard. The sites that experienced high hourly average concentrations did not experience a sufficient number of high hourly average concentrations to have a cumulative 3-month SUM06 value of 24.4 ppm-h. It appears that the sites that violated the ozone standard experienced sufficient scavenging such that there were fewer hourly average concentrations greater than or equal to 0.06 ppm than those that occurred at sites that did not violate the standard. On the other hand, the sites that did experience cumulative 3-month SUM06 values greater than or equal to 24.4 ppm did not experience a second hourly maximum concentration greater than 0.125 ppm.

For some sites listed in the EPA AIRS database, it appears that there is a poor

Table 8.2. Summary of Percentiles for Ozone Monitoring Sites in 1989 (April–October) Using a 3-Month SUM06 Index

AIRS site	Name	Percentiles									Maximum uncorrected SUM06 value (ppm-h)	Number of observations over 7-month period
		Min	10	30	50	70	90	95	99	Max		
220330004	Baton Rouge, LA	.000	.002	.008	.016	.028	.047	.057	.078	.138	8.4	4791
481990002	Kountze, TX	.000	.000	.010	.020	.030	.050	.060	.080	.130	10.6	4630
220770001	New Roads, LA	.000	.001	.011	.021	.033	.052	.062	.083	.141	12.0	4964
220191003	Westlake, LA	.000	.003	.013	.022	.033	.052	.061	.082	.137	12.2	4811
060833001	Santa Barbara Co., CA	.000	.010	.020	.030	.040	.050	.060	.080	.140	12.3	5077
090091123	New Haven, CT	.000	.003	.010	.019	.029	.045	.056	.091	.156	12.9	4502
490353001	Salt Lake City, UT	.000	.002	.014	.029	.041	.053	.061	.079	.140	13.0	4544
060830010	Santa Barbara, CA	.000	.010	.020	.030	.040	.050	.060	.080	.220	13.3	4663
471630009	Kingsport, TN	.001	.001	.005	.017	.032	.054	.062	.078	.125	13.4	4252
060374002	Long Beach, CA	.000	.010	.020	.020	.030	.050	.060	.080	.160	13.6	4876
482011034	Houston, TX	.000	.000	.010	.010	.030	.050	.060	.100	.220	14.0	4595
220331001	E Baton Rouge, LA	.000	.003	.012	.022	.034	.056	.066	.092	.171	14.4	4890
481410027	El Paso, TX	.000	.010	.020	.030	.040	.050	.060	.080	.260	14.9	4484
220470002	Iberville Par, LA	.000	.005	.014	.023	.034	.057	.068	.093	.149	15.9	5040
482011037	Houston, TX	.000	.000	.010	.010	.030	.050	.060	.110	.250	16.3	4729
090010113	Bridgeport, CT	.000	.002	.011	.022	.033	.048	.059	.091	.156	16.5	4865
230052003	Cape Elizabeth, ME	.001	.017	.027	.034	.042	.055	.064	.093	.146	16.7	4627
482010062	Houston, TX	.000	.000	.010	.020	.030	.050	.070	.110	.170	16.8	4600
060190003	Livermore,CA	.000	.000	.010	.030	.040	.050	.060	.090	.140	17.0	5067
060690008	Santa Barbara, CA	.000	.010	.020	.030	.040	.050	.060	.080	.190	17.1	4823
490350003	Salt Lake Co., UT	.000	.001	.008	.029	.042	.056	.062	.083	.125	17.4	4585
220330003	Baton Rouge, LA	.000	.001	.009	.021	.034	.059	.069	.094	.168	17.4	4964
060375001	Hawthorne, CA	.000	.000	.020	.030	.040	.060	.060	.080	.190	18.1	4894
060371301	Lynwood,CA	.000	.000	.010	.020	.030	.050	.070	.100	.140	18.1	4793

Site ID	Location												
482010024	Harris Co., TX	.000	.000	.010	.020	.030	.060	.070	.110	.230	19.2	4728
390030002	Allen Co., OH	.000	.007	.022	.032	.043	.060	.068	.086	.107	24.5	4854
510610002	Fauquier Co., VA	.000	.009	.021	.033	.045	.061	.069	.084	.122	24.6	5050
550270001	Horicon, WI	.002	.019	.029	.037	.047	.062	.070	.088	.111	24.6	4142
310550032	Omaha, NE	.002	.021	.030	.037	.047	.062	.067	.075	.098	24.9	4160
350431001	Sandoval Co., NM	.000	.010	.020	.030	.040	.060	.060	.070	.090	25.1	5059
420770004	Allentown, PA	.000	.003	.016	.028	.039	.060	.070	.087	.102	25.1	5040
240053001	Essex, MD	.000	.002	.010	.024	.038	.059	.069	.089	.121	25.2	5028
170491001	Effingham Co., IL	.000	.009	.023	.036	.046	.063	.070	.081	.104	25.3	4600
180970042	Indianapolis, IN	.001	.006	.021	.034	.046	.063	.070	.085	.103	25.4	4592
240030014	Anne Arundel, MD	.000	.006	.021	.032	.045	.064	.072	.090	.120	25.5	4360
510130020	Arlington Co., VA	.000	.001	.010	.023	.037	.059	.073	.088	.116	25.7	5029
370270003	Lenoir, NC	.000	.007	.019	.032	.045	.062	.071	.078	.092	25.8	4806
391510016	Canton, OH	.000	.008	.019	.030	.042	.060	.067	.088	.110	26.3	4875
371470099	Farmville, NC	.000	.010	.023	.034	.044	.062	.070	.083	.100	26.4	4833
370810011	Guilford Co., NC	.004	.010	.023	.034	.046	.063	.070	.083	.113	27.7	4853
551390007	Oshkosh, WI	.002	.016	.028	.038	.048	.063	.070	.084	.121	27.9	4206
120094001	Cocoa Beach, FL	.002	.017	.024	.032	.042	.059	.068	.077	.094	28.7	5012
061011002	Yuba City, CA	.000	.000	.020	.030	.040	.060	.070	.080	.100	29.0	4623
420070003	New Brighton, PA	.000	.008	.021	.032	.043	.062	.070	.087	.102	29.4	5055
040132004	Scottsdale, AZ	.000	.006	.018	.031	.045	.062	.071	.084	.107	31.7	5070
170190004	Champaign, IL	.000	.008	.020	.029	.039	.065	.072	.078	.088	32.0	5091
511870002	Shen NP (DKY RDG)	.004	.027	.037	.045	.054	.065	.071	.082	.100	33.5	4454
060070002	Chico, CA	.000	.010	.020	.030	.040	.060	.070	.080	.100	33.5	4690
470090101	Smoky Mt. NP, TN	.000	.025	.036	.044	.053	.065	.070	.081	.098	35.9	4764
060430004	Yosemite NP, CA	.000	.008	.022	.035	.049	.065	.072	.083	.111	37.6	4853
060170009	South Lake Tahoe	.000	.020	.030	.040	.050	.060	.070	.080	.100	44.8	4768
360310002	Essex Co., NY	.016	.033	.042	.050	.056	.067	.073	.086	.106	45.6	4070
060710006	San Bernardino Co., CA	.000	.020	.040	.050	.060	.070	.080	.090	.100	70.5	4856

correlation between the 3-month cumulative index SUM06 value and the infrequent occurrence of high hourly average concentrations. The results summarized in Table 8.2 show that a cumulative 3-month SUM06 value at a specific monitoring site will not necessarily relate to the occurrence or absence of high hourly average ozone concentrations. A 3-month SUM06 value of 24.4 ppm-h or greater indicates only that there are a large number of hourly average concentrations greater than or equal to 0.06 ppm; a low SUM06 value indicates that there are a small number of hourly average concentrations greater than or equal to 0.06 ppm. In either case, the occurrence of hourly average concentrations greater than or equal to 0.12 ppm is not necessarily a function of the value of the cumulative SUM06 value.

An important question still remains. If peak exposures are important and the SUM06 index does not correlate strongly with peak concentrations, why were the results of Lee et al. (1991) so consistent? The answer is straightforward. Lee et al. (1991) used NCLAN data, and as discussed by Lefohn et al. (1989), the NCLAN experimental protocol applied incremental and proportional additions that resulted in these treatments experiencing elevated ozone exposures. Thus, the artificial regimes created by the NCLAN addition protocols contained the elevated hourly average concentrations that were reflected in the determination of the absolute values of the cumulative indices that were calculated by Lee et al. (1991). Therefore, the magnitude of the SUM06 index, calculated using NCLAN protocols, contained the peak exposures that correlated well with the observed growth reductions. As discussed above, the strong relationship between peak concentrations and the value of the SUM06 index does not necessarily occur under ambient conditions.

The single-parameter exposure index, such as the SUM06, does not necessarily relate well with the occurrences of elevated hourly average concentrations. The application of a 3-month SUM06 exposure index value of 24.4 ppm-h as a standard to protect vegetation will provide inconsistent results for protecting agricultural crops. Because of the strong correlation between the SUM06 and W126 exposure indices (Lefohn et al., 1989), it appears that the magnitude of the W126 index will also have difficulty in characterizing the upper tail of the hourly average-distribution curve. This does not imply that all currently used cumulative exposure indices are not appropriate for describing ozone exposure. Rather, it appears that cumulative indices, such as the SUM06 index, will have to be combined with other parameters to describe accurately the occurrence of the high hourly average concentrations.

Any ozone exposure index that is used to describe those regimes that cause vegetation effects must be able to characterize adequately the upper tail of the hourly average-distribution curve to avoid the problems summarized in Table 8.2. As indicated, it appears that a combination of indices may be required to adequately describe the upper end of the distribution curve. Krupa and Nosal (1989) have discussed the use of the application of spectral coherence analysis to develop multiple-parameter indices to describe the relationships between ozone exposure and crop growth.

In conclusion, we have seen that the current form of the standard, as well as the long-term seasonal mean and the SUM06 cumulative exposure index, do not necessarily capture the presence or absence of the occurrence of high hourly average concentrations. Although the use of cumulative exposure indices is appropriate, at this time it appears that the use of a single-parameter exposure index will not guarantee that the most important components of exposure have been captured, and that additional combination of indices will be required.

8.2.2 Comparing Experimental Exposure Regimes with Actual Ambient Conditions

8.2.2.1 Establishing Reference Points

In section 8.2.1, we focused on whether indices that are selected as surrogates for "dose" describe adequately the most important exposure characteristics that elicit an adverse effect on vegetation. In this section we will explore those exposure indices that order themselves properly when comparing the absolute values experienced in an experiment with those values calculated under actual ambient conditions. Linking exposures that occur under experimental conditions with those that occur under ambient conditions is an important step in establishing standards to protect vegetation.

Some vegetation researchers have used the seasonal average of the daily 7-h (0900 to 1559 h) average as the exposure parameter in exposure-response models (Heck et al., 1982). For quantifying the effects of air pollution on crops and trees, some investigators have used controlled-environment and field methods with charcoal-filtration systems (Olszyk et al., 1989). In both the design of the experiments and the analysis of the data, the 7-h (0900 to 1559 h) seasonal mean reference point for ozone was **assumed** to be 0.025 ppm. The 0.025 ppm concentration has been used to estimate crop loss across the U.S. (Adams et al., 1985, 1989).

There has been concern that the ozone concentrations experienced in charcoal-filtration chambers may not mimic the levels experienced in "clean" areas. Heuss (1982), commenting on the work of Heck et al. (1982), believed that the 0.025 ppm level of ozone selected by NCLAN was too low and that the resulting agricultural loss estimates derived from the NCLAN models were too high. Heuss (1982) was concerned that the ozone levels experienced in the charcoal-filtration chambers did not mimic the levels experienced in "clean" areas, and therefore, the assumed 0.025 ppm 7-h exposure-period average used by NCLAN investigators was too low.

A key and controversial issue is how reference points should be established. Lefohn et al. (1990a) discussed two possible approaches. The authors stress that one approach is to estimate unpolluted, natural background levels prior to disturbance by human influence. This was the approach adopted by the NCLAN program.

Lefohn et al. (1990a) pointed out that natural background can be defined as the unpolluted conditions in preindustrial times (i.e., absolutely unpolluted air in which there is no human interference). Natural background can also be defined as the condition currently existing at any location that is presently free from human influence. However, as pointed out by the authors, it is difficult, if not impossible, to determine whether any geographic location on Earth is free from human influence (Finlayson-Pitts and Pitts, 1986). For the following reasons, Lefohn et al. (1990a) stated that the definitions of natural background and the use of the information are subject to much uncertainty when characterizing ozone exposures to be used as controls in vegetation research:

- Little is known with certainty about the nature of past unpolluted conditions.
- Even if all anthropogenic emissions of ozone precursors were eliminated, it is unlikely that ozone concentrations in, for example, eastern North America, would return to preindustrial levels. Since preindustrial times, major land-use changes have occurred. As substantial amounts of natural emissions of ozone precursors are derived from soils and vegetation, especially during the warmer months, it is probable that these land-use changes have modified the emissions of ozone precursors and, thus, changed the concentrations of ozone.
- Vegetation is no longer exposed to those ozone levels that may have existed hundreds of years ago; it is possible that vegetation has adapted to changed levels.

Another approach for establishing reference points, which was the one used by Lefohn et al. (1990a), is to examine ozone data from the cleanest sites in the world, including North America, and describe the range of exposures at "clean" sites that can be used for defining reference points. Although it is likely that there is some human influence at these sites, the ozone exposures at these sites were so low that they should be considered as reference-point exposures. The characterization of ozone exposure regimes that occur at "clean" locations can form a base from which one can compare more polluted regimes.

Prior to establishing an ozone standard to protect vegetation, the proposed value should be compared with those exposures that occur in "clean" locations, to identify possible anomalies. If comparisons had occurred using the 7-h seasonal mean exposure index, it would have been clear that there would have been difficulties in applying this exposure index in the standard-setting process. For example, although Lefohn et al. (1990a) reported that most remote sites outside North America experienced seasonal 7-h averages as low as the seasonal 7-h averages used by research investigators (i.e., seasonal 7-h means of 0.025 ppm) (Table 8.3), in the continental U.S. and southern Canada, values ranged from approximately 0.028 to 0.050 ppm. Lefohn et al. (1990a) reported that the Theodore Roosevelt National Park, ND site, in 1984, 1985, and 1986, experienced a 7-month (April to October) average of the 7-h daily average concentrations of 0.038, 0.039, and 0.039 ppm, respectively.

Table 8.3. The Value of the Ozone Season (7-Month) Average of the Daily 7-h (0900–1559 h) Concentration[a]

Site	Elevation (m)	1976	1977	1978	1979	1980	1981	1982	1983	1984	1985	1986	1987
South Pole, Antarctica	2835					0.025	0.027		0.026	0.027	0.024	0.025	
Bitumount, Alberta, Canada	350				0.028								
Barrow, AK	11					0.022	0.025	0.024	0.024	0.022	0.026	0.022	0.026
Theodore Roosevelt N.P., ND	727									0.038	0.039	0.039[b]	
Custer National Forest, MT	1006				0.043	0.044			0.042				
Ochoco National Forest, OR	1364					0.043	0.035	0.038	0.038				
Birch Mtn., Alberta, Canada	850				0.036								
White River Oil Shale Proj., UT U-4	1600								0.045	0.045			
Fortress Mtn., Alberta, Canada	2103											0.041	0.050
Apache National Forest, AZ	2424					0.054	0.039	0.047	0.040				
Mauna Loa, HI	3397					0.035	0.039	0.034	0.038	0.035	0.035	0.034	
Whiteface Mtn., NY	1483				0.049	0.046	0.040	0.034	0.041	0.044	0.043	0.043	0.045
Hohenpeissenberg, Germany	975	0.047	0.040	0.044	0.040	0.037	0.043	0.047	0.040	0.043			
American Samoa	82					0.010	0.010	0.011	0.009	0.012	0.010	0.011	

[a] Units in ppm (Lefohn et al., 1990a).
[b] Collection did not occur during the months of October, November, and December.

The 7-month seasonal averages calculated at the Theodore Roosevelt National Park (i.e., 0.038 and 0.039 ppm) appear to be representative of values that may occur at other fairly "clean" sites located in the U.S. and other locations in the Northern Hemisphere. Lefohn and Foley (1991) reported that the 7-month seasonal values for several Class I national park and national monument sites in the U.S. are in the range described for the "clean" sites reported by Lefohn et al. (1990a). It appears that a 7-h seasonal average of 0.025 ppm, used as a reference point for estimating yield loss, may be too low, and the use of this value may have resulted in an overestimation of predicted agricultural losses. Recognizing the possible problems in selecting too low a value as a reference point for the 7-h seasonal average concentration, the National Acid Precipitation Program (NAPAP) (1990) selected multiple reference points (i.e., 0.025, 0.030, 0.035, and 0.040 ppm) for assessing the possible effects of increases in ozone in the U.S. agricultural sector.

Lefohn et al. (1990a) summarized the ozone exposures at "clean" locations, using several exposure indices. Table 8.4 lists the sites used by Lefohn et al. (1990a) in their analysis. Table 8.5 shows the percentile distributions calculated from the data analyzed by the authors. Although the maximum hourly mean ozone concentrations recorded at the South Pole, Bitumount, and Barrow sites were different (i.e., 0.047, 0.130, and 0.067 ppm, respectively), the annual average concentrations experienced at the sites were similar. The annual average concentrations at the sites were low (i.e., 0.030 ppm and lower); the Bitumount and Barrow sites experienced annual average concentrations slightly lower than did the South Pole site. The infrequent high hourly average concentrations recorded at the Bitumount and Barrow sites were more than offset by low concentrations, and this resulted in the two sites experiencing lower annual average concentrations than the South Pole site. Omitting the Samoa results, these three sites experienced the lowest exposures recorded at the remaining 13 sites.

The sigmoidal exposure (W126) values, calculated over an annual period, are provided in Table 8.6. The lowest W126 values were those calculated for the South Pole, Bitumount, and Barrow sites (i.e., ranging from 2 to 4 ppm-h). Although not experiencing ozone exposures as low as the previously discussed sites, there were four other sites with low W126 values. The W126 values for Theodore Roosevelt National Park, ND were in the range 6.48 to 8.03 ppm-h. The maximum hourly average concentration reported at the site was 0.068 ppm. The W126 values at the Custer National Forest, MT and Ochoco National Forest, OR sites ranged from 5.79 to 22.67 ppm-h (Table 8.6). The maximum hourly average concentrations measured at each site were 0.075 and 0.080 ppm, respectively. The W126 values calculated for the Custer National Forest and Ochoco National Forest sites showed greater variability from year-to-year than did the values calculated for the South Pole, Barrow, and Theodore Roosevelt National Park sites (Table 8.6). The Birch Mountain, Alberta site experienced a W126 value of 19.73 ppm-h. The maximum hourly average concentration was 0.075 ppm.

Table 8.4. Location of Ozone Sites Used in the Analysis[a]

Site	Latitude				Longitude				Elevation (m)
South Pole, Antarctica	90°	0'		S					2835
Bitumount, Alberta, Canada	57°	23'	3"	N	111°	32'	00"	W	350
Barrow, AK	71°	18'		N	156°	36'		W	11
Theodore Roosevelt N.P., ND	47°	36'	07"	N	103°	15'	50"	W	727
Custer National Forest, MT	45°	18'	01"	N	106°	09'	42"	W	1006
Ochoco National Forest, OR	44°	13'	30"	N	119°	42'	25"	W	1364
Birch Mtn., Alberta, Canada	57°	41'	30"	N	111°	49'	30"	W	850
White River Oil Shale Proj., UT U-4	39°	55'		N	109°	44'		W	1600
Fortress Mtn., Alberta, Canada	50°	49'	40"	N	115°	12'	20"	W	2103
Apache National Forest, AZ	33°	45'		N	109°	00'	00"	W	2424
Mauna Loa, HI	19°	30'		N	155°	36'		W	397
Whiteface Mtn., NY	44°	23'	26"	N	73°	51'	34"	W	1483
Hohenpeissenberg, Germany	47°	48'		N	11°	01'		E	975
American Samoa	14°	12'		S	170°	36'		W	82

[a] Lefohn et al. (1990a).

As the W126 values increased, the magnitude of the year-to-year variability also increased. For 2 years of data, the W126 values calculated for the White River U-4 Oil Shale (UT) site were 19.98 and 32.10 ppm-h (Table 8.6). The maximum hourly concentration recorded was 0.079 ppm. Similarly, the Fortress Mountain, Alberta site experienced W126 values of 25.04 and 83.89 ppm-h. The maximum hourly concentration was 0.122 ppm. The W126 values calculated for the Apache National Forest, AZ site ranged from 10.24 to 81.39 ppm-h. The highest hourly average concentration was 0.090 ppm.

The values calculated using the SUM0 parameter are summarized in Table 8.7. With this parameter, it appears that, at times, the Theodore Roosevelt National Park site experienced lower ozone exposures than the South Pole site (Table 8.7). The percentile distributions for the two sites show that this conclusion is incorrect. Lefohn et al. (1990a) pointed out that the SUM0 index did not provide sufficient focus on the higher concentrations to describe the higher ozone exposures experienced at the North Dakota site. Lefohn et al. (1989, 1991) have discussed the performance differences between the SUM0 and other exposure indices and have concluded that the SUM0 index has serious limitations that preclude its use in quantifying ozone exposures and that it should not be used for this purpose.

Using the annual average values of the hourly concentrations, Lefohn et al. (1990a) characterized the diurnal pattern (see Figure 8.1). The diurnal variation in ozone concentrations was absent or small (typically 0 to 0.01 ppm) for the remote sites. A varying diurnal pattern at three of the NOAA sites was almost nonexistent (see Figure 8.1a). The Mauna Loa, HI site experienced its highest exposures during the late-evening and early-morning hours. The Canadian "clean" sites showed slightly varying, diurnal patterns (see Figure 8.1b). The

Table 8.5. Comparison of Percentile Distribution of the Hourly Average Ozone Concentrations[a]

Site	Year	0	10	30	50	70	90	95	99	100	No. hours
South Pole, Antarctica	80	.010	.021	.025	.029	.032	.035	.036	.037	.040	8624
	81	.014	.020	.025	.032	.035	.038	.040	.042	.045	7869
	82	.016	.022	.028	.031	.034	.036	.037	.038	.041	7645
	83	.014	.020	.025	.029	.032	.034	.035	.036	.038	8443
	84	.013	.020	.027	.033	.036	.037	.038	.041	.047	8513
	85	.013	.020	.025	.029	.033	.036	.038	.039	.040	8010
	86	.012	.020	.025	.032	.036	.038	.039	.040	.041	8606
	87	.013	.018	.021	.025	.029	.031	.032	.033	.038	6536
Bitumount, Alberta, Canada	77	.000	.010	.020	.025	.030	.040	.045	.070	.130	6023
	78	.000	.015	.020	.030	.040	.050	.050	.060	.075	6658
	79	.000	.015	.020	.025	.030	.040	.040	.050	.075	7436
Barrow, AK	80	.000	.015	.022	.028	.032	.036	.037	.041	.044	8436
	81	.000	.018	.024	.028	.031	.035	.037	.041	.050	8619
	82	.000	.011	.024	.029	.034	.037	.039	.042	.046	8596
	83	.000	.016	.023	.028	.032	.036	.038	.040	.046	8527
	84	.000	.005	.021	.025	.031	.038	.040	.042	.049	8006
	85	.000	.019	.025	.030	.033	.036	.037	.040	.051	7235
	86	.000	.011	.022	.027	.032	.036	.038	.041	.067	8207
	87	.000	.017	.025	.029	.034	.039	.040	.043	.050	8506
Theodore Roosevelt N.P., ND	83	.000	.015	.022	.027	.032	.041	.047	.061	.080	5691
	84	.000	.016	.023	.029	.034	.043	.048	.056	.068	8544
	85	.000	.020	.026	.031	.035	.043	.047	.053	.061	7809
	86	.001	.018	.026	.032	.037	.045	.048	.054	.062	6425
	87	.003	.016	.026	.032	.038	.046	.049	.054	.064	3726

Site	Year										
Custer National Forest, MT	79	.005	.015	.025	.030	.040	.050	.050	.060	.075	8488
	80	.005	.020	.030	.035	.045	.055	.055	.065	.070	7754
	81	.005	.020	.025	.030	.035	.040	.045	.050	.070	6363
	82	.005	.020	.025	.030	.035	.040	.045	.050	.055	5672
	83	.010	.025	.030	.035	.040	.045	.050	.055	.065	7967
Ochoco National Forest, OR	80	.010	.025	.035	.040	.045	.050	.055	.065	.080	7756
	81	.010	.020	.025	.030	.035	.040	.045	.050	.075	7768
	82	.000	.025	.030	.035	.035	.045	.050	.055	.065	7793
	83	.010	.025	.030	.035	.040	.045	.050	.055	.060	7288
Birch Mountain, Alberta, Canada	77	.000	.025	.030	.035	.040	.050	.055	.060	.120	4657
	78	.005	.020	.030	.040	.040	.050	.055	.065	.080	6563
	79	.000	.025	.030	.040	.045	.050	.055	.065	.075	7649
White River Oil Shale U-4 Utah	83	.013	.030	.036	.039	.043	.049	.053	.061	.072	8545
	84	.011	.030	.036	.041	.047	.055	.058	.067	.079	8317
Fortress Mtn., Alberta, Canada	86	.003	.030	.036	.040	.046	.053	.056	.060	.071	8304
	87	.013	.032	.039	.043	.053	.069	.076	.093	.122	7139
Apache National Forest, AZ	80	.010	.030	.040	.045	.055	.065	.070	.080	.090	8374
	81	.010	.025	.030	.035	.040	.045	.050	.055	.065	8298
	82	.010	.030	.035	.040	.045	.055	.055	.065	.075	7977
	83	.004	.025	.035	.035	.040	.050	.055	.060	.070	7853
Mauna Loa, HI	80	.009	.022	.031	.037	.044	.054	.059	.065	.083	8604
	81	.013	.026	.034	.040	.047	.058	.064	.072	.088	8674
	82	.008	.019	.030	.038	.045	.055	.060	.070	.092	8650
	83	.010	.025	.034	.041	.049	.061	.067	.079	.180	8743
	84	.010	.023	.031	.037	.042	.050	.054	.067	.085	7624
	85	.012	.023	.031	.038	.044	.055	.060	.072	.086	8440
	86	.010	.021	.029	.036	.043	.052	.056	.069	.083	8331

Table 8.5. Continued

Site	Year	0	10	30	50	70	90	95	99	100	No. hours
	87	.010	.022	.030	.036	.042	.049	.052	.061	.068	5795
Whiteface Mountain, NY	78	.001	.026	.035	.042	.053	.072	.080	.095	.115	7077
	79	.000	.023	.032	.041	.051	.068	.079	.095	.136	8045
	80	.000	.023	.031	.038	.046	.064	.074	.094	.123	8017
	81	.000	.019	.027	.034	.041	.056	.064	.081	.101	7418
	82	.000	.015	.025	.034	.040	.054	.061	.081	.114	7969
	83	.000	.017	.027	.033	.040	.057	.067	.083	.115	7491
	84	.001	.023	.030	.036	.044	.059	.065	.078	.122	8079
	85	.003	.025	.032	.037	.043	.057	.065	.080	.096	8158
	86	.003	.025	.032	.038	.043	.055	.062	.073	.120	7641
	87	.005	.026	.034	.040	.048	.062	.069	.083	.104	8191
Hohenpeissenberg, Germany	71	.000	.008	.022	.029	.037	.050	.054	.064	.122	8252
	72	.000	.011	.023	.030	.038	.049	.054	.064	.096	8752
	73	.000	.011	.024	.031	.039	.052	.060	.071	.130	8675
	74	.000	.013	.026	.031	.038	.050	.056	.064	.086	8746
	75	.000	.010	.024	.033	.042	.058	.068	.086	.112	8719
	76	.000	.014	.029	.036	.047	.062	.069	.081	.107	8730
	77	.000	.016	.027	.032	.039	.052	.058	.070	.094	8726
	78	.000	.014	.029	.036	.043	.056	.061	.072	.110	8658
	79	.000	.013	.026	.031	.039	.051	.057	.068	.104	8748
	80	.000	.018	.028	.032	.039	.050	.057	.069	.110	8782
	81	.002	.021	.029	.034	.041	.054	.060	.071	.093	8438
	82	.000	.020	.031	.038	.046	.059	.064	.079	.133	8685
	83	.002	.018	.027	.032	.039	.050	.054	.067	.091	8757
	84	.000	.017	.026	.031	.041	.057	.066	.080	.107	8711

Samoa										
80	.002	.006	.009	.012	.016	.023	.026	.032	.039	7599
81	.001	.007	.010	.012	.015	.021	.024	.027	.030	8670
82	.001	.006	.009	.013	.017	.023	.026	.029	.035	8002
83	.001	.006	.009	.012	.016	.022	.025	.029	.037	8704
84	.001	.006	.008	.013	.018	.023	.026	.029	.042	7513
85	.000	.007	.010	.013	.016	.022	.024	.029	.035	8503
86	.001	.006	.009	.012	.016	.024	.028	.032	.036	7386
87	.000	.005	.009	.013	.020	.026	.029	.033	.036	8445

[a] Concentration in ppm (Lefohn et al., 1990a).

Table 8.6. The Value of the W126 Sigmoidal Exposure Parameter Calculated Over the Annual Period[a]

Site	Elevation (m)	1976	1977	1978	1979	1980	1981	1982	1983	1984	1985	1986	1987
South Pole, Antarctica	2835					2.65	3.72	3.01	2.41	3.54	2.76	4.09	
Bitumount, Alberta, Canada	350				2.99								
Barrow, AK	11					2.60	2.60	3.15	2.36	2.79	2.03	2.46	3.69
Theodore Roosevelt N.P., ND	727									8.03	6.69	6.48[b]	
Custer National Forest, MT	1006				14.08	22.67			12.18				
Ochoco National Forest, OR	1364					19.54	5.79	9.10	8.02				
Birch Mtn., Alberta, Canada	850				19.73								
White River Oil Shale Proj., UT U-4	1600								19.98	32.10			
Fortress Mtn., Alberta, Canada	2103								17.48			25.04	83.89
Apache National Forest, AZ	2424					81.39	10.24	27.18	48.90	19.18			
Mauna Loa, HI	3397	61.28	25.04	35.64	86.50	27.48	45.68	33.68	37.82	42.94	32.66	24.48	
Whiteface Mtn., NY	1483				21.76	68.30	33.75	32.03	19.85	40.43	41.36	32.07	58.33
Hohenpeissenberg, Germany	975					18.53	29.53	49.00					
American Samoa	82					0.28	0.24	0.25	0.28	0.30	0.26	0.30	0.32

a Units in ppm-h (Lefohn et al., 1990a).
b Collection did not occur during the months of October, November, and December.

Table 8.7. The Value of the SUM0 Exposure Parameter Calculated Over the Annual Period[a]

Site	Elevation (m)	1976	1977	1978	1979	1980	1981	1982	1983	1984	1985	1986	1987
South Pole, Antarctica	2835					245.00	233.96	231.55	236.98	234.40	229.27	258.51	
Bitumount, Alberta, Canada	350				175.80								
Barrow, AK	11					219.17	232.82	214.97	204.78	196.51	178.79	192.36	225.94
Theodore Roosevelt N.P., ND	727									248.93	243.41	202.25[b]	
Custer National Forest, MT	1006				273.46	285.20			280.30				
Ochoco National Forest, OR	1364					298.46	242.55	265.37	247.23				
Birch Mtn., Alberta, Canada	850				286.95								
White River Oil Shale Proj., UT U-4	1600								337.68	347.71			
Fortress Mtn., Alberta, Canada	2103					396.64						338.70	338.76
Apache National Forest, AZ	2424					323.68	292.75	328.87	297.40	280.02	324.53	302.88	
Mauna Loa, HI	3397						359.06	326.60	332.01				
Whiteface Mtn., NY	1483				351.60	330.34	265.22	272.03	264.73	310.28	319.26	297.97	344.54
Hohenpeissenberg, Germany	975	330.25	289.83	312.71	282.08	262.42	281.23	338.16	291.50	300.67			
American Samoa	82					95.81	114.72	97.30	113.90	100.13	106.64	99.44	106.36

a Units in ppm-h (Lefohn et al., 1990a).
b Collection did not occur during the months of October, November, and December.

NOAA GMCC Remote Sites

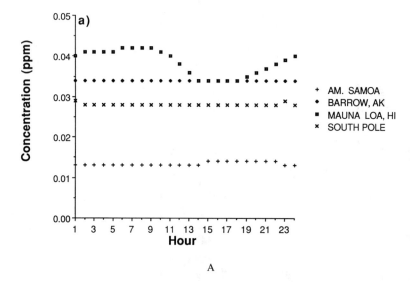

A

Canadian "Clean" Sites

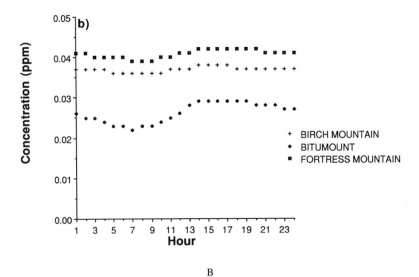

B

Figure 8.1.A-B. The annual average diurnal patterns experienced at the "clean" and remote monitoring sites (Lefohn et al., 1990a).

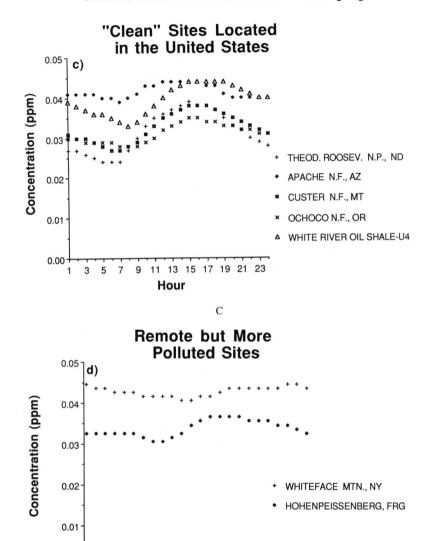

Figure 8.1.C-D. The annual average diurnal patterns experienced at the "clean" and remote monitoring sites (Lefohn et al., 1990a).

patterns for the high-elevation sites at Birch Mountain and Fortress Mountain were almost flat, while the pattern for the site at Bitumount showed small diurnal variations. The diurnal patterns for the five "clean" sites located in the continental U.S. showed some variation in the hourly concentrations.

Historically, hourly average ozone concentrations have been used to generate monthly average concentrations that differentiate the cleanest sites from those influenced by anthropogenic activities. However, the specific month when monthly average concentrations are highest may not correlate well with the timing of the highest hourly average concentrations, which are of particular importance for assessing vegetation effects. If there are many high concentrations, the timing of the highest monthly average concentrations will strongly correlate with peak exposures (Lefohn and Benedict, 1985). On the other hand, if there are few high concentrations, the timing of the highest monthly average concentrations may not strongly correlate with peak exposures (Winner et al., 1989; Lefohn et al., 1990b). Because vegetation effects caused by ozone are linked with the occurrence of high hourly mean concentrations, it is difficult to associate effects with a long-term average concentration such as a monthly mean.

8.2.2.2   Comparing Experimental Regimes with Actual Ambient Conditions

Linking exposure regimes (used in experimental fumigation chambers) that result in vegetation effects with those regimes that are observed under ambient conditions is important for establishing standards to protect vegetation. Lefohn et al. (1990a, 1992) reported results showing that only some indices order themselves properly, when comparing the absolute value experienced in an experiment with the value calculated under actual ambient conditions. Although the use of a long-term seasonal average-concentration index may be appropriate to relate South Coast Air Basin (CA) ozone exposures to vegetation effects, the relationship between the long-term average-concentration index and vegetation effects becomes much weaker when milder exposures are used.

Using two different approaches, Lefohn et al. (1992) compared the performance of four exposure indices and found that the value of the SUM0 index did not adequately characterize the ozone exposure. Charcoal-filter (CF) chambers are normally used as a basis of comparison with other ozone exposure treatments (Chappelka et al., 1990). The authors compared the ozone exposures for the CF chambers with those experienced at two "clean" ozone sites, South Pole and Pt. Barrow (AK). These two sites were used previously by Lefohn et al. (1990a) in their analysis to characterize ozone exposures at "clean" locations. Except for the SUM0 values, the SUM06, W126, and SUM08 indices indicated that the ozone exposures in the CF chambers and at the two "clean" ozone sites were low and similar. In 1989 there were fewer than 11 occurrences of hourly average concentrations greater than or equal to 0.06 ppm in any of the CF chambers. The SUM06 values for the CF treatments reflected more occurrences of hourly average concentrations greater than or equal to 0.06 ppm than at the other two sites. Using the SUM0 index, one might incorrectly have concluded that the ozone exposures measured in the Auburn CF treatments were approximately 50% less than those experienced at either the South Pole or Pt. Barrow.

Furthermore, Lefohn et al. (1992) compared the ambient ozone exposures experienced at the Auburn, AL experimental site with information characterized for a select set of "clean" ozone-monitoring sites that experienced SUM0 values similar to the Auburn site. Although the SUM0 values for all sites were similar, the percentile distribution of the hourly average ozone concentrations and the values for the other three cumulative indices were not. The Auburn site (ambient plot) experienced higher maximum hourly average concentrations, with more hourly average concentrations above 0.07 ppm, than the "clean" sites. The values of these three cumulative indices for the "clean" sites were less than the values for the ambient ozone exposures at the Auburn site. With the focus on identifying those sites that experience high hourly average concentrations, the SUM0 index did not order the exposures properly. The SUM0 index failed because the magnitude of the index was heavily influenced by the lower hourly average ozone concentrations and, thus, was less sensitive than those cumulative indices that focused more on the higher concentrations (Lefohn et al., 1989).

8.3 THE USE OF SINGLE- VS. MULTIPLE-PARAMETER INDICES TO DESCRIBE OZONE EXPOSURES

8.3.1 Introduction

As discussed earlier in this chapter, a major concern about the use of any exposure index (e.g., cumulative or seasonal average concentration) is whether the value of the index is linked to a specific exposure regime. The absolute value of the index reflects only the mathematical calculation performed using hourly average ozone concentrations. Given the observation that the distribution of the highest hourly average concentrations (i.e., the upper tail of the distribution) is an important factor in affecting vegetation, we have shown previously (Section 8.2) that a single-parameter exposure index may not be sufficient to describe those important distributions that cause an ozone-related effect. As we have seen, the absolute value of some single-parameter exposure indices (e.g., SUM06) may represent many distributions: some with high maximum hourly average concentrations, and some with low maximum concentrations.

Lefohn et al. (1989) have discussed the merits of applying any index for the purposes of summarizing exposure and have cautioned that any index used should adequately describe the range of ozone exposure regimes that elicit an adverse effect on vegetation. It is important that the index used adequately focus on the important parts of the ozone exposure regime that are thought to be responsible for affecting crops adversely. In addition, it is important that the exposure index selected be consistent so that a low value indicates relatively low risk to agricultural crops, while a high value represents a high risk. If this consistency does not occur, then the application of the index as a secondary standard will result in either under- or overprotecting the desired target. Although there are difficulties in applying single-parameter indices in linking exposure-response information with ambient air quality data, single-parameter exposure indices have been successfully applied for other purposes.

8.3.2 Using Single-Parameter Indices for Describing Regional Exposure

Single-parameter indices have been useful for describing regional ozone exposure in the U.S. (Lefohn et al., 1987; Lefohn et al. 1990b). Figure 8.2 shows the results of interpolating characterized hourly average ozone data using kriging of the W126, 7-month seasonal ozone exposure index in $1/2 \times 1/2°$ grids for the eastern U.S.. The figure shows that ozone exposures in the East were higher in 1987 than in the two previous years, 1985 and 1986. Trends analysis performed by the U.S. EPA (1991) confirm this observation. Yet, given the fact that we have shown that the magnitude of cumulative exposure indices, such as the W126 or SUM06 exposure index, is not necessarily strongly associated with the occurrence of high hourly average ozone concentrations, why is it possible to successfully describe regional exposures using the single-parameter cumulative indices?

The magnitude of the cumulative index is related to the distribution of the hourly average concentrations. The ozone exposures experienced at each site are influenced by a multitude of factors. The elevation of a specific site, its ground cover (i.e., sorptive capacity), as well as its latitude, may influence ozone production and destruction and the absolute ozone exposure value experienced at a specific site. Many of the ozone monitors used in the kriging analyses were situated near urban-oriented locations (Lefohn et al., 1990b). Thus, the distribution of the hourly average concentrations may have been similar. For example, most of the urban-oriented monitoring sites may experience similar scavenging processes that result in 30% or more of the hourly average concentrations occurring below 0.015 ppm. In addition, the maximum hourly average concentrations experienced at many of these sites were similar. Thus, with similar hourly average distribution curves, it would be assumed that the magnitude of a cumulative exposure index, such as the W126 or SUM06, would order itself properly, with the higher value corresponding to the higher exposure.

Besides using cumulative exposure indices to describe regional ozone exposures, the indices have been used in trends analysis. Lefohn and Shadwick (1991) summarized trends for ozone exposures over 5- and 10-year periods (i.e., 1984 to 1988 and 1979 to 1988) for rural locations in the U.S. The investigators explored the evidence for trends at each monitoring location. **Evidence for**

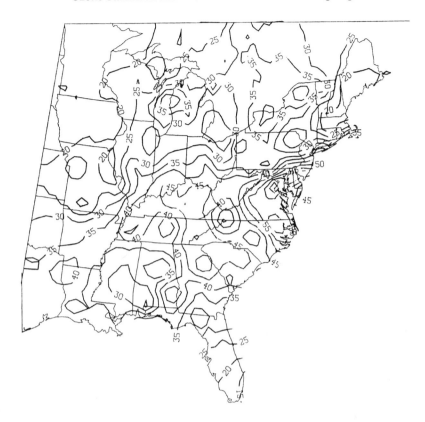

Figure 8.2. Interpolation of April–October ozone exposures for 1987 for the eastern U.S. using the W126 index.

regional trends was based on studying the individual time trends that were observed for each of the sites in the region. The seasonal W126 cumulative exposure index was used to investigate trends. The results reported by Lefohn and Shadwick (1991) were consistent with the findings reported by the U.S. EPA (1990).

The explanation for the successful application of the cumulative index in the trends analysis was similar to the one given for the kriging analysis. Given a specific monitoring site, the hourly average distribution curve was similar over the years studied by Lefohn and Shadwick (1991). The scavenging processes remained the same over time at a specific site. Thus, the difference in magnitude of the W126 index, at any one site over time, was related to the difference in the occurrences of the hourly average ozone concentrations at the upper end of the distribution curve.

8.4 FUTURE RESEARCH LINKING EXPOSURE WITH EXPERIMENTAL RESULTS AND ESTABLISHING A SECONDARY STANDARD TO PROTECT VEGETATION

From the previous discussions in this chapter, it appears that a fine-tuning is necessary when single-parameter exposure parameter indices are applied. Single-parameter indices have been successfully used in describing regional ozone exposures, as well as for characterizing the evidence for regional trend patterns. However, they have not been entirely successful in describing the critical components of exposure that result in adverse effects on vegetation.

A review of Table 8.2 shows that, for many of the sites that experienced the highest hourly average concentrations, the number of occurrences of hourly average concentrations greater than or equal to 0.06 ppm was low. In many cases, the scavenging processes that occurred at these locations were effective and resulted in reducing the number of concentrations above this level. Although the federal primary standard for ozone was exceeded (based on the second-highest maximum hourly average concentration that occurred over a 24-h period), the cumulative effect of the ozone exposures that occurred over a 3-month period was low. As indicated by Lee et al. (1991), the current form of the ozone standard does not relate well with what is known about protecting vegetation from ozone exposures.

For developing a secondary standard to protect vegetation, the combined exposure statistics should be selected so that they are based on the observation that high concentrations are expected to cause greater impact on vegetation than are lower concentrations. The following important factors, summarized by Lee et al. (1991), may be important when selecting an appropriate standard to protect vegetation:

- Peak concentrations are more important than low concentrations in determining plant response
- Ozone effects are cumulative (i.e., increasing the duration of the exposure period is expected to cause greater biological response)
- Exposure cannot be characterized as the unweighted product of concentration and time, because the effect of ozone on vegetation yield depends on the cumulative impact of high concentrations during the growing season
- Plant sensitivity is not constant, but varies according to stage of development.

As discussed by Lefohn et al. (1990b) and Lee et al. (1991), the use of a 7-h seasonal average as a form of a standard to protect vegetation is inappropriate. The long-term average concentration does not correlate well with the occurrence of high hourly average concentrations, except in those areas of the country experiencing frequent occurrences.

Lefohn et al. (1988, 1992) and Lee et al. (1988, 1989, 1991) have shown that, when high hourly average concentrations are present in an exposure regime, single-parameter cumulative indices can be used to relate ozone exposures with growth reductions. However, as indicated in the previous section, the application of a single-parameter exposure index in the form of a standard for protecting agricultural crops will provide inconsistent results. Because of the limitations of the use of these indices, an exposure index will have to be combined with other exposure-related parameters to assure that a strong correlation exists between the combination of these parameters and the occurrence of high hourly average concentrations. The combination of exposure parameters (i.e., multiple indices) that is used to describe those regimes that cause vegetation effects must adequately characterize the upper tail of the hourly average distribution curve.

The estimated ranges of ozone exposures that result in injury and damage effects to vegetation, published by Guderian et al. (1985), provide us with an indication that hourly average concentrations of 0.10 ppm and higher are important in eliciting adverse effects on vegetation. This does not mean that concentrations below 0.10 ppm are not important. In general, however, the recommendations of Guderian et al. (1985) tend to support the hypothesis that hourly average concentrations greater than or equal to 0.10 ppm may have to be experienced before serious injury or damage to vegetation can occur.

Recognizing that some of the lower hourly average ozone concentrations may contribute to adverse vegetation effects, it is important to attempt to define subjectively a lower limit. As discussed earlier, ozone hourly average concentrations of 0.05 ppm routinely occur at many "clean" site locations in the world (Lefohn et al., 1990a), including several Class I areas in the U.S. (Lefohn and Foley, 1991). In addition, occasional occurrences of hourly average concentrations near 0.08 ppm are experienced at these "clean" monitoring locations. Lefohn and Foley (1991) report that in almost all cases, none of the "clean" ozone monitoring sites experienced hourly average concentrations greater than or equal to 0.08 ppm, and the maximum hourly average concentrations were in the range from 0.060 to 0.075 ppm. In addition, the results reported by Lefohn et al. (1988) and Lee et al. (1988, 1991) support the concept that hourly average ozone concentrations greater than or equal to 0.06 ppm are important in the growth reduction of agricultural crops. Thus, when reviewing the possible combination of exposure parameters, the sigmoidally weighted exposure index, as proposed by Lefohn and Runeckles (1987), as well as the SUM06 index, as recommended by Lee et al. (1991), coupled with a combination of indices whose magnitude reflects the presence of high hourly average concentrations, should provide sufficient means to describe the potential for ozone exposures to elicit an adverse effect.

A method used to characterize ozone exposures is the determination of the percentile distributions of the hourly average concentrations. Experience with hourly average-concentration ozone data has revealed both seasonal and daily patterns in time plots of ozone concentrations. Ozone tends to be episodic on a short-time basis (i.e., time frames of days or weeks). The occurrence of "high" ozone values tends to be relatively close in time, as determined by meteorological events. The regularity in the time structure of "high" ozone concentrations gives the appearance of "peaks" in time plots of hourly ozone concentrations. With high confidence, from the percentile distribution of ozone, one can infer that the values in the tail of the distribution represent "peaks" in the time plots of hourly ozone concentrations.

Besides identifying ways to characterize ozone exposures, the results of the retrospective analyses using NCLAN data, reported by Lefohn et al. (1988) and Lee et al. (1988, 1991), provide researchers with the opportunity to better understand the **level** of exposures that result in agricultural-yield reduction. For example, Table 8.8 lists the SUM06 cumulative exposures (in units of ppm-h) that were calculated, using equations supplied by E.H. Lee, for 10%, 20%, and 30% yield losses. However, caution is suggested in applying thes results in the standard-setting process. In almost all cases, NCLAN investigators did not artificially fumigate their crops for 24 h. The results reported by Lee et al. (1991), as well as the SUM06 values listed in Table 8.8, are based on the period of fumigation (i.e., 7, 10, and 12 hours). Ozone hourly average concentrations greater than or equal to 0.06 ppm occur outside of a fixed 7-, 10-, or 12-h window. The crops used in the NCLAN experiments received additional ozone exposures outside of the fixed window, which were not accounted for by Lee et al. (1991). It is possible that the cumulative exposures reported by Lee et al. (1991) and summarized in Table 8.8 may be low. Prior to applying these results, it is recommended that an assessment be made to identify whether the 7-, 10-, and 12-h windows adequately captured the SUM06 cumulative exposures to the crops or whether the equations derived for exposure response, as used by Lee et al. (1991), have to be modified to reflect the actual exposures received by the plants.

Future research efforts associated with NCLAN retrospective studies should involve the characterization of the W126 cumulative exposure index and the distribution of the hourly average concentrations used in the NCLAN experiments. Although the W126 and SUM06 indices are strongly correlated, it may be possible that adverse effects, other than damage, occur below 0.06 ppm and that an alternative to the SUM06 index will be required (Lefohn et al., 1992). By adequately characterizing the distribution of the hourly average concentrations, an evaluation of the upper end of the distribution curve can be made. This additional information would be available for use in the multiparameter exposure index approach that was described previously.

The modification of the results reported by Lee et al. (1991), as well as appropriate additional experimental information, can be linked with characterized ambient air quality data to develop an appropriate secondary standard to protect vegetation. Lefohn et al. (1989) pointed out that any attempt to link experimental results with actual ambient air quality data must be based on identical time periods. Because ozone monitoring data are recorded for a 24-h period, the ozone exposure information that is monitored in the experimental work should be characterized for the same period. This is important when applying exposure-response models, developed from experimental data, to predict the effects of ambient exposures on vegetation.

Table 8.8.   **SUM06 Cumulative Exposures for the Experimental Period, Using the SUM06 Lee et al. (1991) Equations (Assuming SUM06 = 0 for "Clean" Sites) that Predict 10%, 20%, and 30% Yield Losses[a]**

| | Percent Yield Loss | | | |
Crop	10%	20%	30%	Exposure days
Soybean				
A80SO Corsoy	12.7	20.8	28.4	56
A83CV Amsoy_71	29.3	39.5	47.7	83
A83CV Corsoy_79	38.1	44.4	48.9	83
A85SO Corsoy_79 Dry	65.4	79.6	89.9	87
A85SO Corsoy_79 Wet	65.9	82.0	94.0	87
A86SO Corsoy_79 Dry	93.9	99.2	102.6	94
A86SO Corsoy_79 Wet	63.0	78.8	90.7	94
B83SO Corsoy_79 Dry	5.2	14.9	28.7	63
B83SO Corsoy_79 Wet	10.0	18.6	27.3	63
B83SO Williams Dry	21.2	34.9	47.8	63
B83SO Williams Wet	12.6	22.8	33.1	63
I81SO Hodgson	9.9	20.9	33.4	70
R81SO Davis	20.5	38.2	56.2	110
R82SO Davis	18.9	31.0	42.3	89
R83SO Davis Dry	26.8	56.8	90.7	110
R83SO Davis Wet	24.4	46.5	69.6	110
R84SO Davis Dry	31.2	45.9	58.5	106
R84SO Davis Wet	31.2	45.9	58.5	106
R86SO Young Dry	33.3	58.4	83.0	107
R86SO Young Wet	33.3	58.4	83.0	107
Sorghum				
A82SG Dekalb	65.7	91.3	112.2	84
Wheat				
A82WH Abe	20.3	28.9	36.0	56
A83WH Abe	31.1	34.6	36.8	53
A82WH Arthur_71	21.0	29.7	37.0	56
A83WH Arthur_71	24.6	35.4	44.4	53
I82WH Vona	2.9	6.1	9.7	71
I83WH Vona	5.5	9.6	13.7	36
Corn				
A81MA PAG 397	54.3	64.9	72.5	82
A81MA Pioneer 3780	40.1	52.7	62.5	82
Kidney Bean				
I80KB Cal Lt. Red	14.1	17.2	19.5	21
I82KB Cal Lt. Red	16.6	22.7	27.5	43
Potato				
P85PO Norchip	9.1	19.3	30.7	69
P86PO Norchip	16.9	29.4	41.7	61
Cotton				
C81CO Acala SJ-2 Dry	34.8	49.5	61.7	71
C81CO Acala SJ-2 Wet	21.4	32.5	42.2	71
C82CO Acala SJ-2 Dry	20.5	32.1	42.5	99
C82CO Acala SJ-2 Wet	10.2	19.8	29.9	99

Table 8.8. Continued

Crop	Percent Yield Loss			Exposure days
	10%	20%	30%	
C85CO Acala SJ-2 Dry	94.9	201.0	321.3	73
C85CO Acala SJ-2 Wet	60.2	127.5	203.8	73
R82CO Stoneville	23.7	36.7	48.2	119
R85CO McNair Dry	74.2	96.4	113.6	126
R85CO McNair Wet	26.9	44.0	59.8	126
Lettuce				
C83LE Empire	34.7	40.1	43.9	52
Peanut				
R80PN NC-6	34.7	49.0	60.9	112
Tobacco				
R83TQ McNair 944	24.5	46.9	70.5	96

[a] Adapted from data supplied courtesy of E. H. Lee, U.S. EPA, Corvallis, OR.

8.5 FURTHER CONSIDERATIONS

In this chapter a strong case has been made for selecting multiparameter exposure indices for establishing a secondary standard to protect vegetation from ozone exposures. However, caution is urged. A consistent relationship between multiparameter exposure indices and vegetation effects may not always occur. As indicated in other chapters in this book, besides the importance of the occurrences of elevated hourly concentrations, (1) the amount and chemical form of the pollutant that enters the target organism, (2) the length of the exposure within each episodic event, (3) the time between exposures (i.e., the respite or recovery time), and (4) the sensitivity of the target organism are important factors that affect our ability to predict ozone effects on vegetation. As pointed out in Chapter 3, Showman (1991) reported indications that sensitivity may be an important factor. For field surveys in the midwestern U.S., in 1988, ozone levels were high, but injury to vegetation was low due to drought stress. In 1989 ozone exposures were much lower than in 1988, and optimum growing conditions resulted in greater foliar injury.

Overall, it is unclear how important these four factors are in an overall weighting scheme when predicting vegetation effects. If both sensitivity and the actual dose that enters the organism are as important as ambient concentration in the weighting scheme, then a given pollutant exposure will elicit varying biological responses at different times for the same crop, as conjectured by Krupa and Teng (1982). As additional information becomes available, it may be possible to modify the use of exposure indices to take into consideration such concerns as the sensitivity of the target organism at the time of exposure. However, given the current state of knowledge, for the standard-setting process, concentration should be weighted more heavily than either sensitivity or actual dose. Preliminary support for this assumption is built upon the work described in the South Coast Air Basin, where extremely high ozone exposures occur (Oshima, 1975; Oshima et al., 1976; Thompson et al., 1976; Lefohn and Benedict, 1982).

It is important to note that no matter how good the exposure index, a point will always be reached where all indices begin to fail because of the "noise" in the experiment. Natural biological variations and experimental design will affect the exposures at which the "noise" of the experiment will hinder our ability to identify perceivable effects. Researchers and policymakers should not simply give up and say that all exposure indices are bad and, therefore, only **dose** is important. Those who worry about standard-setting protocols would have an impossible task of establishing "dose" instead of "exposure" standards. Although it is important that we continue research on developing dose-response relationships, we should also keep in mind that the link between dose and exposure needs to be quantified so that relevant exposure-response relationships can be established. The exposure-response relationships will play an important role in helping to establish appropriate secondary ozone standards to protect vegetation.

REFERENCES

Adams, R. M., Glyer, J. D., Johnson, S. L., and McCarl, B. A. (1989) A reassessment of the economic effects of ozone on U.S. agriculture. *JAPCA* **39**, 960–968.

Adams, R. M., Hamilton, S. A., and McCarl, B. A. (1985) An assessment of the economic effects of ozone on U.S. agriculture. *JAPCA* **35**, 938–943.

Chappelka, A. H., Lockaby, B. G., Mitchell, R. J., Kush, J. S., and Jordan, D. N. (1990) Growth and physiological responses of loblolly pine exposed to ozone and simulated acid rain in the field. In *Proc. 83rd Annual Meeting of the Air and Waste Management Association.* Air and Waste Management Association, Pittsburgh.

Finlayson-Pitts, B. J. and Pitts, J. N. (1986) *Atmospheric Chemistry: Fundamentals and Experimental Techniques.* John Wiley & Sons, New York, 1098.

Guderian, R., Tingey, D. T., and Rabe, R. (1985) Effects of photochemical oxidants on plants. In *Air Pollution by Photochemical Oxidants* (ed., Guderian R.). Springer-Verlag, Berlin, 129–333.

Heck, W. W. and Brandt, C. S. (1977) Effects on vegetation: native crops, forests. In *Air Pollution*, Vol. II. Academic Press, London, 157–229.

Heck, W. W., Taylor, O. C., Adams, R. M., Bingham, G. E., Miller, J. E., Preston, E. M., and Weinstein, L. H. (1982) Assessment of crop loss from ozone. *JAPCA* **32**, 353–361.

Heuss, J. M. (1982) Comment on "Assessment of crop loss from ozone." *JAPCA* **32**, 1152–1153.

Jacobson, J. S. (1977) The effects of photochemical oxidants on vegetation. *VDI-Ber* **270**, 191–196.

Krupa, S. V. and Nosal, M. (1989) Application of spectral coherence analysis to describe the relationships between ozone exposure and crop growth. In *Proc. 82nd Annual Meeting of the Air Pollution Control Association.* Air Pollution Control Association, Pittsburgh, 89-89.4.

Krupa, S. V. and Teng, P. S. (1982) Uncertainties in estimating ecological effects of air pollutants. In *Proc. 75th Annual Meeting of the Air Pollution Control Association,* New Orleans. Air Pollution Control Association, Pittsburgh.

Lee, E. H., Hogsett, W. E., and Tingey, D. T. (1991) Efficacy of ozone exposure indices in the standard setting process. In *Transactions of the Atmospheric Ozone and the Environment Specialty Conference* (ed., Berglund, R., Lawson, D., and McKee, D.). Air and Waste Management Association, Pittsburgh, 255–271.

Lee, E. H., Tingey, D. T., and Hogsett, W. E. (1988) Evaluation of ozone exposure indices in exposure-response modeling. *Environ. Pollut.* **53**, 43–62.

Lee, E. H., Tingey, D. T., and Hogsett, W. E. (1989) Interrelation of experimental exposure and ambient air quality data for comparison of ozone exposure indices and estimating agricultural losses. Contract No. 68-C8-0006, U.S. Environmental Protection Agency, Corvallis Environmental Research Laboratory, Corvallis, OR.

Lefohn, A. S. and Benedict, H. M. (1982) Development of a mathematical index that describes ozone concentration, frequency, and duration. *Atmos. Environ.* **16**, 2529–2532.

Lefohn, A. S. and Benedict, H. M. (1985) Exposure considerations associated with characterizing ozone ambient air quality monitoring data. In *Evaluation of the Scientific Basis for Ozone/Oxidants Standards* (ed., Lee, S. D.). Air Pollution Control Association, Pittsburgh, 17–31.

Lefohn, A. S. and Foley, J. K. (1991) Estimated surface-level ozone exposures in selected Class I areas in the United States. In *Proc. 84th Annual Meeting of the Air Pollution Control Association*. Air Pollution Control Association, Pittsburgh, 91-144.2.

Lefohn, A. S. and Runeckles, V. C. (1987) Establishing a standard to protect vegetation — ozone exposure/dose considerations. *Atmos. Environ.* **21**, 561–568.

Lefohn, A. S. and Shadwick, D. S. (1991) Ozone, sulfur dioxide, and nitrogen dioxide trends at rural sites located in the United States. *Atmos. Environ.* **25A**, 491–501.

Lefohn, A. S., Krupa, S. V., and Winstanley, D. (1990a) Surface ozone exposures measured at clean locations around the world. *Environ. Pollut.* **63**, 189–224.

Lefohn, A. S., Laurence, J. A., and Kohut, R. J. (1988) A comparison of indices that describe the relationship between exposure to ozone and reduction in the yield of agricultural crops. *Atmos. Environ.* **22**, 1229–1240.

Lefohn, A. S., Runeckles, V. C., Krupa, S. V., and Shadwick, D. S. (1989) Important considerations for establishing a secondary ozone standard to protect vegetation. *JAPCA* **39**, 1039–1045.

Lefohn, A. S., Knudsen, H. P., Logan, J. A., Simpson, J., and Bhumralkar, C. (1987). An evaluation of the kriging method to predict 7-h seasonal mean ozone concentrations for estimating crop losses. *JAPCA* **37**, 595–602.

Lefohn A. S., Benkovitz C. M., Tanner R. L., Shadwick D. S. and Smith L. A. (1990b) Air quality measurements and characterization for terrestrial effects research, Report 7. In *NAPAP State of Science and State of Technology*, National Acid Precipitation Assessment Program, Washington, D.C.

Lefohn, A. S., Shadwick, D. S., Somerville, M. C., Chappelka, A. H., Lockaby, B. G., and Meldahl, R. S. (1992) The characterization and comparison of ozone exposure indices used in assessing the response of loblolly pine to ozone. *Atmos. Environ.* **26A**, 287–298.

National Acid Precipitation Program (1990) Integrated Assessment: questions 1 and 2. External Review Draft, August 1990, Washington, D.C.

Olszyk, D. M., Bytnerowicz, A., and Takemoto, B. K. (1989) Photochemical oxidant pollution and vegetation: effects of gases, fog, and particles. *Environ. Pollut.* **61**, 11–29.

Oshima, R. J. (1975) Development of a system for evaluating and reporting economic crop losses caused by air pollution in California. III. Ozone dosage — crop loss conversion function — alfalfa, sweet Corn. IIIA. Procedures for production, ozone effects on alfalfa, sweet corn and evaluation of these systems. California Air Resources Board, Sacramento, CA.

Oshima, R. J., Poe, M. P., Braegelmann, P. K., Baldwin, D. W., and Van Way, V. (1976) Ozone dosage-crop loss function for alfalfa: a standardized method for assessing crop losses from air pollutants. *JAPCA* **26**, 861–865.

Runeckles, V. C. (1987) Exposure, dose, vegetation response and standards: will they ever be related? In *Proc. 80th Annual Meeting of the Air Pollution Control Association*. Air Pollution Control Association, Pittsburgh, 87–33.2.

Runeckles, V. C. and Wright, E. F. (1988) Crop response to pollutant exposure — the "ZAPS" approach. In *Proc. 81st Annual Meeting of the Air Pollution Control Association*. Air Pollution Control Association, Pittsburgh, 88-69.4.

Showman, R. E. (1991) A comparison of ozone injury to vegetation during moist and drought years. *JAPCA* **41**, 63–64.

Thompson, C. R., Kats, G., and Cameron, J. W. (1976) Effect of photochemical air pollutants on two varieties of alfalfa. *ES&T* **10**, 1237–1241.

Tingey, D. T., Hogsett, W. E., and Henderson, S. (1989) Definition of adverse effects for the purpose of establishing secondary national ambient air quality standards. In *Proc. 82nd Annual Meeting of the Air Pollution Control Association*. Air Pollution Control Association, Pittsburgh, 89-38.2.

Tingey, D. T., Hogsett, W. E., and Lee, E. H. (1991) An evaluation of various alternative ambient ozone standards based on crop yield loss data. In *Transactions of the Atmospheric Ozone and the Environment Specialty Conference* (ed., Berglund, R., Lawson, D., and McKee, D.), Air and Waste Management Association, Pittsburgh, 272–288.

U.S. EPA. (1988) Review of the national ambient air quality standards for ozone assessment of scientific and technical information. U.S. Environmental Protection Agency, Office of Air Quality Planning and Standards, Research Triangle Park, NC.

U.S. EPA. (1990) National air quality and emissions trends report, 1988. EPA-450/4-90-002, U.S. Environmental Protection Agency, Office of Air Quality Planning and Standards, Research Triangle Park, NC.

U.S. EPA. (1991) National air quality and emissions trends report, 1989. EPA-450/4-91-003, U.S. Environmental Protection Agency, Office of Air Quality Planning and Standards, Research Triangle Park, NC.

Winner, W. E., Lefohn, A. S., Cotter, I. S., Greitner, C. S., Nellessen, J., McEvoy, L. R., Jr., Olson, R. L., Atkinson, C. J., and Moore, L. D. (1989) Plant responses to elevational gradients of O_3 exposures in Virginia. *Proc. Natl. Acad. Sci. U.S.A.* **86**, 8828–8832.

INDEX